T0133645

Client-Centered Software Development

The CO-FOSS Approach

Client-Centered Software Development

The CO-FOSS Approach

Allen B. Tucker

CRC Press is an imprint of the
Taylor & Francis Group, an **informa** business
A CHAPMAN & HALL BOOK

CRC Press
Taylor & Francis Group
6000 Broken Sound Parkway NW, Suite 300
Boca Raton, FL 33487-2742

Printed on acid-free paper

International Standard Book Number-13: 978-1-138-58384-9 (Hardback)

Library of Congress Cataloging-in-Publication Data

Names: Tucker, Allen B., author.
Title: Client-centered software development : the CO-FOSS approach / Allen B. Tucker.
Description: Boca Raton, FL : CRC Press/Taylor & Francis Group, [2019] | Includes index.
Identifiers: LCCN 2019010378| ISBN 9781138583849 (hardback : acid-free paper) | ISBN 9780429506468 (ebook)
Subjects: LCSH: Application software--Development. | Computer software industry--Customer services. | Consumer satisfaction.
Classification: LCC QA76.76.D47 T839 2019 | DDC 005.3--dc23
LC record available at https://lccn.loc.gov/2019010378

Visit the Taylor & Francis Web site at
http://www.taylorandfrancis.com

and the CRC Press Web site at
http://www.crcpress.com

To Meg,
my inspiration
and lifelong partner

Contents

List of Figures xvii

List of Tables xxiii

Foreword xxv

Preface xxix

Acknowledgments xxxv

About the Author xxxvii

CHAPTER 1 ▪ The Journey 1

1.1 SOFTWARE 1

1.2 SOFTWARE DEVELOPMENT MODELS 3

1.2.1 Serial Development 3

1.2.2 Agile Development 4

1.2.3 CO-FOSS Development 5

1.2.4 Software Customization: A Continuum 7

Custom Software 7

Off-the-Shelf Software 8

Custom Software with Off-the-Shelf Components 9

1.3 SOFTWARE LICENSING 9

1.3.1 Proprietary Licensing 9

1.3.2 Open Source Licensing 10

1.3.3 FOSS Origins and Impact 13

FOSS Worldwide 16

Terminology: OSS, FOSS, FLOSS, H/FOSS, and CO-FOSS 18

1.4 SOFTWARE ARCHITECTURES 19
1.4.1 Software Frameworks 19
1.4.2 Web Servers and Bundles 21
1.5 NEW VS MATURE OPEN SOURCE PROJECTS 22
1.5.1 Maturity Assessment 23
1.5.2 Incubation 24
Community 25
Bug Tracking 27
1.6 INTO THE WEEDS 28
1.6.1 To the Instructor 29
1.6.2 To the Student 31
1.6.3 To the Client 32
1.6.4 To the Developer 33
1.7 SUMMARY 33
1.8 MILESTONE 1 34

SECTION I Organization Stage

CHAPTER 2 ■ Finding a Client and a Project 37

2.1 CLIENT ACTIVITIES AND SOFTWARE NEEDS 39
2.1.1 The Current Process and Existing Software 41
2.1.2 New Software to Fit a New Need 44
2.2 DOMAIN ANALYSIS 45
2.2.1 Requirements Gathering 48
2.2.2 User Stories 49
2.2.3 Use Cases 50
Unified Modeling Language 52
Writing an Effective Use Case 53
2.3 SOFTWARE DESIGN 55
2.3.1 System and Performance Requirements 55
2.3.2 Software Architecture 57
Layering, Cohesion, and Coupling 57
Domain Class Layer 61
Database Layer 61
User Interface Layer 63
2.3.3 Software Security 63
2.3.4 Encouraging Code Reuse 65

2.4 THE DESIGN DOCUMENT 66
2.4.1 Overall Structure 67
2.4.2 Variations 68
2.5 THE SANDBOX 69
2.6 SUMMARY 70
2.7 MILESTONE 2 70

CHAPTER 3 ▪ Defining the Course 71

3.1 SOFTWARE PROJECT ELEMENTS 71
3.1.1 Collaboration Tools 72
3.1.2 Development Platform 73
3.1.3 Project Hosting 74
3.1.4 The Version Control System 75
3.1.5 Sandbox and Live Versions 77
3.1.6 Reading, Writing, and Documenting Code 79
3.1.7 Unit Testing 82
Unit Testing Tools 84
3.1.8 User Help 85
3.2 THE COURSE 86
3.2.1 The Classroom 87
3.2.2 Team Formation and Dynamics 88
3.2.3 Scheduling and Milestones 90
3.2.4 Ensuring Progress 92
3.2.5 The Syllabus 93
3.2.6 Assignments and Grading 95
3.2.7 Alternatives: The Two-Semester Software Projects Course 97
3.3 SUMMARY 98
3.4 MILESTONE 3 98

SECTION II Development Stage

CHAPTER 4 ▪ Project Launch 101

4.1 THE TEAM 101
4.1.1 Team Dynamics 103

4.1.2 Asynchronous Communication 105
Aside: Mature FOSS Projects 106
4.1.3 Synchronous Communication 107
4.1.4 Shared Documents 108
4.2 THE DEVELOPMENT TOOLS 109
4.2.1 Programming Languages 109
JavaScript 110
Python 110
Java 111
Ruby 111
PHP 111
HTML and CSS 111
Other Languages 112
4.2.2 Software Platforms 112
The Apache/MySQL/PHP Server 113
Server-Side Java 114
Python 114
Ruby 114
4.2.3 IDEs for Development 114
Eclipse IDE 115
Python IDEs 116
Ruby IDEs 116
Java IDEs 116
Choosing and Installing an IDE 117
4.2.4 Working with the VCS 117
4.3 THE PRODUCT 122
4.3.1 Reading the Design Document 122
Identify Classes and Modules 124
Identify Instance Variables 124
Identify Methods and Functions 124
4.3.2 Reading the Code 126
Start from the Top 126
Look for Classes with Unique Keys 127
Avoid the Temptation to Edit the Code 128
4.3.3 Reading and Writing Code 129
4.3.4 Code Reuse 130
4.3.5 Licensing 131

4.4 SUMMARY 132
4.5 MILESTONE 4 132

CHAPTER 5 ■ Domain Class Development 133

5.1 CODING THE DOMAIN CLASSES 134
5.1.1 Reusing External Legacy Code 134
5.1.2 Reusing Internal Legacy Code 136
5.1.3 Coding a Domain Class from Scratch 137
5.1.4 Adding Functionality: Constructor and Getters 138
5.2 SOFTWARE TESTING 139
5.2.1 Test Case Design 141
5.2.2 Unit Testing Frameworks 142
5.2.3 Unit Testing the *Homeroom* Domain Classes 146
5.2.4 Unit Testing the *Homebase* Domain Classes 147
5.2.5 Code Synchronization and Integration Testing 151
5.3 DEBUGGING AND REFACTORING 154
5.3.1 Debugging 154
5.3.2 Identifying Bad Smells 156
Aside: Using Software Metrics 158
5.3.3 Refactoring 159
5.4 CLIENT REVIEW AND ISSUE TRACKING 162
5.4.1 Client Review 162
5.4.2 Issue Tracking 163
5.5 SUMMARY 164
5.6 MILESTONE 5 165

CHAPTER 6 ■ Database Development 167

6.1 DATABASE PRINCIPLES 168
6.1.1 Relations and Tables 169
Table Naming Conventions 170
6.1.2 Queries 172
6.1.3 Normalization 173
6.1.4 Keys 175
6.1.5 Concurrency Control 176

6.2 DATABASE ACCESS 177
6.2.1 Connecting the Program to the Database 178
6.2.2 Table Creation and Dropping 179
6.2.3 CRUD Functions 181
Create: Inserting Rows into a Table 182
Retrieving Rows from a Table 182
Update: Altering Rows in a Table 184
Delete: Removing Rows from a Table 185
6.2.4 Database Security 185
6.2.5 Database Integrity 187
6.2.6 Adding a Database Abstraction Layer 190
6.3 DATABASE TESTING 191
6.3.1 Testing the dbShifts.php Module 191
6.3.2 Testing the dbPersons.php Module 193
6.3.3 Testing the dbBookings.php Module 195
6.3.4 Testing the dbRooms.php Module 196
6.3.5 Integration Testing: Persons, Bookings, and Rooms 197
6.4 CLIENT REVIEW AND ISSUE TRACKING 200
6.4.1 Client Review 200
6.4.2 Issue Tracking 201
6.5 SUMMARY 205
6.6 MILESTONE 6 205

CHAPTER 7 ■ User Interface Development 207

7.1 PRINCIPLES 208
7.1.1 Model-View-Controller Pattern 209
MVC Example 1: Editing a Shift in *Homebase* 211
MVC Example 2: Editing a Person in *Homeroom* 212
MVC Example 3: Editing a Stop in *Homeplate* 213
7.1.2 Linkages among MVC triples 214
7.1.3 User-Level Security 216
User Login and Password Encryption 216
User Access Levels 218
Enforcement of Access Levels 218

7.1.4 Protection against Outside Attacks 219
Avoiding SQL Injection Attacks 219
Avoiding Cross-Site Scripting Attacks 220
7.2 PRACTICE 221
7.2.1 Sessions, Query Strings, and Global Variables 221
7.2.2 Working with Scripts and HTML 223
Scripting Example 1: Editing a Shift 224
Scripting Example 2: Managing a Sub Call List 226
7.2.3 Reading Deeply 227
7.2.4 Using JavaScript and jQuery UI to Improve the User Interface 231
7.2.5 Responsive User Interfaces 234
Responsive user interface design 236
7.3 TESTING, DEBUGGING, AND REFACTORING 238
7.3.1 Testing a User Interface 240
Organizing the Testing Process 243
7.3.2 Refactoring: Removing a Layering Violation 243
7.4 ADDING A NEW FEATURE: ALL LAYERS IMPACTED 246
Changing the Edit Person MVC Triple 247
Changing the Search for Persons MVC Triple 248
Changing the Schedule Person MVC Triple 249
Changing the Edit Shift MVC Triple 250
Changing the Sub Call List MVC Triple 251
7.5 CLIENT REVIEW AND ISSUE TRACKING 252
7.5.1 A User Interface Bug 253
7.5.2 A Multi-Layer Bug 256
7.6 SUMMARY 258
7.7 MILESTONE 7 259

CHAPTER 8 ▪ Preparing to Deploy 261

8.1 TECHNICAL WRITING 261
8.1.1 Writing for an Audience 262
8.1.2 Standards for Writing Quality 264
8.2 USER DOCUMENTATION 267
8.2.1 User Manuals, FAQs, and Demo Versions 267
Example: Firefox User Manual 269

Example: OpenMRS FAQ and Demo 270
Example: Homebase Demo 270
8.2.2 On-Line Help 271
8.2.3 Example: Homebase On-Line Help 273
Context-Sensitive Help 273
Help Table of Contents and Navigation 274
Help System Architecture 275
8.3 OTHER USER SUPPORT 278
8.3.1 User Training 278
8.3.2 Feedback Surveys 279
8.3.3 Final Presentations 280
8.4 CLOSURE FOR STUDENTS 281
8.4.1 Self-Assessment 281
8.4.2 Leveraging the CO-FOSS Experience 281
8.5 SUMMARY 282
8.6 MILESTONE 8 282

SECTION III Deployment Stage

CHAPTER 9 ▪ Continuing the Journey 287

9.1 TRANSITIONING TO PROFESSIONAL SUPPORT 287
9.1.1 The Hand-Off 288
9.1.2 Case Studies 289
Homebase Hand-Off and Support 289
RMHP-Homebase Hand-Off and Support 289
Homeroom Hand-Off and Support 290
Homeplate Hand-Off and Support 290
BMAC-Warehouse Hand-Off and Support 290
9.2 PROJECT EVALUATION AND CODE RELEASE 291
9.2.1 Potential New Clients 291
Volunteer and Resource Scheduling 291
Food Rescue and Redistribution 292
Agricultural Operations 293
9.2.2 Licensing Choices 293
9.2.3 Project Hosting Alternatives 294
GitHub 294

GitLab 294
Bitbucket 295
SourceForge 295
9.2.4 Maturity Assessment 296
9.3 SOFTWARE MAINTENANCE AS A COMMUNITY ACTIVITY 298
9.3.1 Fixing Bugs: A Case Study 298
User-Developer Discussion 299
Debugging Activities 299
Developer-Developer Discussion 301
Closure 303
9.3.2 Software Maintenance: A Multi-Year Developer Perspective 304
Homebase Maintenance: 2010-2018 304
Homeplate Maintenance: 2012-2018 305
Homeroom Maintenance: 2013-2018 306
BMAC-Warehouse Maintenance: 2015-2018 307
RMHP-Homebase Maintenance: 2015-2018 308
9.4 CREATING A FORUM 308
9.4.1 Example: Wordpress Support Forums 309
9.4.2 Example: Firefox Forums 311
9.4.3 An Example Forum Exchange 312
9.5 EVOLVING INTO A DEMOCRATIC MERITOCRACY 312
9.5.1 Incubation 313
9.5.2 Organization 314
9.5.3 Task-Specific Roles 316
9.5.4 Oversight 317
9.5.5 Decision Making and Conflict Resolution 318
9.5.6 Domain Constraints 319
9.5.7 FOSS Project Foundations 320
9.6 SUMMARY 320
9.7 MILESTONE 9 321
9.8 ENDING THE JOURNEY 321

BIBLIOGRAPHY 323

INDEX 327

List of Figures

1	The Triad.	xxxi
2	The Three Stages of CO-FOSS Development.	xxxi
1.1	The serial (waterfall) software development model.	4
1.2	An agile software development cycle.	5
1.3	The CO-FOSS software development model.	6
1.4	Relationships among common FOSS licenses.	12
1.5	Stand-Alone Computing.	19
1.6	Client-Server Framework.	20
1.7	Cloud Computing Framework.	20
1.8	Life cycle of a bug, from Bugzilla documentation, p 9.	28
2.1	RMH guest referral form (prior to 2011).	46
2.2	RMH guest registration card (prior to 2011).	47
2.3	RMH guest room log (prior to 2011).	48
2.4	*Homeroom* use cases.	53
2.5	Layered Architecture ($\leftrightarrow$ denotes information flow and $\rightarrow$ denotes control flow).	58
2.6	Layered architecture of *Homeroom*.	59
2.7	Some of the initial domain classes for *Homeroom*.	61
2.8	dbRooms table structure in *Homeroom* database.	62
2.9	Room view screen draft for *Homeroom*.	64
2.10	Login Form for Restricting *Homeroom* Access.	64
3.1	The sandbox version: client-developer interaction.	78
3.2	Example code from *Homeroom*.	79
3.3	Output of the example code in Figure 3.2.	80
3.4	Inserting comments into the 2015 version of the *Homebase* Shift class.	81

3.5 PHP documentation generated for the 2008 version of the *Homebase* Shift class. 83
3.6 Some of the functions in the Shift class for unit testing. 84
3.7 Elements of a unit test for the Shift class. 85
3.8 Results of running the TestShift unit test. 86
3.9 Form for filling a vacancy on a shift. 87
3.10 Help screen for filling a vacancy. 88
3.11 Assignment 3 in the BMAC-Warehouse project. 96

4.1 Developing *Homeroom* with the Eclipse IDE. 116
4.2 The code synchronization problem. 119
4.3 Resolving the problem: Copy-modify-merge. 120
4.4 Git Menu Options (on right) from within an Eclipse IDE. 121
4.5 Documentation practice using indented blocks and control structures. 130
4.6 Showing the open source license notice in the user interface. 131
4.7 Displaying the open source license notice in the source code. 131

5.1 Reusable *Homebase* Code in 2008. 136
5.2 Adapting the Code for Reuse in *Homeroom* in 2011. 137
5.3 Original Booking Class for *Homeroom* in 2011. 138
5.4 Revised Booking class for *Homeroom* in 2013. 139
5.5 Room class constructor and getters for *Homeroom*. 140
5.6 Test Suite in the *Homeroom* `tests` Directory. 143
5.7 Results of running a Test Suite. 143
5.8 A Unit Test for the Room Class in *Homeroom*. 144
5.9 Reporting a Unit Test Failure. 145
5.10 Setter Functions for the Room Class in *Homeroom*. 146
5.11 Partial unit test for the Booking Class in *Homeroom*. 148
5.12 The 2013 unit test for the Shift class. 149
5.13 The 2015 unit test for the Shift class. 150
5.14 New ApplicantScreening Class Added to *Homebase* in 2015. 151
5.15 New ApplicantScreening Unit Test added in 2015. 152
5.16 Interdependencies among Classes for Integration Testing. 153
5.17 A Recent GitHub Issue List for the *Homeplate* Project. 155
5.18 Example bad smell—duplicate code. 156
5.19 Example bad smell removal. 157

5.20 Searching the code base for all references to the `get_address` function. 160

6.1 A few rows in the dbDates table. 170
6.2 *Homebase* Shift class instance variables. 171
6.3 Attribute names and types in the dbShifts table. 172
6.4 The entries in the dbShifts table for August 6, 7, and 8, 2018 in Portland. 172
6.5 Connecting to the *Homebase* database. 179
6.6 Template for MySQLi table creation. 180
6.7 Creating the dbDates table in the *Homebase* database. 181
6.8 The phpMyAdmin tool for managing a MySQLi database. 181
6.9 Deleting a date from the dbDates table. 188
6.10 Retrieving a person from the dbPersons table in *Homeroom*. 189
6.11 A unit test for the dbShifts module. 192
6.12 Instance variables for the Person class in *Homeroom*. 193
6.13 A unit test for the dbPersons module. 194
6.14 Instance variables for the Booking class in *Homeroom*. 195
6.15 Portions of a unit test for the dbBookings.php module. 196
6.16 Instance variables for the Room class in *Homeroom*. 197
6.17 A unit test for the dbRooms.php module. 198
6.18 An integration test for dbPersons.php, dbBookings.php, and dbRooms.php. 199
6.19 The first 6 issues posted for the 2015 *Homebase* project. 201
6.20 Simple framework for posting a new issue. 202
6.21 Form for posting a new issue on a GitHub project. 203

7.1 The Model-View-Controller pattern. 210
7.2 The Edit Shift view in *Homebase*. 211
7.3 The Person Edit view in *Homeroom*. 212
7.4 The Stop view in *Homeplate*. 213
7.5 The main menu views in (a) *Homebase*, (b) *Homeroom*, and (c) *Homeplate*. 214
7.6 Part of the view and controller for the main menu MVC in *Homebase*. 215
7.7 The View and Controller for the *Homebase* login form. 217
7.8 Ensuring security in *Homebase* using `$_POST` and `$_SESSION` variables. 219

7.9 Controlling navigation using $˙POST variables. 224
7.10 Excerpts from editShift.php view and controller module. 225
7.11 Underlying view and controller for managing a SubCallList. 227
7.12 Using the SubCallList form. 228
7.13 Code snippet for removing a person from a Shift. 230
7.14 Essential steps for deleting a Shift from the dbShifts table. 231
7.15 Essential steps for inserting a Shift into the dbShifts table. 232
7.16 Coding calendar date using HTML selects. 233
7.17 Coding calendar date using a jQuery UI datepicker widget. 234
7.18 A Responsive user interface. 235
7.19 The *Homeplate Mobile* home screen. 236
7.20 A responsive user interface view. 238
7.21 HTML code underlying part of the view in Figure 7.20. 239
7.22 The Calendar view inside *Homebase* Use Case 4. 241
7.23 Layering Violation: a user interface module directly querying the database. 244
7.24 Layering Violation fixed and bad smell removed. 246
7.25 Showing a person's status in the Edit Person view. 248
7.26 Coding to show a person's status in the Edit Person view. 248
7.27 Updating a database entry with the new status field. 249
7.28 Searching for "applicant" status. 249
7.29 Search results for status = "applicant". 250
7.30 searchPeople.php code for selecting a person's type. 250
7.31 Listing only "active" volunteers when filling a vacancy. 251
7.32 Changing editMasterSchedule.php to list "active" volunteers. 251
7.33 Selecting only active volunteers for filling a calendar vacancy. 252
7.34 Code for selecting only active volunteers. 252
7.35 Issues 7-16 posted for the 2015 *Homebase* project. 254
7.36 Locating a bug in the calendar.php module. 255
7.37 The `process_edit_notes` function inside calendar.inc. 257
7.38 Locating a bug in the dbDates module. 258

8.1 First page of the *Firefox user manual*, including Help link. 269
8.2 The Introductory OpenMRS FAQ List. 270
8.3 The OpenMRS on-line demo. 271
8.4 The Homebase on-line demo. 272
8.5 Context-sensitive help for the search page. 273

8.6 The first two steps in the **Searching for People** help page. 274
8.7 Enlarged thumbnail in **Step 2** of **Searching for People**. 274
8.8 The on-line help table of contents in *Homebase*. 275
8.9 Integrating help pages within the code base. 276
8.10 HTML code for Step 2 in the help file searchPersonHelp.inc.php. 277

9.1 Reproducing the bug. 300
9.2 Locating the defect. 300
9.3 Designing the fix. 301
9.4 Testing the fix: editing a person. 302
9.5 Points of access to the Wordpress forums. 310
9.6 Snapshot of the Installing Wordpress Forum. 310
9.7 Accessing the Firefox user forum. 311
9.8 Organizational levels in the *Sahana* project. 317

List of Tables

2.1 Process a Referral. 54
2.2 Overall Structure of a Design Document. 67

3.1 A few PHPDoc Tags and their Meanings. 82
3.2 Example Course Syllabus Schedule: Spring 2015 Semester. 94

4.1 Overall Structure of the *Homeroom* Code Base. 127

6.1 Relations in SQL Queries. 173
6.2 Redesigning the dbShifts table to improve normalization. 175
6.3 Programming Language Database Extensions for SQL. 178
6.4 Common Attribute Types in MySQLi Tables. 180

7.1 CRUD Functions in the dbShifts module. 230
7.2 The three views in the Editing the Calendar use case. 240
7.3 MVC steps for adding a new feature. 247

8.1 *Homebase* User Questionnaire and Results. 279
8.2 Agenda for a Final Presentation. 280

List of Tables

[illegible]

Foreword

Client-Centered Software Development: The CO-FOSS Approach provides a much needed guide and resource for undergraduate software development or capstone courses that seek to engage students in a real-world software-development project.

Such a course offers unique and daunting challenges. As someone who has taught such a course intermittently over my 30+ year career, the goal was always to give students a real sense of what software development is like. But the challenges are many. How do you identify a project that can be done and done well in a 14-week semester? How do you manage teams of undergraduate CS majors, with different skill sets and motivations? What combination of platforms and software tools can be used effectively under such constraints? How do you evaluate student effort and contributions? What happens to the "product" once the semester ends? These are just some of the issues.

In this book, Allen Tucker has laid out a well-tested and practical model for addressing these challenges. The development approach is called *CO-FOSS*, which stands for *client-oriented* software development using *free and open source software*. The class project involves developing and deploying a software product for an actual client, which is typically a local non-profit organization that needs mission critical software but cannot afford to hire a professional software-development company. The software platform and tools used in the project are all freely available and openly licensed. The book is full of instructive examples that cover all of the parts and stages of a substantial software-development project. It ends with a practical and innovative model for supporting the student-built product after the semester has ended. This is very important – many of the software-development projects one finds in undergraduate courses end up sitting on shelves.

It's great to see the evolution of the type of FOSS-development course that this book describes. Ten years ago or so, I and other faculty tried to organize such courses under the banner of the *Humanitarian Free and Open Source Software* project (HFOSS). The idea then was to get students involved with existing FOSS projects, particularly those that served "humanitarian" purposes. The goal was to teach students about FOSS development – something that was not typically part of the CS curriculum at the time – by getting them engaged as contributors to some real FOSS projects. We collaborated with the Sahana project (a disaster management system), OpenMRS (a medical records system), GNOME (Linux-based accessibility software), TOR (privacy-based browser software), the Mozilla project, and others. While we had many successes, and while many students made significant contributions to these projects, the logistics of managing collaboration with such projects within a one-semester course proved difficult. The CO-FOSS model addresses the challenges that the HFOSS approach faced in creative and practical ways. This book shows that you really can get students involved in meaningful FOSS development in a one-semester course.

The book is organized into three main parts. The *Organization Stage* section is written primarily for the instructor and provides practical advice on identifying a client and creating a plan for a doable software product that would help that client, as well as constructing the syllabus for the course. A key part of the syllabus is a carefully thought-out sequence of milestones that, if followed, will lead to successful completion of the project.

The *Development Stage* section is written primarily for students and is meant to be read and followed during the semester. It concisely covers all of the main elements of software development with numerous practical examples: creating development teams, object-oriented design, database design, user-interface design and development, software documentation, and support. Among other things, this section has brief but authoritative discussions of:

- FOSS licensing
- The LAMP, MAMP, and WAMP server stack – i.e., Linux, Apache, MySQL, and PHP
- Software hosting (e.g., GitHub) and issue tracking
- Communication software such as Skype and Slack
- Creating and using unit tests for all parts of the software
- Principles of Model-View-Controller design
- Effective debugging tools and strategies
- IDEs for various programming languages, including PHP, Python, Ruby, and Java
- Database essentials, including normalization and CRUD functions
- Principles of software security
- Writing useful documentation and user-help features

Each of these topics is supported with helpful examples, including many code snippets, taken from successful CO-FOSS projects that Allen and others have conducted at various undergraduate institutions, including Bowdoin College, Dickinson College, University of New Hampshire, Whitman College, and others. Importantly, the projects created at these schools are hosted on

GitHub, and available to be used as models or even templates, depending on the type of software product a client needs.

The *Deployment Stage* section describes how to transition the product from the classroom to professional support so that the product can live on. An important feature of this section is the role played by the *Non-Profit FOSS Institute (NPFI)*, an organization started by Allen that provides help in identifying and supporting professionals who can realistically be expected to host the software and manage its ongoing debugging and support. When no such professionals can be found, NPFI shoulders some of these tasks itself. This is an incredibly powerful resource, which has the potential to make all the difference between a class project that dies once the class ends and a software product that truly adds value to the client's mission.

Some other important features of the book include:

- Milestones: Each of the nine chapters include a short list of milestones. These serve both as a means of keeping the project on track, and also as assignments that can serve as the basis for evaluating student work.
- Course organization: In addition to providing a template that can be used to model the course syllabus, the book provides helpful ideas on how to evaluate student work. Like other parts of the book, these have the benefit of being based on courses that have tried and tested many of the ideas in the book.

Designing and implementing a software-development course in an undergraduate CS program can be intimidating. It exposes the instructor to risks not found in other courses: Will he or she be able to manage the relationship with the client? Will the students be able to create a quality piece of software, and will they see it as an important education experience? Will the instructor receive credit for taking such risks and going beyond the usual course expectations? On this last point, it is worth noting that more and more schools seem to be encouraging "community involvement," and many have set up centers designed to serve as interfaces between town and gown. The model described in this book would fit in well with such institutional initiatives.

This book provides a workable model that helps mitigate some of these worries. The projects used as examples throughout the book serve as a proof of concept for what can be done, and the book itself serves as a step-by-step guide to getting it done. If you are considering an undergraduate software-development course that teaches the principles of FOSS development, you won't go wrong by starting with this book.

Ralph Morelli
Professor of Computer Science (Emeritus)
Trinity College
Hartford, CT
April 20, 2019

Preface

Software development is a complex and dynamic field. Its complexity appears in many forms – the sheer variety of software clients and applications, the rapid evolution of software development tools, the wide range of skills among professional developers, the rapid evolution of computing platforms, and the diversity of strategies used to develop the software itself.

This book is about one particular strategy for software development called the "CO-FOSS approach." The term CO-FOSS is short for "Client-Oriented Free and Open Source Software." A project using the CO-FOSS approach aims to develop a customized software product for a single client, either from scratch or (more likely) by reusing open source components from prior projects.[1]

The client for a CO-FOSS project is typically a non-profit humanitarian, educational, or public service organization, such as a Ronald McDonald House, a local school system, a Habitat ReStore, a food distribution organization, or a senior center. The key here is that the client has a genuine need for new software that will streamline a mission-critical operation, such as volunteer calendar scheduling, inventory management, donation tracking, or room scheduling.

The CO-FOSS approach has been evolving since 2008. It has been used in intermediate and capstone undergraduate computing courses where teams of students learned the principles of software development while they gained practical experience implementing a real-world software product. The key distinction for CO-FOSS in this setting is that the software product itself is real: the students are developing it for a real client, so both the risks and the rewards are high in comparison with a more traditional software development course with no real product.

Organizing such a course requires an unusual effort by the instructor. Because some of this effort may be unrewarded by typical institutional measures for excellence in teaching, the instructor must view the benefits of taking this "outside the box" approach as worthwhile. Additional support for making this

[1] The term "CO-FOSS" was coined in a 2014 study by MacKellar, Sabin, and Tucker [25], which discusses the results of using this approach in courses at three different types of institutions. The original idea of "client-oriented FOSS" was presented in a 2011 book by Tucker, Morelli, and de Silva [43], where it was contrasted with the idea of "community-oriented FOSS." While both ideas engage students with FOSS development, the latter creates a more generic product that is not customized for a single client.

extra effort may come from the instructor's home institution or from various outside sources such as the `Non-Profit FOSS Institute`.[2]

Our experiences with the CO-FOSS approach have yielded the following benefits:

1. Undergraduate computing students are uniquely motivated by community service experiences that are embedded within their formal academic training. Uniformly, they report great satisfaction when using their computing skills to develop software that serves the larger community (e.g. the page `https://npfi.org/student_evaluations/` shows the complete student evaluations for the software development course taught at Whitman College in 2015).

2. Client organizations benefit by receiving free customized software that directly supports their mission-critical activities. For example, the Ronald McDonald House in Providence, RI received volunteer database and scheduling software called *Homebase* developed by a 5-student team in that 2015 Whitman College course (see `https://npfi.org/projects/the-rmhp-homebase-project/`). That software is still in use today.

3. The fact that a CO-FOSS product is free and open source software allows any of its source code to be reused and refitted to suit the needs of a future project. For example, the *Homebase* software mentioned above was adapted from `an earlier version` developed in 2013 by Bowdoin College students for the Ronald McDonald House in Portland, ME. Thus, an open source license like the GNU General Public License or the Mozilla Public License is an essential element of the CO-FOSS approach.

4. Students gain experience learning about key elements of the software development process, including coding, testing, refactoring, and writing user documentation, as they would in a conventional software development course. However, these students also gain practical experience by working within a team, communicating with a client, using a client-centered development model, sharing a code base, and reusing legacy code – experience that prepares them well for entry into the modern software industry.

This book aims to provide instructors, students, clients, and professional software developers with a roadmap to guide them through the development of a new CO-FOSS product from conceptualization to deployment. We use

[2]Throughout this book, any word or phrase that appears in `typewriter font` represents a link to a Web page that provides more details. Of course, those links work only for the e-book version. Readers using the print version should be able to locate most of these Web pages by doing a Google search for that word or phrase.

our own experiences with this approach to illustrate each step in the process, detailing its technical elements, its methodologies, and its outcomes.[3]

The CO-FOSS approach views each project as having three connected elements that form a *triad*, as pictured in Figure 1. The *student team* is one element of the triad, the *client* is the second, and the *professional developer* is the third. The goal of a triad is to design, implement and deploy a customized software product that supports a specific mission-critical activity of the client.[4]

The instructor is involved in all three elements of a CO-FOSS project. The instructor organizes it, leads the student team through project development, and delivers the completed software product to the professional. The students, who are intermediate-level computing majors, develop the software using both the requirements document and a client-centered approach. The professional installs the completed software on the client's server, and then provides ongoing support thereafter.[5]

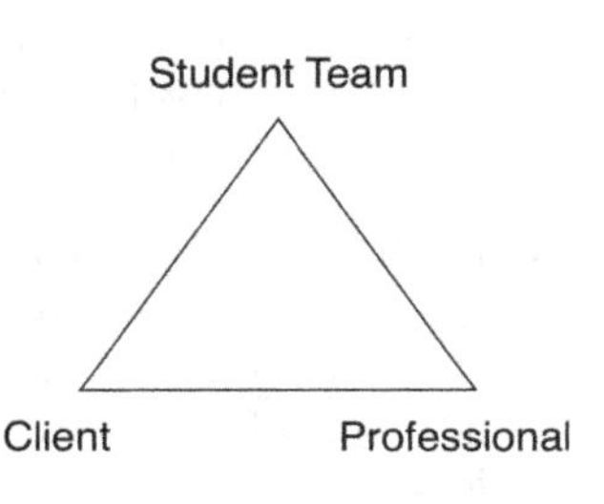

FIGURE 1 The Triad.

The project has three stages: a 2-month *organization stage*, a 3-month *development stage*, and a 1-month *deployment stage*, as shown in Figure 2.

Month 1	2	3	4	5	6
Instructor/Client		*Instructor/Students/Client*			*Instructor/ Professional*
Organization Stage		Development Stage			Deployment Stage

FIGURE 2 The Three Stages of CO-FOSS Development.

So the instructor provides the glue that holds these three stages together, as outlined below:

[3]All the examples in this book use the PHP/Javascript/MySQL/HTML platform, since that is the platform on which our own CO-FOSS projects were built. So while instructors may find the organizational aspects of this book to be useful, this book may be supplemented by reference materials that uses a different platform, such as Django or Rails.

[4]Absent the student team, a CO-FOSS product can always be designed, implemented, and deployed by a professional developer working directly with the client.

[5]Because the requirements and the design document are prepared during the organization stage, the course itself should be properly labeled "software development" rather than "software design."

1. During the *organization stage*, the instructor identifies the client's mission-critical software need, and then develops the requirements for a software product that will fulfill that need. This includes eliciting an initial set of use cases from the client, developing an initial design document and course syllabus that has an implementation timeline embedded, and forming the student team.

2. During the *development stage*, the instructor, student team, and client representative use a client-centered process to create a software product that fulfills the requirements. That is, in a 1-semester course, the team iteratively develops the product and refines the requirements, taking into account the client's feedback at each iteration.[6]

3. During the *deployment stage*, the instructor turns the product over to the developer who installs the product on the client's server (website). Here, the developer and the client also collaborate to iron out any lingering issues for the product and agree on a long-term support plan going forward.[7]

To introduce the CO-FOSS approach, this book has an introductory chapter followed by three Parts. The introductory chapter provides an overview of software, open source licensing, and major software development methodologies. Everyone should read this chapter first and complete **Milestone 1** at the end of the chapter before continuing.

Each Part thereafter explores one of these three stages by sharing our knowledge of designing, developing, and deploying a CO-FOSS product using the *triad* as an organizational framework.

> Part I is written mainly for the instructor and secondarily for the client. It covers the details of finding a client and a CO-FOSS product to be developed, defining that product's requirements, and organizing a course in which students can develop the product. The instructor should complete **Milestones 2** and **3** before continuing to Part II.
>
> Part II comprises the bulk of the book and is written mainly for the instructor and the students. It covers the principles and practice of client-centered software development, with many examples from CO-FOSS projects that our students have completed in recent years. The chapters

[6]We know of several CO-FOSS courses that spread the project's development stage over two semesters, either with the same group of students or with two different groups of students. In one case, two different groups of students worked on the same project in successive offerings of the same course. In another case, the same group of students worked on the project over a unified 2-semester capstone "Software Projects" course. Both of these approaches are viable when the institutional setting allows that flexibility.

[7]The recent rise of Web application hosting services, often called `''platform as a service"` or `PaaS`, may reduce or eliminate the need for a professional developer to be involved in the deployment stage. In this case, the instructor should be willing to maintain the software after the project has been deployed.

in Part II should normally be taken in order, and each chapter's own **Milestone** should be completed before continuing to the next chapter.

Part III is written mainly for the instructor and the professional developer, providing guidance on deploying a new CO-FOSS product, supporting it, and disseminating it to the larger open source community. The last **Milestone** appears at the end of this chapter and its completion signals completion of the entire project.

The Table of Contents shows how the chapters are laid out in each of these three Parts. Of course, the devil is in the details, so let's get started!

Acknowledgments

The CO-FOSS approach to software development is people-intensive. I am fortunate to have worked with many extraordinary people who have contributed to the CO-FOSS projects described in this book. I gratefully acknowledge:

The student developers at Bowdoin and Whitman Colleges, for their willingness to make the connection between academic work and community service by completing these projects successfully: Adrienne Beebe, Hartley Brody, James Cook, Johnny Coster, Moustafa El Badry, Felix Emiliano, Connor Hargus, Jerrick Hoang, Richardo Hopkins, Noah Jensen, Phuong Le, Alex Lucyk, Dylan Martin, Ruben Martinez, Nolan McNair, Jackson Moniaga, Jesus Navarro, Luis Munguia Orta, Maxwell Palmer, David Phipps, David Quennoz, Oliver Radwan, Sam Roberts, Luis Rojas, Taylor Talmage, Xun Wang, Nicholas Wetzel, Ivy Xing, and Judy Yang.

The non-profit clients, for providing real-world settings in which the software could be developed, customized, and deployed: *The Blue Mountain Action Council of Washington* (Kathy Covey and Jeff Mathias); *Ronald McDonald House Charities of Maine* (Gabrielle Booth, Robin Chibroski, Georgia Doucette, Whitney Linscott, Ashley MacMillan, Alicia Milne, Gretchen Noonan, Karla Prouty, and Raymond Ruby); *Ronald McDonald House Charities of Rhode Island* (Susan Czekalski, Michelle LePage, and Joanna Powers); and *Second Helpings of South Carolina* (Bruce Algar, Lili Coleman, and Jon Peluso).

The professional developers, for supporting the software on the clients' servers: Artopa LLC (David Tripp), Coursevector LLC, The Non-Profit FOSS Institute, Pragmatics, Inc. (Dr. Long Nguyen), and Vivio Technologies, Inc.

My faculty colleagues, for helping me understand the challenges of teaching FOSS development as an academic and humanitarian enterprise: Jim Bowring, Grant Braught, Janet Davis, Greg Hislop, Steve Huss-Lederman, Bonnie MacKellar, Craig McEwen, Ralph Morelli, and Mihaela Sabin.

The reviewers of this manuscript, for providing a wealth of conceptual and detailed suggestions for improving it: Jim Bowring, Janet Davis, and Steve Huss-Lederman.

Dr. Jennifer Tucker, for helping me develop the idea of the Non-Profit FOSS Institute, and then serving as its first Executive Director. And finally my wife, Meg, for her lifelong commitment to education and humanitarian volunteerism, and especially for introducing me to the first CO-FOSS client at the Ronald McDonald House in Portland, ME in 2007.

Allen B. Tucker, February 2019

About the Author

Allen B. Tucker is the Anne T. and Robert M. Bass Professor Emeritus in the Department of Computer Science at Bowdoin College. He held similar positions at Colgate and Georgetown Universities. He is currently a professional software developer and President of the Non-Profit FOSS Institute (NPFI), a 501(c)(3) organization that supports the development of free open source software for non-profits by students and professionals.

Allen earned a BA in mathematics from Wesleyan University and an MS and PhD in computer science from Northwestern University. He is the author or coauthor of several books and articles in the areas of programming languages, software design, natural language processing, and computer science education. He co-authored the 1986 Liberal Arts Model Curriculum in Computer Science, served as Editor-in-Chief of the *Handbook of Computer Science*, and co-authored the textbooks *Programming Languages* and *Software Development*. He also served as Fulbright Lecturer at the Ternopil Academy of National Economy in Ukraine, a visiting Erskine Lecturer at the University of Canterbury in New Zealand, a Visiting Lecturer at ESIGELEC in France, and a Visiting Professor at Whitman College.

Allen has been a member of NSF's CISE Advisory Board, the Association for Computing Machinery (ACM), the IEEE Computer Society, Computer Professionals for Social Responsibility, and the Liberal Arts Computer Science (LACS) Consortium. In 1991, he received the ACM Outstanding Contribution Award and shared the IEEE Meritorious Service Award. He is also a Fellow of the ACM and a recipient of the ACM SIGCSE Award for Outstanding Contributions to Computer Science Education.

From 2008 to 2012, Allen served on the Advisory Board for the NSF CPATH grant that supported Trinity, Wesleyan, and Connecticut College's Humanitarian Free and Open Source Software (HFOSS) initiative. That experience inspired him to begin engaging his own Bowdoin students in HFOSS and developing a curricular model called CO-FOSS (client-oriented FOSS) with his colleague Ralph Morelli at Trinity College.

From 2008 to 2015, he taught several software-development courses at Bowdoin and Whitman Colleges using the CO-FOSS model with different student teams. As a byproduct of this work, he developed strong working relationships with non-profits such as the Ronald McDonald Houses in Maine and Rhode Island, and food distribution organizations in South Carolina and Washington.

In 2013, with the belief that the CO-FOSS model would be viable in a large number of undergraduate settings, Allen co-founded NPFI. NPFI's mission is to spread the development and deployment of open source CO-FOSS products, teaching methods, grants, and other resources to other computing faculty, so that they can engage their own students with real-world HFOSS development for many more non-profit organizations in the future.

CHAPTER 1

The Journey

"Change your opinions, keep your principles;
Change your leaves, keep intact your roots."
—*Victor Hugo*

THis chapter provides an overview of software — its nature, its development models, its licensing alternatives, its architectures, and its maturity. Thus it offers a useful perspective within which the development of a new software product can be viewed.

CO-FOSS is a model for developing new software. It is a particularly valuable model, both for learning about the software process and for developing an actual software product for a real client. To provide a broader context, Section 1.2 discusses three different software development models: the serial model, the agile model, and the CO-FOSS model.

Fundamental to CO-FOSS development is the free and open source license that accompanies the software itself. Without such freedom, CO-FOSS development would not be possible. This idea is discussed more carefully in Section 1.3.

A key characteristic of any software product is its underlying architecture, or organization. A coherent architecture is always an essential component of all but the most simple software products. Section 1.4 introduces the client-server family of software architectures that underly the organization of many CO-FOSS products.

Different software products also vary in their maturity. The idea of CO-FOSS applies mainly to the development of new software, often from pre-existing components. However, most software is more mature, having evolved through various levels of maturity over its lifespan. Section 1.5 looks at this larger temporal context in which CO-FOSS development lies.

1.1 SOFTWARE

Simplistically, "software" can be viewed as all the programming in a computer that is not hardware. But the very idea of "software" is a complex one. Even the software on a single computer exists at two distinct levels – the operating system/network level (think Linux, Windows, MacOS, or Apache Server) and the application level (think Microsoft Office, OpenOffice, the Chrome browser, or the Google Maps application).

Professional software developers have skills that reflect the level and type of software that they develop. For example, Linux and network software developers work at the operating system/network level. Their skills allow them to work with such tools and techniques as C programming and process synchronization.

Other professional software developers work at the application level, such as Web programming or database design, which requires a different set of skills. The application level alone spans a wide range of distinct areas, each of which has its own community of developers. Here's just one taxonomy of software application areas that appears in `Wikipedia`:

> **Information systems software** supports corporate payroll, accounting, and inventory management, and individual word processing, spreadsheet, and visual presentation needs.
>
> **Entertainment software** includes video games, mobile games, and social networks.
>
> **Educational software** includes course management, survey management, and language learning support.
>
> **Enterprise infrastructure software** includes project management, database systems, document management, and content managed websites.
>
> **Simulation software** simulates social networks, battlefield scenarios, airline flight control, and vehicle driving control.
>
> **Media development software** includes computer graphics and animation, graphic art, image galleries, audio and video editing, and digital music generation.
>
> **Product engineering software** includes compilers, interpreters, virtual machines, computer aided design tools, integrated development environments (IDEs), version control systems (VCS), and debuggers.

How big is the software industry? The number of professionals in this industry is large and growing. `A recent study` estimated that there were 21 million software developers worldwide in 2016. Of those, nearly 4 million worked in the United States, and they comprised 2.5% of the total US workforce. At the same time, `demand for software professionals` greatly exceeds supply, creating a favorable job market for new developers who are completing computer science, IT, and computer engineering degree programs.

1.2 SOFTWARE DEVELOPMENT MODELS

Software is also complex in the sense that a software product can be developed using different methodologies, or "development models." On the one hand, it can be developed serially, starting from a fixed set of requirements, proceeding to a design specification, followed by writing the code and finally testing the code. On the other hand, it can be developed from the "bottom up," starting with a small prototype and incrementally adding new requirements and functionality with each iteration.

Additionally, some software can be developed as a generic product for a large (real or imagined) market, while other software can be developed as a customized product for a single client. The former approach is potentially more profitable, while the latter approach is useful for an organization that has unique software needs that are unmet by commercially-available software.

Finally, software can be developed from scratch (sometimes called a "greenfield" project), or it can be developed incrementally using pieces of code borrowed from other software with similar features (a "brownfield project").

This section briefly addresses three different software development models, their constraints, and their tradeoffs.

1.2.1 Serial Development

The serial approach to developing software originated as the so-called "waterfall model," and it was the predominant approach to developing software throughout the 1970s and 1980s. It is based on the assumption that a software product's functional requirements can be fully specified at the outset, and that subsequent stages in the development process can be carried out more-or-less serially. These stages are called "requirements analysis," "design," "coding," "testing," and "delivery."

Each stage in this process is viewed as a single discrete event. One stage typically does not begin until the previous stage is completed. Typically, the client is involved in the beginning and ending stages, but not in the crucial middle stages. This is illustrated in Figure 1.1.

If the requirements can be fully specified at the outset, the serial model can work. For example, an embedded software module that measures and reports the altitude of an airplane in real time can be designed and implemented using this model.

However, this serial approach to software development has had a poor record of success in completing software products for customers. For example, the 2015 Chaos Report [19] surveyed 50,000 software projects around the world to learn how well they met the following three criteria:

1. completed on time,
2. completed on budget, and
3. completed with all features implemented.

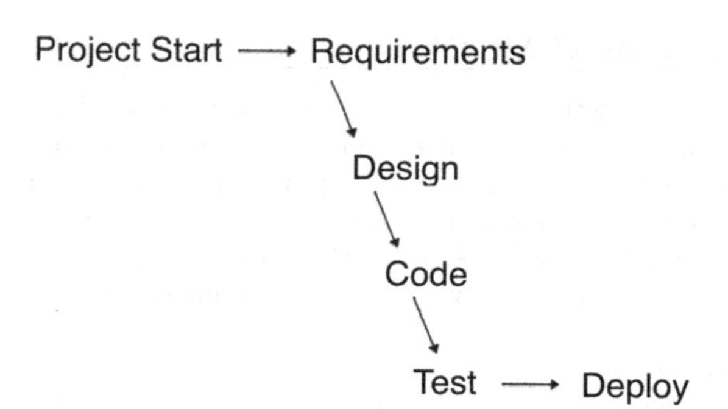

FIGURE 1.1 The serial (waterfall) software development model.

The Chaos Report found that only 11% of all projects using the traditional serial model met all three criteria, while 60% were "challenged" (that is, they were completed but did not meet all three criteria), and the remaining 29% failed (that is, they were never completed).

So in many situations, the serial development model does not work well. Its main problem lies in the assumption that the requirements of a software product can be fully specified at the outset, and that those requirements will not vary throughout the development process. In reality, various outside factors (such as changing user needs or the emergence of a new computing platform) can alter the requirements. For example, the 2015 Chaos Report [19] confirmed that not incorporating end users' feedback throughout the development process was a frequent cause for software project failure.

1.2.2 Agile Development

Since the 1990s, and in response to these problems, software development methodologies have been gradually evolving away from the serial model. Newer methodologies known as "rapid application development," "dynamic systems development," "scrum," "extreme programming," and "feature-driven development" have been shown to be more effective in settings where changing user requirements or computing platforms had become the norm. These newer methodologies all led to the 2001 publication of the *Manifesto for Agile Software Development* [5], which crystallized them into a coherent statement of principles and a development model.

In recent years, the agile model and its variants have yielded significant improvements over the traditional serial model. For example, the same 2015 Chaos Report [19] found that 39% of all projects that used the agile model met all three of the above criteria for success, while 52% were challenged and only 9% failed.

The main reasons for its improved success rate are explained by the nature of the agile process itself. In an agile project, the software product starts with a minimal set of requirements and iterates several times through a 6-stage development cycle, as pictured in Figure 1.2. The process is fluid, in the sense that each cycle improves the requirements and develops new code in response to client feedback from reviewing the results of the previous cycle.

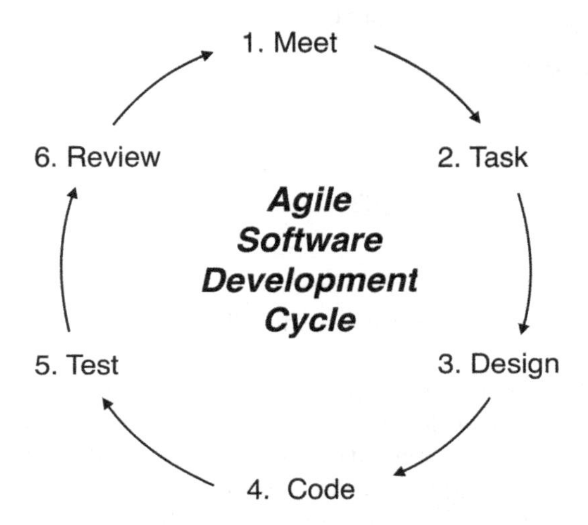

FIGURE 1.2 An agile software development cycle.

Let's look at some of the details in the agile cycle. In stage 1, the developers **Meet** with the client and discuss the client **Review** of the partially-completed software from stage 6 of the previous cycle. In stage 2, the developers and client assume new **Task**s for making progress by adding new functionality and incorporating the client's feedback from stage 1. Developers then independently complete their respective **Design**, **Code**, and **Test** stages, thus preparing the next version of the partially-completed software for client **Review**.

1.2.3 CO-FOSS Development

The CO-FOSS model for developing software is a hybrid of the serial and agile models for software development. It borrows pieces from both, as summarized in Figure 1.3.

To enable students to develop a useful piece of software for a single client in one semester, the software **Design** is organized by the instructor and the client before the semester begins, which is reminiscent of the serial model. This activity is described in detail in Chapters 2 and 3.

Then the software **Development** is completed by students and the client through a series of meet-task-code-test-review cycles. Each 1-2-week cycle is repeated 5 or 6 times throughout the semester, each repetition achieving a pre-determined milestone that ensures successful project completion.

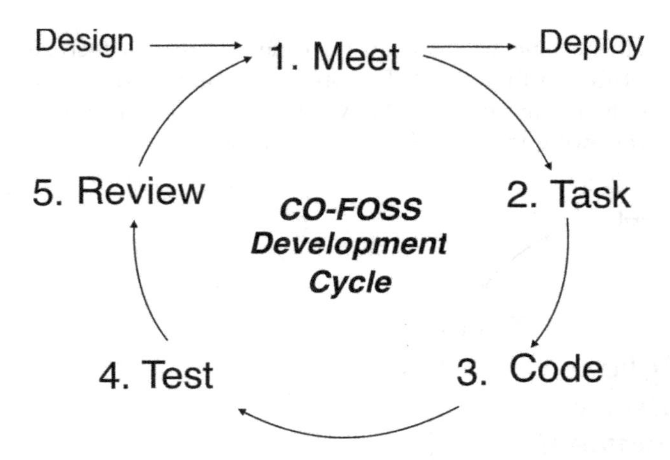

FIGURE 1.3 The CO-FOSS software development model.

The **Code** stage relies on the fact that developers work with free and open source software (FOSS). In the FOSS world, mature well-tested code can be freely downloaded for reuse in any application with a similar functional need. Thus, developers need to read and work with code written by others. Real software is less often developed from scratch by a single individual. Instead, it is usually developed incrementally by a team, each member adding and refining parts of an existing "code base" written by others.

The **Test** stage in each iteration of the cycle provides a new opportunity for *debugging* and *refactoring* the code base in preparation for client **Review** and the next iteration.

> *Debugging* means finding and correcting errors in the program. Bugs, or instances of incorrect behavior, result from programming errors. Such errors can often be notoriously difficult to find and correct, even when working with a small code base. So as an aid to finding bugs we use an aggressive strategy called "unit testing," where individual units (classes and modules) of code are individually tested at each repetition of a CO-FOSS cycle.
>
> *Refactoring* a program means reading the code, finding instances of poor programming practice (from either a readability or an efficiency standpoint), and reorganizing the code so that it performs the same functions in a more readable and/or efficient way.

Coding, testing, debugging, and refactoring are discussed in detail in Chapters 5, 6, and 7.

Clients further test and evaluate the software during the **Review** stage of each cycle. They play a key part in debugging, since they are the ones who most often identify bugs and provide feedback to developers during each cycle, ensuring that the final product meets their particular expectations. A more careful treatment of the client review process and its close relationship to the Test stage appears in Chapters 5, 6, and 7.

Finally, **Deployment** of the CO-FOSS product takes place at the end of the semester, and is coordinated between the instructor, the client, and a professional software developer. This is described in detail in Chapter 9.

1.2.4 Software Customization: A Continuum

A final consideration for software developers and clients involves the alternatives that are available in selecting/developing a software product to help improve a particular mission-critical activity within the organization. These choices form a continuum – from developing a completely customized software product to obtaining a completely off-the-shelf product, with many other choices in between. Let's take a look at the trade-offs among three key choices in this continuum:

Custom software,

Off-the-shelf software, and

Custom software with off-the-shelf components.

Custom Software

Custom software is just what its name suggests. The developer designs and implements a unique piece of software that can improve a client's mission-critical activity. The software is tailored to match all that activity's needs, processes, and security requirements. Importantly, the client's staff can assimilate that software easily because it uses existing organizational vernacular that the staff already knows.

Custom software may be *open source* or *proprietary* (see Section 1.3), but the client must rely on the developer to keep it up to date with changing organizational needs. So a strong working relationship between the developer and client is essential for custom software to remain effective. Custom open source software is ideally *client-centered*, allowing the client to be involved continuously in the development process.

Custom software is not without its downsides. First, its original development cost can be higher than the alternatives, if there are any. Second, asking the developer to add new features as requirements change may also be billable. Third, custom software has no peer user community outside the client's organization to provide advice on usability issues, though this downside is somewhat mitigated by the developer's ongoing availability.

All the software projects discussed in this book have developed custom open source products. Each one is fitted to satisfy the requirements of a single customer. For example, *Homebase* was originally developed in 2008 for the Ronald McDonald House in Portland, ME. Enhancements were made by different student teams in 2012 and 2013. A single developer returned in 2015 to add more features to *Homebase*. These results would not have occurred if *Homebase* were not *open source* and developed using a *client-centered* approach.

When weighing whether to use custom software, an organization should be sure that there is no satisfactory off-the-shelf product available that can satisfy its requirements at an affordable cost. It should also find a developer that can produce that custom software and provide ongoing support, all at an affordable cost. For example, all the software products discussed in this book were developed and are supported at no cost to their clients. However, since each of these products was developed using the CO-FOSS model, it did require client participation (averaging about 2 person-hours per week) throughout its development process.

Off-the-Shelf Software

Off-the-shelf software is a (proprietary or open source) product developed for a large number of customers. Examples include Microsoft Word, Apache OpenOffice, Google Sheets, and various smartphone- and tablet-based computer games. Off-the-shelf software is aimed at addressing a specific shared need of a mass market audience, such as the need to play a game of Sudoku on a smartphone while waiting for an airplane.

The per-user cost of off-the-shelf software can vary greatly; some products are free, others are costly, and still others are available as both free "introductory" versions and paid "full" versions. The full versions of off-the-shelf software usually come with a preponderance of features, most of which are not needed by the average user. These features are there in order to satisfy the one-size-fits-all requirement. However, their presence can make the software more difficult to learn and use.

Off-the-shelf software can be deployed quickly, usually with a simple download and install step. Another advantage of popular off-the-shelf software products is that they have large, often international, communities of users and forums that provide self-help support. So the user doesn't need to hire a developer to fix a bug or customize the software to fulfill a specific need.

On the downside, off-the-shelf software typically will not match all of an organization's needs, either lacking needed features or providing superfluous features. If customization is even possible, that typically comes at an additional cost. Routine upgrades may also come with additional costs.

Finally, off-the-shelf software can be obsolete or slow to evolve with the industry to which it is targeted. Moreover, its vernacular is invariably out of sync with the user organization's vernacular, requiring users to assimilate a new vocabulary before becoming comfortable with the software. Off-the-shelf software may also require technologies that do not conform to the organization's current computing platform. Moreover, it often comes with the subtle inability for an organization to change to a different vendor in the future when a better alternative emerges (this is sometimes called "vendor lock-in").

Custom Software with Off-the-Shelf Components

There is a middle ground between custom software and off-the-shelf software, which is becoming an increasingly popular solution for organizations. The idea of "custom software with off-the-shelf components" is that an organization finds software that matches most of its specific needs but requires a few additional functions in order to match the rest. Typically, this approach uses open source software at its core, though some of the add-on components can be proprietary as well.

A good example is `Wordpress`, which is free open source software out of the box for building websites. A Wordpress website can be customized by adding "plugins" which are modules that provide specialized functionality so that the website can provide specific functionality that matches an organization's peculiar requirements. The Wordpress plugin library is huge, and it covers a wide range of functionalities, such as membership management, on-line application form processing, and on-line product catalogs for e-commerce.

The advantages of this approach to software development are mainly that it leverages pre-existing libraries of reliable modules to help reduce up-front costs, especially those associated with writing and testing new code. Other advantages are derived from its basic open source nature: the organization owns the software and its attendant database (avoiding vendor lock-in), the software can be continuously updated to meet changing needs, and there are no licensing fees.

The disadvantages of this approach are higher upfront costs vs. off-the-shelf software and the requirement for an ongoing relationship with a developer to make changes and upgrades (which may be billable). Like custom software, this approach has no attendant user community to provide self-help (though the relationship with the developer compensates for this).

1.3 SOFTWARE LICENSING

A software product can be licensed in one of two general ways, *proprietary* or *open source*. The differences between these two types of licenses are significant, especially in regard to the software development process and environment in which the software is created and maintained.

1.3.1 Proprietary Licensing

Proprietary software is that which is licensed and sold as a binary executable program to individual and corporate customers. The source code is the private property of the developer and is kept hidden from the customer. A proprietary software license typically limits the number of computers on which a user can install the software – installing the software on more than one computer costs more money. So a proprietary license prevents the user from copying the software, modifying it, or sharing it freely with associates and friends.

From the 1970s to the mid-1980s, nearly all software was developed and sold with a proprietary license. Proprietary software is developed and maintained by an in-house programming staff of a large organization or by a vendor targeting a specific market. All developers of a proprietary software product must sign a non-disclosure agreement (known informally as an NDA) which binds them to secrecy about the product's source code and architecture.

For example, Word was developed by Microsoft's own programmers to meet the needs of the word processing market. Today it can be bought by a single user either stand-alone (for $110) or as part of Microsoft's "Office 365" bundle, a cloud-based subscription service that includes Word, Excel, PowerPoint, OneNote, Outlook, Publisher, Access, OneDrive, and Skype (for a $70 yearly subscription). The license for a single-user version of Word is a 30-page document "Microsoft Software License Terms," which spells out that the user has the right to install and use a single copy of the software on a single computer, but cannot copy it to a second computer or pass it to a friend.

Google Docs is a proprietary word processor that runs on a web server and is a free alternative to Microsoft Word. While Google Docs is less feature-rich than Word, many users prefer that because of its intuitive functionality and its interoperability with other aspects of cloud computing. Most importantly, Google Docs' cloud-based functionality supports smooth simultaneous editing of a shared document by several persons. Microsoft's cloud-based version of Word, when it is bundled within Office 365, also supports this kind of collaboration.

1.3.2 Open Source Licensing

Free and open source software (FOSS) is software whose source code and binary executable code are freely available for download by any individual or organization. Most significantly, "freely" means that downloaders are free to use the software on any computer, to modify the source code and binary, and to share the modified software with associates and friends. Because of this freedom, FOSS is accessible in markets where proprietary software has no interest and little leverage—non-profit organizations, developing countries, and individuals and businesses who are either unwilling or unable to pay the cost of purchasing proprietary software.

Most proprietary software has a FOSS alternative. For example, a FOSS alternative to Microsoft Word is called "Writer" and is part of the "OpenOffice" bundle, maintained and distributed by the Apache Foundation. OpenOffice allows any individual or organization to freely download and use it on any number of computers. It runs on Windows, Linux, and Macintosh platforms. OpenOffice is distributed under an open source license called the "Apache License Version 2.0," which describes the rights of clients to download and freely use, copy, modify, and redistribute this software. The Android operating system also carries the Apache license [24].

Besides the Apache License, three slightly different types of licenses are used for FOSS products:

`The MIT License` [27] was developed by the Open Source Initiative to provide a totally unrestricted vehicle for reworking and redistributing the source code.

`The GNU General Public License` [14], or GPL for short, was developed by the GNU Foundation to provide a vehicle for reworking and redistributing the source code, but with the caveat that any redistribution must be GPL-licensed FOSS as well. This caveat effectively keeps all derivatives of the product in the FOSS domain for other developers to freely use and refine.

`The Mozilla Public License`, or MPL for short, was developed by the Mozilla Foundation for its Firefox browser and is used by many other software products today.

Many popular FOSS products (Linux and Wordpress, for example) are licensed under the GPL, preventing them from ever being commercialized or embedded inside a proprietary product. Version 2 of the GPL was released in 1991. The GPL has been repeatedly upheld in courts around the world as an enforceable license [2]. Since 1991, a variety of FOSS licenses have evolved alongside the GPL to satisfy different needs within the open source community. GPL version 3 (GPLv3) was released in 2007 to address a wide range of issues, especially its compatibility with these other FOSS licenses.

Unlike the GPL, neither the Apache License nor the MIT license protects a software product from having one of its derivative products converted into a proprietary product and sold for profit. For example, Apple's MacOS operating system is proprietary software derived from the FOSS product BSD Unix, which carries an MIT-like (permissive) license.

The LGPL and MPL represent a middle ground between the permissive MIT license and the protective GPL license. That is, while they protect the FOSS software and its derivatives from becoming fully proprietary, they allow the software to be embedded in a larger proprietary product.

Today, there are dozens of different FOSS licenses. The Free Software Foundation's own list of other licenses cites rulings on which ones are compatible with the GPL as well as guidance on how to define customized FOSS licenses [15]. The Open Source Institute maintains a similar list as part of its effort to define *open source software* [27].

One of the most difficult questions for FOSS developers is how the various licenses relate to each other. Figure 1.4 [44] provides an overview of the more widely used licenses and their inter-relationships. Each box represents a particular kind of license.

A license is more or less *protective* depending on how strongly it protects the freedoms listed in the FOSS definition, particularly the freedom to redistribute derivatives of the software. The left-to-right arrangement of the

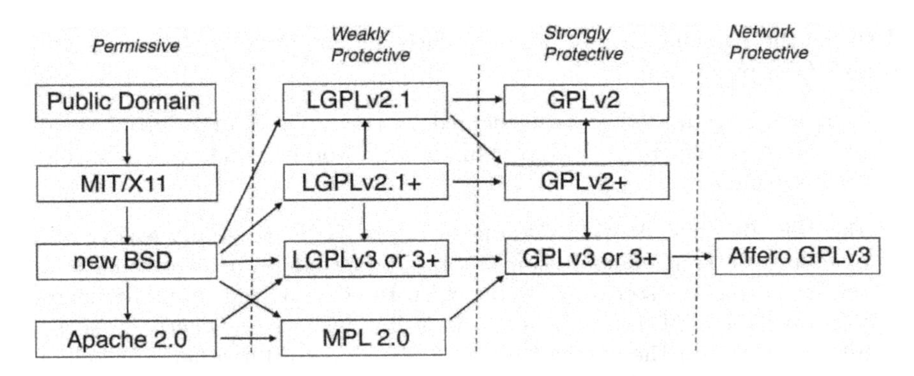

FIGURE 1.4 Relationships among common FOSS licenses.

licenses in Figure 1.4 shows how they progress from least to most protective in this sense. Here are the main distinctions among the columns in Figure 1.4:

Permissive Software in the *public domain* is, strictly speaking, unlicensed and therefore completely unprotected. Thus, someone can take a piece of public domain software and re-distribute it under another open source or proprietary license. The other three licenses in the *permissive* column also allow derivative products to become *proprietary*. For example, software with an Apache license can be turned into a proprietary product.

Weakly protective *Weakly protective licenses* are often used for source code libraries or modules. They protect the software from becoming proprietary but allow it to be used as part of a larger proprietary package. The Lesser General Public License (LGPL) and the Mozilla Public License (MPL 2.0) are the most widely used licenses of this type.

Strongly protective Licenses in the *strongly protective* column in Figure 1.4 require that derivative works must also be licensed, as a whole, under the GPL. This effectively prevents derivatives from becoming proprietary software.

Network protective The Affero GPLv3 expands the reach of the GPL so that users of a Web application can receive its source code. This also applies to network-interactive software, including programs like game servers.

The arrows in this figure represent *compatibility*, indicating where two different FOSS-licensed products can be merged into one and share a common license. To determine this sort of *compatibility* between two licenses, we trace the arrows from the two products' licenses in Figure 1.4 to a common license. For example, two software products distributed under an Apache 2.0 license and an MPL 2.0 license, respectively, can be combined into a new product and distributed under a GPLv3 license.

Licensing a complex FOSS product is sometimes a complex matter. For example, one of the most challenging issues Mozilla faced was to respect the prior licensing of Firefox's many embedded third-party modules. Agreements with the owners of these modules had to be worked out so that their code could be shipped either as open source or as binary code. The alternative would be to remove these modules from the code base altogether.

Finally, we note that all software licenses, whether they be proprietary or open source, carry some sort of warning notice that the software is provided in "as-is" condition, thus attempting to free the developer from being sued should the software cause harm or inconvenience to a client who downloads and uses it. There are many exceptions to the force of this disclaimer, such as `a 2017 lawsuit` (see `https://npfi.org/iphone-lawsuit/`) filed by iPhone users against Apple for intentionally slowing their iPhone software as it got older.

1.3.3 FOSS Origins and Impact

The free software movement was started in the early 1980s by Richard Stallman [38] and his colleagues. Stallman was a programmer at MIT's Artificial Intelligence lab and learned to program as part of the hacker culture that was thriving in much of the programming community during the 1960s and 1970s.

Having grown frustrated with the directions that the proprietary software industry was taking, Stallman started the GNU (read "GNU is Not Unix") project in 1983. This project was an effort to build an entirely free and open operating system [36]. It is clear from Stallman's original announcement about GNU that his motivations were ethical and humanitarian [18]:

> I consider that the golden rule requires that if I like a program I must share it with other people who like it. I cannot in good conscience sign a nondisclosure agreement or a software license agreement. ... I'm looking for people for whom knowing they are helping humanity is as important as money.

In 1985 Stallman founded the `Free Software Foundation` to help support this new movement. He developed the definition of free software along with the concept of *copyleft*, which uses software licensing to protect the freedom of software users and developers to share their work [5]. Under a copyleft license, *free software* guarantees users the freedom to:

1. Run the software for any purpose,
2. Study and modify the software (which requires access to the source code),
3. Distribute copies of the software to help their neighbors, and/or
4. Improve the software and release those improvements to the public so that the whole community benefits [5].

Notice that these four freedoms imply "open source" as well, especially considering items 2 and 4. So using the term *free software* as defined here is equivalent to using the term *free and open source software*, or FOSS.

Despite the ambiguity of the English word "free," Stallman's definition of *free software* has nothing to do with the *price* of the software; in his own words, it means "free as in 'free speech' not as in 'free beer.'" As a byproduct, however, most software that is licensed under this definition is also distributed free of charge.

In 1989, to help protect programs developed as part of the GNU project, Stallman created the GNU General Public License [14]. The GPL is widely regarded as the strongest copyleft license, since it requires that all derivative works be made available under the four freedoms listed in the above definition of copyleft licensing.

By 1991, Stallman and his collaborators had developed an entire UNIX-based operating system, minus the kernel program. It was in this context that Linus Torvalds, working with a broad international community of programmers, developed the Linux kernel program [3]. Linux became licensed under the GPL and became the core of the GNU/Linux operating system [42]. GNU/Linux, or Linux as it is popularly called, is one of the best and most widely known examples of FOSS.

Following the dramatic success of Linux, the Open Source Initiative (OSI) was founded with the purpose of making the FOSS development process acceptable to the software industry itself [29]. In his formulation of the *open source definition* Bruce Perens and other founders hewed closely to Stallman's principles, preserving the basic freedoms that Stallman articulated. Despite this effort, for many the OSI provided a means to distance the movement from what they saw as Stallman's anti-business stance. As a result the OSI has focused more on the practical benefits of the FOSS development model.

As open source gained popularity within the software industry, a schism developed between free software and open source proponents. Perens eventually resigned from OSI [28], saying:

> Most hackers know that Free Software and Open Source are just two words for the same thing. Unfortunately, though, Open Source has de-emphasized the importance of the [four] freedoms involved in Free Software. It's time for us to fix that. We must make it clear to the world that those freedoms are still important, and that software such as Linux would not be around without them.

However, despite the efforts of Perens and others to emphasize the moral dimension, the gap between the two branches of FOSS continued to grow. Stallman himself has continued to emphasize the moral motivation behind the free software movement and has repeatedly emphasized the fact that it is the commitment to software freedom, not the temporary practical advantages, that make the FOSS movement viable [37]. In July 2009 Stallman was still encouraging the FOSS community to place its emphasis on software freedom [39]:

> As the advocates of open source draw new users into our community, we free software activists must work even more to bring the issue of freedom to those new users' attention. We have to say, "It's free software and it gives you freedom!" more and louder than ever. Every time you say *free software*, rather than *open source*, you help our campaign.

In recent years, the FOSS movement has been highly successful and has grown to encompass a significant share of the software market. Two important events came together to contribute to this success, the emergence of the Red Hat business model and the transformation of the Netscape browser into Mozilla Firefox.

The Red Hat Business Model Red Hat Corporation was the first to show that FOSS development can be sustained by an effective business and economic model [45]. In 1993, prior to Linux's surge in popularity, Red Hat's founder, Robert Young, was running a software distribution company specializing in Unix applications. As sales of Linux distributions began to pick up, he and Marc Ewing founded Red Hat Software, Inc. in January 1995.

Red Hat's business model is to work with Linux development teams from around the world to put together the hundreds of modules that make up a Linux (or, more accurately, a GNU/Linux) distribution. Rather than selling a license for the software, as a proprietary software vendor would do, Red Hat sells service. In the 1990s, selling service, rather than branding the software as *intellectual property* and selling it, was a revolutionary concept. The Red Hat model thus provides convenience, quality, security, and service to its customers. By 2017, Red Hat Linux had gained a 67% share of the Linux market [34].

Following in Red Hat's footsteps, many other companies have discovered that rather than owning a proprietary software product, a successful business can be built around the concept of supporting and servicing a FOSS product. This fact contradicts the programmers-need-to-eat skepticism that had greeted the GNU Manifesto when it appeared in 1983[1]. That is, the GNU Manifesto defended open source as an idea that is compatible with that of financial viability in the software industry, and the Red Hat model independently verified that idea.

From Netscape to Firefox The creation of the Mozilla community was another watershed event in the history of the open source movement.

[1]Stallman originally wrote the GNU Manifesto [36] to help gain financial support for the development of the GNU operating system.

Unlike its successful FOSS predecessors (e.g., Linux and Apache) that mainly benefit professional programmers, the Firefox browser became the first FOSS product to be successfully distributed to all computer users. Here's how Firefox came into being.

In 1994, Netscape began providing unrestricted distributions of its Navigator browser. In January 1998 Netscape announced that, in addition to freely distributing its browser, it would also freely distribute the source code for its browser software, known as *Mozilla* [20]. Thus, Netscape became the first large corporation to open-source its proprietary software in the interest of widening corporate development of open source environments. This event forever changed the way software is distributed on the Internet.

Mozilla's current open source bowser, called *Firefox*, has become a major combatant in the so-called "Browser Wars," alongside Google's Chrome, Microsoft's Internet Explorer, and Apple's Safari. Of these, only Chrome and Firefox are open source browsers and they combined to command 72% of the desktop browser market in 2017. The proprietary software alternatives, Internet Explorer and Safari, command only 19% of the same market [33].

Today, the Mozilla Foundation (mozilla.org), originally formed to manage the Mozilla development effort, has evolved to become a model open source community. Mozilla has only about 300 paid employees. Another 1,500 or so volunteer programmers from a broad international community contribute to its most recent software releases. In addition to programmers and developers, the Mozilla community includes tens of thousands of testers and users, who work to promote the browser and have helped to translate it into more than 70 languages worldwide (see [16] for more details).

As Red Hat, Mozilla, and many other open source projects have demonstrated, the FOSS development model is compatible with the idea of commercial success in the software business. Today, many major software companies, including IBM, Google, Hewlett-Packard, and others, support open source development in various and substantial ways. Companies that rely on the success of systems such as Linux and Apache assign members of their own software development staffs to work, more or less full time, as contributors to these projects.

FOSS Worldwide

At this writing, the FOSS movement has spread far beyond its origins in GNU, Linux, and Mozilla. Many new FOSS communities have emerged to develop important consumer-related software products. Because of its accessibility, affordability, transparency, and association with freedom for the user, FOSS has become a major force in the software industry. Here are three notable examples:

GIMP The `GNU Image Manipulation Program` (GIMP) provides a software suite for photographic and other image manipulation. It is a free alternative to the proprietary Adobe Photoshop software.

OpenOffice `OpenOffice` is an office productivity suite that includes word processing, spreadsheets, and presentation modules. As such, it is a free alternative to the proprietary Microsoft Office software, which includes Word, Excel, and PowerPoint.

Wordpress `Wordpress` is an open source platform for developing content management systems (CMS) in websites. As of 2017, Wordpress had a commanding lead in CMS market share over Drupal and Joomla, which are also open source platforms.

Businesses, governments, academic institutions, and non-profit organizations throughout the world are increasingly turning to FOSS for their software needs. Here are a few examples:

> The 5th consecutive International Conference on Open Innovation 2.0 took place in Romania in June 2017. It is attended by innovation experts, policy-makers, academic scholars, practitioners and individuals who are engaged in various aspects of open source development. The 2017 conference site has a strong IT and innovation ecosystem with 11 Universities and several innovation and technology parks [11].

> The Brazilian government was one of the first to experiment with FOSS, beginning shortly after the election of Luiz Inacio Lula da Silva in 2002. Following Brazil's leadership, the entire Latin American Region started initiatives to promote FOSS usage and development, including many grass roots efforts. In April of each year since 2005, free software festivals are held in 200 cities and 18 Latin American countries [12].

> The French Gendarmerie is reported to have saved an estimated 50 million Euros since 2004 in moving from Microsoft to the Ubuntu/Linux desktop [22].

> In 2009, the Amsterdam city government made OpenOffice and Firefox their default systems on all its desktops [21].

> Also in 2009, the United Kingdom government announced an effort to avoid vendor lock-in by considering FOSS alternatives equally when deciding IT procurements [7].

> Throughout the last several years, U.S. government organizations have made major commitments to FOSS, including the Library of Congress,

the U.S. Postal Service, the U.S. Census Bureau, the Department of Defense, the FBI, and many state governments (see [32] p. 182 and [30]).

In addition, some of the most dramatic FOSS progress has occurred in developing nations, where governments have seen FOSS as a way to save money, avoid the vendor lock-in problem, and bridge the technology gap.

The emergence of FOSS is fueled by many forces, including the world's need for affordable computing, the effectiveness of agile and related software development methodologies (see Section 1.2), and the increasing worldwide sense of public ownership of the Internet and its resources. A 2015 survey estimated that 78% of all companies were using open source software, up from about 42% five years earlier [40]. According to that survey, the two main reasons cited by companies for preferring open source over proprietary software are:

1. Open source delivers better security than proprietary software.

2. Open source scales better and is easier to deploy than proprietary software.[2]

Considering its current momentum, popularity, and openness to self-forming development communities, the free and open source software movement promises to remain a healthy and prominent part of the software industry for the foreseeable future.

Terminology: OSS, FOSS, FLOSS, H/FOSS, and CO-FOSS

Overall, the terms OSS and FOSS cover the broad scope of open source software. These two have somewhat different licensing variations, from more permissive (OSS) to less permissive (FOSS) restrictions on reuse. But three other nearly-equivalent terms – FLOSS, H/FOSS, and CO-FOSS – have also come into use. Let's quickly sort them out.

FLOSS was coined in 2001 as an acronym for "free/libre and open-source software" [13]. Proponents of this term point out that parts can be translated into other languages, for example the "F" representing free (English) or frei (German), and the "L" representing libre (Spanish or French), livre (Portuguese), or libero (Italian), and so on. So FLOSS is essentially equivalent to FOSS.

A significant sub-concept within the FOSS umbrella is called "Humanitarian FOSS" (H/FOSS for short), which is open source software designed for use by global relief organizations, non-profit organizations, and society at large.

[2]In this book, you will explore some of the security challenges and solutions associated with FOSS development. You will also gain hands-on experience with the complete process of open source development and deployment. So in the end, your own experiences will enable you to corroborate these two particular survey results.

Proprietary software developers largely ignore the particular needs of these organizations, since they usually lack technology budgets. Excellent examples of H/FOSS development can be found at `http://hfoss.org/`. [31].

CO-FOSS is that subset of H/FOSS in which the client is a single organization and the software is customized to fit the needs of that client [25]. Whether or not the software has broader uses beyond that single client is not an immediate concern to a CO-FOSS project. In fact, any temptation to prematurely broaden the reach of a CO-FOSS project beyond the specific needs of its client tends to work against the goal of project success. This broadening is sometimes called "mission creep."

1.4 SOFTWARE ARCHITECTURES

Software is the code that resides in a computer and enables a person to use it in a way that facilitates their work or enhances their lifestyle.

The computer may be a laptop or desktop device, a smartphone or tablet, or a server running a complex collection of applications for a large organization.

1.4.1 Software Frameworks

Stand-alone computing is characterized by the picture shown in Figure 1.5. That is, a single computer runs the software for the user working in isolation from any outside services. An example is a person preparing a document using word processing software on a single computer and then printing it on a printer directly connected to that computer. No network services are required to complete this task.

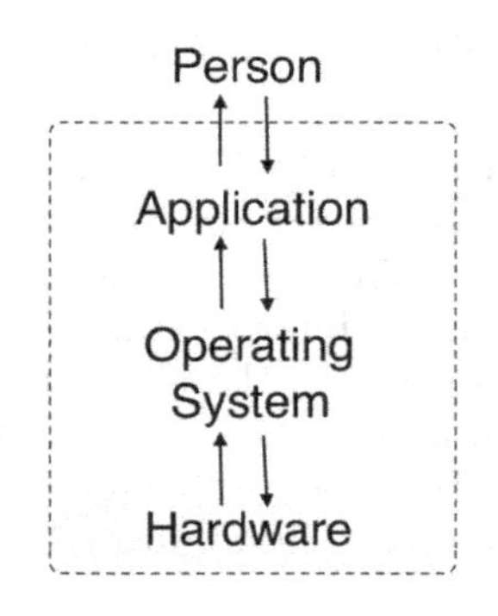

FIGURE 1.5 Stand-Alone Computing.

However, today's computer users generally accomplish their computing tasks by acquiring software services through an Internet connection, provided either by a local wi-fi signal or by a cell tower. When computers connect directly to software in this way, they are using a "client-server" framework, as shown in Figure 1.6. Here, each application is split into a so-called "client side" and a "server side," which interact with each other via the Internet. The server side of an application usually includes interaction with an application-dependent database, as suggested in Figure 1.6.

Examples of client-server computing include apps embedded within a Web browser for purchasing goods at an on-line store or paying bills from a bank account. The browser and the application client sit on the person's computer or smartphone. The on-line store or bank account's server resides on the other side of the connection. The person typically gains secure access to the full application by providing a Username and a Password. SSL encryption is a

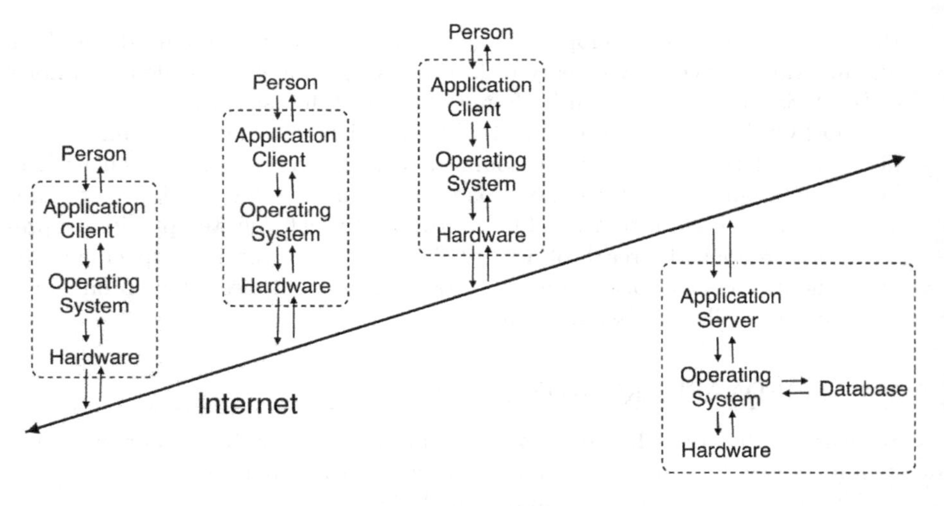

FIGURE 1.6 Client-Server Framework.

common tool for connecting the application's client with its server, ensuring that the information traveling on the Internet between the two remains secure.

When several clients can each connect to a *network* of remote servers hosted on the Internet, this is called "cloud computing." The metaphor here is that, for a user, the individual servers providing different services are invisible, as if obscured by a cloud. Thus, cloud computing is an extension of the notion of client-server computing – it just adds more interrelated servers to the mix, as shown in Figure 1.7.

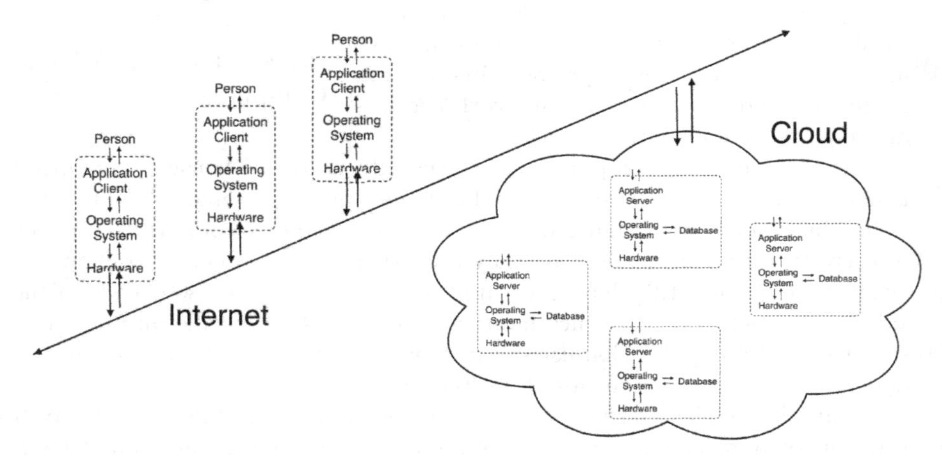

FIGURE 1.7 Cloud Computing Framework.

An example of cloud computing occurs when a user connects to a collection of related application servers through a single secure login. Once logged in, the user can use several different services at once, such as Google Drive, Google Contacts, GMail, Google Calendar, and Google Meet, while engaging in a videoconference with co-workers and team members.

1.4.2 Web Servers and Bundles

A *web server* is the hardware-software configuration that provides web services on the Internet. A *bundle* or *stack* includes the web server, a host operating system, a database system, and one or more programming (scripting) languages. Each of the servers in Figures 1.6 and 1.7 depicts such a bundle. Two popular web servers/bundles are the Apache Web Server and Microsoft IIS.

> The Apache web server is currently the world's most widely-used web server (44% of all active sites in January 2018).[26] One of its bundles is called the "LAMP stack" and it contains an Apache server, a MySQL database, and PHP language scripts that run on a Linux operating system. This is a popular combination for developing software embedded in a web site, partially because it is open source and partially because it runs equivalently on Windows (called WAMP) and MacOS (called MAMP), in addition to Linux (LAMP).
>
> Another Apache-based open source bundle is called AMPPS, which adds Perl and Python to the above collection of programming tools for web apps. AMPPS also runs equivalently on Windows, MacOS, and Linux servers.[35] You can find more detailed information on the different Apache bundling alternatives by visiting `the Apache site` itself.
>
> The NGINX bundle, like Apache, is open source and runs on a variety of platforms. However, NGINX also runs faster than Apache. These two factors are mainly responsible for NGINX market share growing steadily (21% of all active sites in January 2018), compared with Apache.[6]
>
> Microsoft's IIS web server has a smaller market share (7% of all active sites in January 2018). Its application bundle includes the ASP .NET scripting language and Microsoft's MSSQL database. One reason for its limited impact is that IIS is proprietary software and runs only on Windows machines.

At another level is the more modern notion of a "web framework," which differs from a bundle because it hides from the programmer much of the interface between the client, the server, and the database. Two prominent web frameworks in use today are called "Django" and "Rails:"

> `Django` is an open source web framework with the Python programming language at its core.
>
> `Rails` is an open source web framework based on the Ruby programming language.

Choosing from among the many bundles and frameworks available for developing a new web-based CO-FOSS product is governed by several factors. Prominent factors include the nature of the application, the availability of open source code for reuse in the new application, and the experience/preferences of the instructor and the student developers themselves.

To facilitate code reuse and to capitalize on prior student preparation, all our CO-FOSS projects are client-server Web applications. They all use an Apache web server bundle with PHP, JavaScript, and MySQL. So all the examples in this book use this bundle. If we were to develop a new CO-FOSS product with different code-reuse constraints, we would consider using a different web framework such as Django or Rails. In this case, our students would need a different book to supplement this one as a source of examples.

1.5 NEW VS MATURE OPEN SOURCE PROJECTS

While the project and course design discussed in this book can yield both a functioning software product and a successful hands-on learning experience for students, it is important to distinguish this experience from the experience of contributing to a more mature FOSS product.

Mature FOSS products, such as `Eclipse` or `Wordpress`, are far more complex, contain millions of lines of code, and are designed to serve many clients, not just one. Such software is thus not customized (though pluggable architecture encourages customization in a different way), and its development team is larger and more fluid (team members come and leave) than the small team that develops a new CO-FOSS product.

In a mature project, team members have different *levels* of membership, depending on their prior experience and familiarity with the software itself. Cockburn (e.g., [9], p. 9) equates the three levels of team participation—novice, apprentice, and expert—with the three levels of mastery in *Shuhari*, a Japanese martial art concept.[3]

Achieving a particular level of contribution—novice, apprentice, or expert—is defined by a meritocracy, in which merit is measured by the quality and quantity of an individual's prior contributions to the code base. What constitutes merit can include any of the following:

- Code contributions
- Infrastructure support contributions
- Mentorship contributions
- Documentation contributions
- Testing and bug reports

[3]In Shuhari, "shu" means *follow tradition*, "ha" means *break with tradition*, and "ri" means *leave tradition behind*. For more information, see `https://en.wikipedia.org/wiki/Shuhari/`.

With this overview, we can see that students in a 1-semester software projects course will likely gain little hands-on experience as developers by engaging a mature FOSS software project as novices. That is, the complexity of mastering that product to a level where they can become a code contributor is far greater than what most students can expect to achieve in a 1-semester course. Nevertheless, it is useful to explore these mature projects in a bit more detail, so as to inform students interested in entering the software field about the projects and professional communities they can expect to encounter.

1.5.1 Maturity Assessment

Models have been proposed for assessing the maturity of an open source project.[4] Maturity models provide clear quantifiable measures for evaluating a project.

Several factors are considered when evaluating the maturity of an open source project and its community. These include quality assurance, scalability, security, performance, adoption, community strength, community governance, support, and IT management.

A critical measure of an open source project's maturity is the strength of the community that surrounds its development. A strong community can provide a wealth of input from around the world. In comparison, a proprietary product can only benefit from the input of its employees.

A second measure of maturity evaluates the licensing terms and intellectual property management policies and controls in the project. As we saw in Section 1.3, several popular open source licenses have proven to be effective.

Finer-grained methodologies that can assess an open source software project are defined in products like `QSOS` (Qualification and Selection of Open Source Software) and OpenBRR (Open Business Readiness Rating, developed by `Carnegie Mellon University West` and others).

A simpler way to make a quick judgment about the maturity of an open source project is to ask whether each of its core developers has had at least 3 months experience on the project, the project has many more users than developers, and the following actions have been taken:

- the source code is in a repository for public download,
- the project has a public forum where users can post bug reports and queries about the system's adaptability to new users, and
- the development team is open to taking on new members who may volunteer.

This is different from a new CO-FOSS project, which has no immediate intention of broadening either its client base or its developer team, but may

[4] In fact, maturity models are not new in the software engineering industry. The Capability Maturity Model Integration (CMMI) is one such example [1]. CMMI includes best practices for planning, engineering, and managing product development and maintenance.

become more mature in the future. That is, after project completion it may be desirable to consider broadening the developer team or reaching out to other clients who have needs for a similar product. An example of this type of project is the *FarmData* project, which was originally developed by students at Dickinson College for a single client but is now more mature and has a larger user and developer community.

A newly-completed CO-FOSS project that promises to have broader impact may transition to a more mature phase which has been called the *democratic meritocracy* phase. Democratic meritocracy is an ideal form of governance for a young FOSS project, in the sense that all the project's participants are representatives from the meritocracy of contributors. Typically, the sponsors of these types of projects are non-profit foundations whose boards of directors are also selected by the membership. An excellent example of a democratic meritocracy is the Debian community, which has now grown to include thousands of voluntary developers and a very representative process of voting and selecting project leaders.

In the next section, we discuss the long-term transition of a young CO-FOSS project into a more mature project. We will return to that discussion in Chapter 9, which considers the initial steps for making that transition after the project has been completed.

1.5.2 Incubation

The formation of a vibrant community of users and developers marks a critical stage in a CO-FOSS project's transition from origination to maturity. This stage is sometimes called *incubation*, and its purpose is to establish a self-sustaining open-source project with a long lifespan. Both the Eclipse Foundation and the Apache Software Foundation have created *incubators* that invite young open source projects to join.

Two key activities govern how successfully an open source project can pass through its incubation phase and become healthy and sustainable for the long run: building a vibrant community and establishing a viable bug tracking process. At its beginning, a project has only a single (lead) developer, a sponsoring client, and a single user. As the code base evolves, a core group of developers can emerge alongside a handful of "bleeding edge" new users.

How can a CO-FOSS project transform its fragile community into one that has a significant number of developers and users, who are actively reporting issues, and whose developers are contributing bug fixes and new features as the product evolves? Three inter-related groups are vital to this transition: users, contributors, and committers.

Users know and use the software actively. They provide feedback to developers (contributors and committers) when they find bugs or other difficulties when using the software. They also suggest new features that could improve the software's usability or applicability.

Contributors are users who also contribute bug fixes and minor features to the software, but don't have the right to alter the code base itself. Contributions can also be in the form of documentation, administrative support, and testing.

Committers are developers who review contributions and install them in the code base. In this activity, the committer ensures that the code base keeps its integrity—i.e., that the new features are correctly implemented and that the bugs are actually fixed.

Attracting new contributors and committers to the project requires active recruiting, not just passive "openness" for outsiders to join. For example, a certain amount of professional and social networking must be directly associated with the project. The lead developers especially must make reasonable efforts to recruit promising new contributors.

An active Web presence, including easily accessible developer and user forums and project wiki, are valuable catalysts that encourage the successful incubation of a new FOSS project. The establishment of effective on-line forums allows developers and users to discuss specific issues related to the usability of the software itself.

These forums provide an immediate avenue of expression through which a user can report a bug or other technical issue related to using the software. They also provide timely information to developers about the status of all active bugs and other new features that are being considered for the next release of the software.

Finally, these forums provide documentary evidence that can be used when a contributor applies "promotion" to committer status. That person's contributions can be retrieved from the forum's discussion threads and then used by project leaders to evaluate that application.

Community

Abstractly, Jensen [23] characterizes FOSS project organization as a socio-technical interaction network, or *STIN* for short. STINs are always in flux; they are self-organizing networks of activities, people, and tools, and often all these parts are geographically distributed around the world.

More concretely, mature FOSS projects tend to have three main organizational distinctions from proprietary projects. These are:

1. **Self-organizing** Mature FOSS projects allow participants to find their own level and project activity with which to become engaged, based on their interests and skills. Proprietary project leaders assign each participant to a project activity.

2. **Egalitarian** Mature FOSS projects openly invite contributions from everyone. Proprietary projects are hierarchically organized and closed in this regard.

3. **Meritocratic** Mature FOSS projects organize their work around public discussions, and decisions about future directions are based on merit. Proprietary projects organize around the results of private discussions among project leaders, and decisions about future directions are highly influenced by cost and profit.

A FOSS community itself is quite fluid—most users of the software are, in fact, passive users. Jensen [23] (rather harshly) calls these users "free-riders," since they give nothing back in return for the privilege of using the software at no cost.

Users who do provide feedback to the developers do so in an entirely voluntary spirit. Feedback typically occurs through the software's user forum, which is prominently accessible at the project's Web site. Some users may go a step further by providing bug fixes or suggestions for new features in the form of code patches. Engaging in this activity self-promotes the user to *contributor* status. Whether or not a contributor's suggestions are accepted and become part of the code base, however, is decided by a committer.

Promotion to *committer* status is done on the merits of a person's collected contributions to the project over time. In this sense, the contributions become a portfolio of work that can be evaluated to assess the merits of that person's case for assuming the responsibilities of a committer. Who decides on the promotion of a contributor to committer status? This is often done by a core project leadership group. In Apache, for example, the *Project Management Committee* (*PMC* for short), is the group of committers that oversees the project's organization, including making promotions.

While users and contributors are most likely volunteers, many committers become paid employees of the project. What particular skills are required to attain committer status? Generally, an applicant's portfolio contains two types of contributions: those that illustrate technical competence and those that exhibit social skills.

Finally, research [23] has shown that successful FOSS projects must maintain a critical mass of contributors and committers, in relation to "free-riding" users, in order to remain vibrant over the long run. Too many free-riders can kill the project.

Policy, procedural, and technical decision-making in an open source project aims to be fully transparent to all community members. When a complex policy, procedural, or technical issue arises, a member of the PMC typically posts the issue on a public discussion forum, where it is debated and either ratified or rejected by consensus or majority vote. Here, *consensus* means that at least two other developers support a particular solution and no other developers post strong disagreements.

Sometimes, of course, achieving consensus on a contentious issue is not possible. Often conflicts arise during discussions about community infrastructure, technical direction, expectations about developer roles, or interrelationships among roles. These kinds of *conflicts* can be resolved by a process involving a small PMC made up of prominent members of the community. This group has the job of ensuring fairness throughout the community by solving persistent disputes.

In addition to maintaining the code base, the participants in a mature FOSS project play roles that accomplish several other important project tasks. Here is a summary of these roles and their respective activities:

The **project leader** maintains the project's release plan and current status, and moderates the developer forum.

The **expert user** maintains the software's actors, use cases, requirements, and user roles.

The **lead developer** maintains the software architecture.

Other **developers** maintain the user interface design, domain classes, database design, code base, unit test suite, build package, and build schedule.

Testers manage bug reports and the user forum.

Writers maintain on-line help text.

Bug Marshalls oversee opened bugs and pass them on to developers.

Release Managers overlook the packaging and releasing of new versions of the software for general public download and use.

We also note that it is not unusual for an individual to play two or more of these roles simultaneously, depending on his/her particular interests and the size of the developer and user community.

Bug Tracking

As the code base becomes complex and its community grows, an open source software project should establish a viable process for identifying and tracking the status of bugs that are reported by users and developers. Bug tracking is an important process, especially in a development culture where Linus' Law, "given enough eyeballs, all bugs are shallow," is held in such high esteem.

Bug tracking is a formalized process that governs how bugs are identified and how their resolution is managed by the developers who work with the code base. A particularly interesting bug tracking process is the one established by the Mozilla Foundation, based on its open source bug-management tool called `Bugzilla`.

The "life cycle of a bug" refers to a bug's progress through a series of discrete states. Such a life cycle can be described in the form of a state diagram, as shown in Figure 1.8 for Bugzilla. There, we see that a bug can be in any of five states: "unconfirmed," "confirmed," "in progress," "resolved," and "verified." A bug can be introduced by any user. The management of a bug

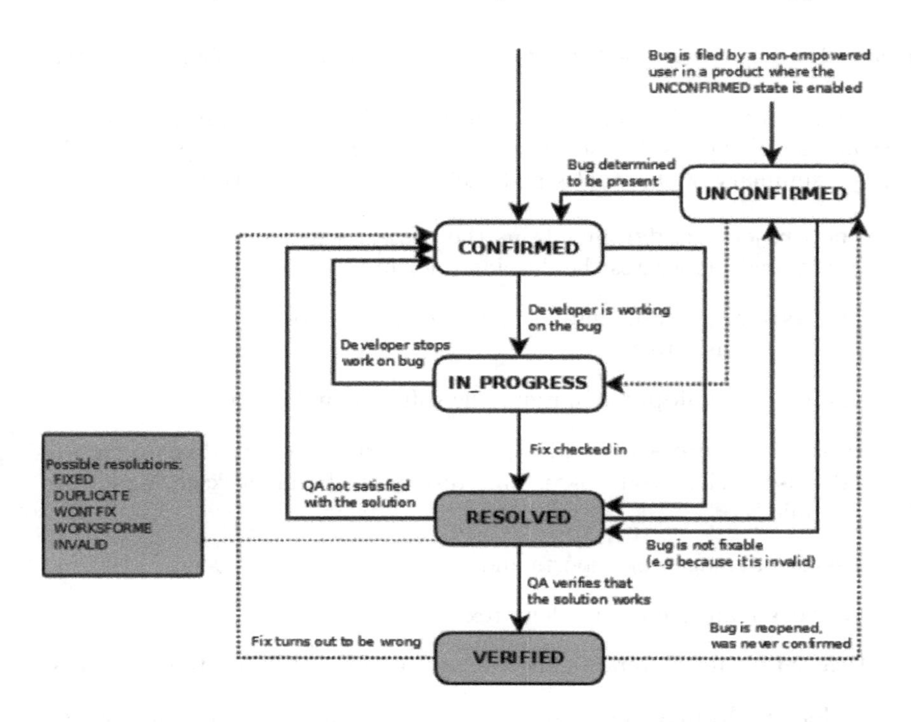

FIGURE 1.8 Life cycle of a bug, from Bugzilla documentation, p 9.

is done by developers with commit privileges, and the whole process can be quite complex.

For example, the Eclipse community uses `Bugzilla` (see `https://npfi.org/bugzilla/`). A quick look at the `Bugzilla User Guide` shows that the process of bug management is not trivial. The process works only if the project has both an active user community for detecting and reporting bugs, and an active developer/committer community for fixing, resolving, and verifying the fixes.

For projects with large user and developer communities, this summary provides insight into how to implement Linus' Law in a practical setting. It also shows that the process of fixing a bug and verifying that fix relies strongly on both the openness of the source code and the atmosphere of trust that exists among users and developers.

1.6 INTO THE WEEDS

This section provides guidance on how to best navigate the rest of this book, depending on whether you are an instructor, a student, a client, or a professional software developer.

1.6.1 To the Instructor

You are about to organize a software development course that has ambitious goals and outcomes for your students.

As in a traditional course, you can expect that your students will learn basic principles of software development, including the stages in the development cycle – requirements, design, coding, testing, deployment, and maintenance. You can also expect that they will master the steps in a client-centered development cycle – meet, plan, code, test, and evaluate – as they develop their CO-FOSS product with the help of a real client.

Two key ideas will help ensure the success of your course: 1) the idea of open source software and 2) the idea of client-centered development. An important goal of your course is to convince students that software development can be a successful and productive enterprise, and that each one of them can play an essential role to achieve that success.

To accomplish this goal, you will need to introduce students to specific software tools to facilitate their work – a system stack, a "sandbox server," an integrated development environment (IDE), a software repository with version control, unit testing, face-to-face collaboration, and open source licenses. Working as a team, you, your students, and your client will be engaging in a client-centered process that will help ensure successful project completion at the end of the course.

For the projects discussed in this book, our students used a system stack with an Apache server, MySQL, PHP, and JavaScript running on Linux, Mac OSX and Windows machines. In one project, students also used Android development tools to develop an app for an Android tablet using Java. They used Mercurial version control with Google Code repositories. Because Google Code is no longer supported, these projects have migrated over to GitHub repositories and they use the Git version control system. For a peek at the current versions of these projects and their code bases, see `https://github.com/megandalster/`.

For coding, testing, and committing code to the repositories, our students used an Eclipse IDE with either PHP/JavaScript or the Android Java programming language. For unit testing their PHP code, they used SimpleTest. At this time, however, PHPUnit would be the preferred unit testing tool, and Android Studio would be the preferred IDE for developing an Android Java project.

The tools you select for your project will depend upon several factors: your own experience, your client, your application, other open source code bases that you can reuse, and your students' skills coming into the course. In our experience, students were intermediate or advanced CS majors, with moderate-to-excellent programming skills, Java or Python language skills, modest IDE experience, and data structures familiarity. They had little or no prior experience with large software projects, databases, version control, or team programming.

Finding a client and a software project that you and your students can complete in this course requires particular care and attention. While this work goes beyond an instructor's normal teaching load, it is key to successfully organizing and teaching the software development course in this way.[5] For more detailed guidance on finding a client and a project, see Chapter 2.

It is no accident that the 3-month project implementation stage is the length of a 1-semester course. This requires that you design the software project carefully so that your students can complete it by the end of the course. We think that student success in this regard is the most important outcome. Failure to complete the project would not only be a disservice to the client; it would also leave your students with a negative view of software development as a profession. For more detailed guidance on organizing your project and your course around it, see Chapter 3.

To support your teaching the course itself, this book provides detailed guidance on the development process, including links to code bases, "sandbox" databases, assignments, and mini-lectures drawn from our own experiences. In particular:

Chapter 4 introduces students to the fundamentals of the client-centered process, open source licensing, team roles, using the code repository, and communicating with team members.

Chapter 5 covers principles of programming languages, IDEs, and coding the domain classes. It also introduces the idea of test case design, unit testing, code synchronization with a "sandbox" server, issue tracking, and client review.

Chapter 6 covers the principles and practice of database design, tables, queries, CRUD functions, testing, security, and client review.

Chapter 7 introduces design and development of the user interface, including the model-view-controller pattern, usability testing, and client review.

Chapter 8 offers suggestions that will help your students finalize their project and prepare the software for productive use. We offer advice about technical writing, user documentation, developer documentation, and user training.

Chapter 9 addresses issues that you will confront after the semester is finished. It discusses finalizing your project's public repository with the key artifacts that your students have developed – code base, issue tracking, Wiki pages, "sandbox" database, requirements document, and developer documentation. They also provide guidance on finding a developer and passing these artifacts to that developer for deployment and ongoing support.

---

[5] Because this preparation normally takes up to 2 months prior to the beginning of the course, you may want to seek outside support, either as release time from your home institution or as a grant from an outside source like NPFI.

1.6.2 To the Student

You are about to begin a course where you will learn the modern principles and practices of software development. As a central activity in this course, you will help create a real software product that fulfills an important need for a real client. (You may view this as a service learning opportunity embedded within a regular academic course.)

Your success in this course will be determined by many factors, including the quality of your contributions to the software itself (programming, testing, and documentation) and your contributions to the team effort (sharing code and collaborating).

In this course you can expect to use what you have learned about programming and data structures in earlier courses. But you can also expect to gain new understanding about the principles and practice of software development, including the following:

Chapter 4 introduces you to the fundamentals of the client-centered process, open source licensing, team roles, using the code repository, and communicating with team members.

Chapter 5 covers principles of programming languages, IDEs, and coding the domain classes. It also introduces the idea of test case design, unit testing, code synchronization with a "sandbox" server, issue tracking, and client review.

Chapter 6 covers the principles and practices of database design, tables, queries, CRUD functions, testing, security, and client review.

Chapter 7 introduces design and development of the user interface, including the model-view-controller pattern usability testing, and client review.

Chapter 8 offers suggestions that will help you finalize your project and prepare the software for productive use. We offer advice about technical writing, user documentation, developer documentation, and user training.

Unlike other courses, this one is driven by three major elements:

1. teamwork
2. a real-world software product
3. self-education

First, you will work with a team to develop a new software product. The team will include a few other classmates, your instructor as "benign dictator," and a client as the evaluator and recipient of your completed software. So you will learn to work with a team and communicate with teammates and the client as your project develops.

Second, the software product you will be developing is real; you may complete it by completing a series of assignments (called "milestones") with your teammates. Your team will be working with a real client who expects to receive a viable product at the end of the semester. Rather than a final exam or homework grades, you will be evaluated on the basis of one outcome – did your and your team's work result in a viable software product that the client can begin using to help improve his/her work experience?

Third, to achieve a successful result, you will be expected to work like a software professional. That is, you should plan to educate yourself about a variety of topics in programming, database development, and user interface programming. The instructor will not spoon-feed any of this to you. In a broader sense, this course provides you with an opportunity to really embrace the idea of yourself as a "lifelong learner," regardless of your future profession.

If you do enter the software field after graduation, you will quickly learn that it evolves rapidly—new languages, new software methodologies, and (especially) new applications will appear throughout your professional lifetime. To stay at the forefront of the field, you will learn most of what you need to learn on-the-job, and much of this knowledge does not even exist today! So life-long learning is a key element of survival in the software field.

1.6.3 To the Client

You are about to take part in a project that will develop new software for your organization. When completed, this software will help improve one of your organization's mission-critical activities. Your role in this project is to participate in regular meetings with student developers and provide them with critical feedback on the quality of their progress at each stage of the project. Your time commitment for this work may be 1-2 hours per week.

Before the project begins, you will work with the instructor to help him/her understand your organization's software need and compile a "requirements document" that the developers will use as guidance for developing the software to fulfill that need. For more concrete guidance about what to expect from this initial step, please take a look at Chapter 3.

Soon after the project begins, and every 1-2 weeks thereafter, the developers will provide you with a new working prototype of your software. Each time you exercise it, you will see new features that you can test, evaluate, and make suggestions for improvements. Later in the project, as you encounter issues with the software, you will be able to post them on the project's "issues board" so that the developers can address them in a timely manner. For more guidance on what to expect during this central stage of the project, please take a look at the "Client Review" sections of Chapters 5 through 7 .

Especially important to the success of this software is your feedback on the quality and ease of use for the features introduced in Chapter 7. Toward the end of the project, your constructive suggestions on the "user help" pages and other documentation (see Chapters 8) will also be very important.

1.6.4 To the Developer

You are about to receive a newly-developed software product to install on the client's server or web site. You should expect to receive the source code, database, and all documentation from the instructor as a complete package.

You should test the software for robustness and work with the instructor to fix any new issues you discover before installing it on the client's server. After installing the software, you should expect to provide ongoing support as new issues arise, especially during the first few weeks of active use by the client. For more guidance on interfacing with the instructor and the client and deploying the product, please take a look at Chapter 9.

In some CO-FOSS projects, a professional developer like yourself has been involved much earlier in the project, assisting the instructor and client with project requirements and/or meeting regularly with the student team to provide professional advice on various technical challenges of the project as they occur. For example, this approach was taken successfully in 2016 by an instructor at Green River College whose students developed software with continuing support and mentoring from a professional developer (for more information, see `https://npfi.org/the-2016-npfi-grant-award/`).

You may see other benefits from your joining the project earlier than the time when the instructor hands you the final product for deployment on the client's server. For example, you may want to brush up your own skills with client-centered development using platforms and tools you haven't used before. Or else you may want to engage actively with the class as an indirect recruiting tool for your own company. Or you may simply want to give back to your community in a way that utilizes your technical skills more directly than what you do in your "day job."

Whatever your motivation, you should talk with the instructor about engaging more actively in the project and work out an arrangement that works for both of you. Once you make such an arrangement, you can use the rest of this book as a resource guide for working alongside the student developers, selecting from among Chapters 4 through 7 the ones that are appropriate to your needs.

Finally, we should mention that the Non-Profit FOSS Institute (NPFI) exists solely to provide *pro bono* support for instructors, clients, and developers who would like to start a new CO-FOSS project or get involved with a current one. Many aspects of NPFI support are mentioned throughout this book. For information about a particular current or future project that interests you, please feel free to `contact NPFI` at `https://npfi.org/contact/`.

1.7 SUMMARY

This chapter introduces the larger ideas behind software development, focussing on a model for open source development which we call CO-FOSS. This model is particularly suitable for inclusion in a course where students gain real-world experience developing a new software product for a local client. Software architectures, frameworks, and licensing are presented with a particular focus on their influence on open source development.

The distinctions between a new CO-FOSS product and the mature FOSS products that students will encounter if they enter the software profession after graduation are also presented. After all, most mature FOSS products are the result of incubation from earlier CO-FOSS projects.

This chapter concludes with some advice to the instructor, the student, the client, and the professional developer on how to best use the material in the remainder of the book.

1.8 MILESTONE 1

1. If you are an instructor, what are the risks and rewards for your launching a new CO-FOSS project with a student team?
2. If you are a student, what did you learn that you did not already know about the software world or the process of software development?
3. If you are a non-profit representative, what critical operations in your organization would be well-served by the addition of new customized free and open source software?
4. If you are a professional software developer, how would engaging in a new CO-FOSS project improve your resume or your personal well-being?

I

Organization Stage

CHAPTER 2

Finding a Client and a Project

"Empowerment of individuals is what makes open source work,
since innovations tend to come from small groups,
not from large, structured efforts."
—*Tim O'Reilly*

Starting an open source project to develop a custom software product for a single client works well with a small team of developers. The client-centered approach begins when the instructor identifies a staff member who will provide direct feedback throughout the project and will "own" the software product when it is completed.

What should the instructor look for when seeking a client for a new CO-FOSS project? We have found the following client characteristics to be essential in all of our projects:

1. **Desire** The client must want to participate in a 6-month effort that produces customized software for one of its mission-critical activities.

2. **Need** The client must have a clear need for that software. That is, one of its current activities is encumbered because the supporting technology is either obsolete, too costly, or badly-fitted to that activity.

3. **Maturity** The client must be a mature organization, with a stable core staff dedicated to fulfilling the organization's mission. Often a young non-profit may have a great desire and need, but other growth priorities may prevent it from effectively supporting a new project.

4. **Participation** The client must commit a staff member to actively participate in the project throughout its development period. This participation amounts to 1-2 hours a week for 3-5 months.

How can an instructor find a client that fulfills these four requirements? Each institutional and geographical setting provides unique circumstances and opportunities, but here are two suggestions for getting started:

1. Most undergraduate institutions support a community outreach center that identifies local organizations where its students, faculty, and staff can volunteer. That center often interacts with faculty so that students in courses can engage in a curricular or co-curricular community activity that supplements their formal coursework. For example, at Bowdoin College this center is called the `McKeen Center for the Common Good`, at Whitman College it is called the `Student Engagement Center`, and at Beloit College it is called the `Campus Community Outreach Center`.
2. Instructors who are inclined to teach courses like this often have connections with local non-profits where they or their family members volunteer when they are not working. For example, the author's spouse, a retired teacher, was an active volunteer at the Ronald McDonald House in Portland, ME in 2008 when she introduced the author to the House Manager and her frustration with keeping a paper volunteer calendar up-to-date. That introduction began a working relationship that spawned the creation of two successful CO-FOSS products, *Homebase* and *Homeroom*, that have been in productive use at the House ever since.

In our overall CO-FOSS experience, eight different student teams have developed five successful software products for five different clients: three Ronald McDonald Houses, one food warehouse, and one food rescue organization.

By "successful" we mean that the software was fully implemented, installed on the client's server, and is still in productive use today. In each case, the client met all four of the above requirements.

These products are (see `https://npfi.org/projects/` for details):

`Homebase` (2008, 2011, 2013) - online volunteer database and calendar scheduling software, securely accessed from a browser.

`Homeroom` (2011, 2013) - online room scheduling software, securely accessed from a browser.

`Homeplate` (2012) - online food pick-up and delivery scheduling software, securely accessed from a browser and from an Android tablet.

`RMHP-Homebase` (2015) - a variant of Homebase (above) developed for a different client with similar volunteer scheduling needs.

`BMAC-Warehouse` (2015) - online food warehouse shipping, receiving, and inventory management software, securely accessed from a browser.

In addition to these, two other student teams developed software products that were not successful in this sense. One client was a senior center and the other was a soup kitchen and food pantry. Although the students completed the software, neither product was installed for productive use on the client's server. In one case, the client decided not to deploy the software for organizational reasons. In the other case, the client was a new organization with other priorities for growth that became more pressing than deploying our software.

The main lesson that we learned from these experiences was that finding the right client and the right project must be done carefully, keeping in mind the four requirements for success listed above. Even with such care, there is no guarantee that the resulting software will be successful. However, careful project and client selection should normally lead to a complete and useful software product at the end of the development period.

2.1 CLIENT ACTIVITIES AND SOFTWARE NEEDS

Once a client is found, a software project needs to be identified and clearly defined in preparation for the 1- or 2-semester development effort. That definition requires learning about the current process that the new software will support, and then laying out the project itself. Project layout includes identifying specific functional requirements and domain-specific use cases, an overall software architecture, and an initial code base upon which the product can be built. These steps are described later in this chapter.

The kinds of activities for which a client may need software are many and varied. In our experience, we know of the following very common activities:

1. *Donor database and donation management.* All non-profits rely on charitable donations to support their on-going mission-critical activities. Organizing fund-raising events, managing the donations that result from those events and other activities, and keeping track of donors' contact information for future fund-raising activities are key elements of this activity. Modern tools are important for helping these organizations manage their donor and donation databases.

2. *Volunteer database and shift scheduling.* Many non-profits rely on a dedicated corps of volunteers to help them fulfill their missions. For example, hundreds of Habitat for Humanity ReStores staff their operations with volunteers who typically work in groups of 3 or 4 on half-day shifts throughout the week. For another example, each of the 300+ Ronald McDonald Houses worldwide relies on volunteers who typically work in groups of 2 or 3 on 3-hour shifts throughout the week to help manage the house. A modern calendar and database is a key asset to help these organizations manage their volunteer schedules and contact information.

3. *Scheduling of resources, such as overnight rooms.* Many non-profits own resources, such as meeting rooms, overnight sleeping rooms, or other

facilities that their clients use in their day-to-day activities. For example, every Ronald McDonald House provides rooms for families to stay overnight while their children are being treated at nearby hospitals. These rooms are nearly fully booked, so that a modern scheduling and database system is an important asset for keeping track of room availability and family contact information.

4. *Inventory management, including receipts and shipments.* Non-profit food warehouses and food banks throughout the country must keep track of all the food items that they receive from donations and other sources, as well as the items that they ship to soup kitchens and other charities for feeding the hungry. Modern tools for managing this data in a timely way are essential. For example, Feeding America has over 200 Food Banks in the United States. Local food warehouses, partnering with Feeding America, require software that helps manage their inventory, shipments, receipts, and client database on a daily basis.

5. *Truck scheduling for pick-ups and drop-offs of food or other merchandise.* Local food rescue organizations and thrift shops also run trucks daily to pick up and deliver food and merchandise to various agencies that distribute it to people in need. These organizations often rely on volunteer drivers and a modern record-keeping system for recording the weights of food and other items picked up and delivered, as well as volunteer schedules. One such agency, called Second Helpings, rescues and delivers almost 3 million pounds of food annually with the help of over 200 volunteers. In another case, a Habitat ReStore runs trucks daily to pick up household merchandise that can be resold in the store to raise money for building low-cost housing.

Beyond these activities, we know of many others in which local organizations, both public and private, have significant needs for new cost-effective software. For example, many public organizations conduct surveys of their clients to help them clarify their missions and tailor their resources to better meet clients' needs. For another example, many institutions, like colleges and universities, are making efforts to reduce their carbon footprints and their impact on the environment. These institutions need software tools and devices to help monitor their progress in these efforts. One such tool is the `Open Energy Dashboard` (OED) that has been developed by computer science students and faculty at Beloit College for several different organizations in the Midwest.

So this discussion identifies several distinct types of organizations and activities that exist in urban, suburban, and other municipalities throughout the United States and the world that could become clients of future CO-FOSS projects. Many of them have specific software needs like the ones identified above. We expect that these types of projects, because they are open source and can reuse existing code, can be replicated in the future by other teams of computer science students to help their own local organizations modernize their mission-critical activities.

2.1.1 The Current Process and Existing Software

After finding a local client organization that has the above four characteristics and a mission-critical software need, what happens next? Three important steps must be taken to create a viable foundation for a new project.

Step 1 Identify a staff member with whom you and the team will work to develop the project.

Step 2 Work with that staff member to review existing software that might already fit the client's need and budget.

Step 3 Absent such software, work with that staff member to develop an initial design for a new CO-FOSS product that will fit the need.

Before Step 3 begins, Step 2 must conclude that there is no existing software that fits the organization's need and budget – it would make no sense to reinvent the wheel. For example, we know of the following existing software products that address the above five non-profit activities with software needs:

1. *Donor database and donation management.* Many proprietary software products are available for non-profits in this area. However, many non-profits do not have the budgets to pay for these products. A few FOSS products are also available for donor and donation management.

   We have experience with `CiviCRM`, a highly-regarded and fully functional FOSS product that can be configured to fit the specific fundraising needs of most non-profit organizations. Moreover, CiviCRM is smoothly integrated with Wordpress (also a FOSS product), so this combination can be attractive for a non-profit that already has a Wordpress website.

   CiviCRM, however, comes with a significant technical learning curve for the non-profit. The level of technical support to configure CiviCRM to fit the organization's fundraising model and donor database can be significant. The non-profit can hire a tech-savvy consultant to do this work. Alternatively, this sort of project provides an opportunity for a student class to develop CiviCRM and database expertise and then customize CiviCRM to fit that non-profit's fundraising and donor database model.

2. *Volunteer database and shift scheduling.* There are a few software products that support volunteer scheduling. Some are proprietary and others are open source. In our experience, these products either are costly, promote vendor lock-in, or provide uneven user training and support.

   Another temptation is to use a free calendar app, such as Google Calendar, for the scheduling activity. However, Google Calendar has several limitations that make it not fit this need particularly well. For example, volunteer scheduling comes with volunteer database and reporting components, neither of which is available with Google Calendar.

A CO-FOSS alternative for volunteer database and shift scheduling is called *Homebase*, originally developed by Bowdoin students in 2008 for the Ronald McDonald House in Portland, ME. The current version of *Homebase* is in use today at three different Ronald McDonald Houses in Maine and Rhode Island. *Homebase* is highly regarded by clients for its ease of use and its customized fit to their specific scheduling models. Here is some recent feedback on Homebase from one House Manager:

> We are very fortunate to have worked with [the] student team that created *Homebase*, our entire volunteer management database. I can't say enough about how smooth the development process was ... . We are very grateful for the system and it has streamlined so much for us.

Homebase runs on a server with a database and is accessible from a browser. It is embedded within the client's Web site. So the overall operating cost of *Homebase* is negligible, since bug fixing and feature enhancements are provided by NPFI.

3. *Scheduling of resources, such as overnight rooms.* There are several facilities scheduling software products available, both proprietary and open source. A highly-rated open source choice in this category is called `Booked`, which has been used by some colleges and other organizations for room scheduling.

   Our experience is with *Homeroom*, a CO-FOSS product developed in 2011 by Bowdoin College students for the Ronald McDonald House in Portland, ME to schedule its 21 rooms and manage its database of families who stay overnight in these rooms. Here is some feedback from the House Manager about a recent bug fix and the overall effectiveness of *Homeroom* in meeting both houses' mission-critical needs:

   > Thank you so much! I just tested out running a report in Data for 1/1/2017-1/1/2018 for room 126 and the page layout and content look great. I really appreciate you updating this in both Homeroom databases. It's going to save us SO much time in the future when pulling stats on specific rooms.

   Since it is open source, *Homeroom* was customized again in 2012-13 by a team of students at St. John's University for use at the Ronald McDonald House in Manhattan, NY.

   Homeroom runs on a server with a database and is accessible from a browser. It is embedded as an app within the client's website hosting plan. So the overall operating cost of *Homeroom* is negligible, since bug fixing and feature enhancements are provided by NPFI.

4. *Inventory management, including receipts and shipments.* Most existing inventory management software is proprietary, although some products provide modest free versions for getting started. However, these are all oriented toward selling products, which is not in line with the missions of non-profit organizations. There are also some free alternatives, as discussed at `https://npfi.org/free-inventory-software/`.

   Our experience is with *BMAC-Warehouse*, a CO-FOSS product developed in 2015 by 5 Whitman College students for the Blue Mountain Action Council's food warehouse in Walla Walla, WA. The software provides online tracking of food inventory, receipts, deliveries, providers, and customers. The 500+ providers are local food donors as well as local, state, and federal food agencies like the USDA. The 90+ customers are local soup kitchens, churches, and other charitable non-profit organizations in the Walla Walla area.

   BMAC-Warehouse runs on a server with a database and is accessible from a browser. It is installed as an app within the client's website hosting plan. So the overall operating cost of *BMAC-Warehouse* is negligible, since bug fixing is provided by NPFI.

5. *Truck scheduling for pick-ups and drop-offs of food or other merchandise.* Most software in this category is aimed at medium and large fleets of trucks with sophisticated scheduling and routing needs. Software for smaller fleets, say 3-5 trucks, is also available. For example, ClearPathGPS "is a fleet management system that specializes in working with small to midsize local fleets in construction, HVAC, electrical, delivery, moving and storage, landscaping, tree service, solar and many others." Others include Fleet Manager and Titan GPS. All these products, however, are proprietary.

   Our experience is with *Homeplate*, a FOSS product developed in 2012 by three Bowdoin College students and an instructor. This software supports the scheduling and reporting of food rescue and delivery operations carried out by several refrigerated trucks on a daily basis. The trucks pick up food at local markets and restaurants and deliver it to soup kitchens, churches, and other organizations that feed the hungry. *Homeplate* has two components, a scheduling component and a data capturing component. The project required an academic semester and a summer internship for the students to complete both components.

   The *Homeplate* scheduling component runs on a server with a database and manages truck pick-up, drop-off, and driver scheduling as well as data reporting. *Homeplate*'s data capturing component runs on Android tablets in the trucks and is used by drivers to record the weights of food picked up and dropped off at each stop. It also synchronizes that data with the server via a simple file transfer protocol from a wi-fi hotspot (think Dunkin' Donuts), thus avoiding the monthly cost of a telecommunications data plan for the tablets. The scheduling component runs as an app within the client's website hosting plan.

The overall operating cost of *Homeplate* includes only the occasional technical support needed for adding new features and fixing bugs, which is provided by NPFI. Here is some recent feedback to the developer from the Second Helping Office Manager about the quality of that support and the ongoing value of *Homeplate* to their mission-critical operations:

> I have now closed all issues for the current release. These modifications will be major improvements to the effectiveness and accuracy of our *Homeplate* system. They allow us to record and report additional information about our agencies. They will also allow us more accurate weight entry and reporting, by allowing us to enter weights for donors or recipients that were not on the standing schedule. Thanks for the good work. Happy Holidays.

These bug fixes and improvements required about 60 hours of *pro bono* work from the developer. This could not have been accomplished had *Homeplate* not been a FOSS product.

2.1.2 New Software to Fit a New Need

Assuming that both a client and an unfulfilled software need have been found, we are ready to begin thinking about an initial design for the software itself. This step is a collaborative process between the instructor and the client representative, and its goal is to achieve a viable design for a new CO-FOSS product that is customized to satisfy that need.

Since the product will be open source software, the design should incorporate where possible similar components from existing FOSS products that satisfy specific requirements, such as secure login access or intuitive volunteer database maintenance. Incorporating existing components into new software is a skill that saves significant programming and development effort and will help students understand that avoiding the "not invented here" syndrome is essential for effective software development.

The design should also anticipate completing the product in a single semester or academic year, depending on the nature of the course in which the students will be working. A common temptation here is to over-estimate the capabilities of the student team within a 3-month development period, and then discover late in the process that the project cannot be completed on schedule. Thoughtful design can help avoid this very negative outcome, both for the students and for the non-profit client. The design process should create the following three elements:

1. A Domain Analysis
2. A Description of the System and Performance Requirements
3. A Description of the Software Architecture

Together, these elements combine to make up the initial Design Document, which will be used by the student developers throughout the development phase for the software.

The following sections describe these elements in more detail.

2.2 DOMAIN ANALYSIS

Domain analysis creates a non-technical description of how the current system functions. The term "domain" is a software developer's word for the client's own professional setting where the software will be used. So the domain analysis itself must be written using terminology with which people who will use the software are already familiar. It must also be written at a sufficient level of detail to give developers a degree of familiarity with the domain itself.

A good domain analysis introduces developers to all the individual artifacts—both digital and manual—that are in use to support the current system that the new software will eventually replace or enhance. It also identifies the major roles for different types of users who will use the system.

For example, here is part of the 2011 domain analysis that guided the development of *Homeroom*:

> Prior to installation of the new software, guests are normally referred to the House Manager by a Social Worker in a local hospital, using the guest referral form shown in Figure 2.1.
>
> Once a room is located for the required dates, the House Manager fills out a guest registration card for the family (see Figure 2.2). If the family has stayed at the House in the past, their previous guest registration card is retrieved from a Rolodex-style file and the information for their new stay is entered onto that card.
>
> At the same time, the House Manager "reserves" the room by placing a "sticky" on the room schedule for the date requested (see Figure 2.3). On the day of their arrival, the family must call to confirm their reservation, the number of guests who will be staying in the room, and the time of their arrival. When they arrive, the House Manager checks them in by removing the "sticky" from that entry on the schedule and writing in the names of the guests.
>
> In addition to the names of the persons staying in a room, the following information is recorded:

Ronald McDonald House Referral Form

250 Brackett Street, Portland, Maine 04102 Phone: (207)780-6282

Please Fax to RMH @ (207)780-0198

Families should call on the day of arrival for room availability and confirmation.

RMH USE for SUBSEQUENT SELF-REFERRALS

Volunteers: Check Box When Confirmed (Same Day Conf. Only)

Consent to Release Information

I consent to the release of the following information from the referring hospital / care provider to the Ronald McDonald House of Portland, Maine, Inc.

Please print neatly.

Name of Patient ______ **DOB** ______

Name of Primary Guest(s) ______

Relationship(s) of above to Patient ______

Home Address ______

______

Phone # ______ County (if Maine) ______

Name(s) of Additional Guests (Immed. family or support person(s) / Limit to 4 total please or call):

______

Name of **Referring Social Worker** / Other ______

Hospital / Dept of Tx ______ Phone ______

Payment Arrangement: $10 per night ______ or Other ______

Check Priority: ___ 1. SCU, NICU, High-Risk Prenatal

___ 2. In-Patient Pediatrics, MCCP Out-Patient Treatment Stays

___ 3. Other: Specialist Appointments, Spring Harbor Hospital, Mercy Westbrook Eating Disorder Program

Check Approved Use: ___ Overnight ___ Day ___ Both Overnight/Day Use

Date of Initial Visit or Will Call (W/C) ______ ☐ (check when confirmed)

I will read, abide by, and respect the rules of the Ronald McDonald House. I will be responsible for my family/party and supervise my children at all times. I enter the House at my own risk and understand that the Ronald McDonald House is not responsible for personal injuries. I understand that information concerning my child or family may be released by or communicated to the Hospital by the Staff or Weekend Manager, and that my room may be entered at any time for maintenance or safety issues. Any infraction of the guidelines may result in the loss of the privilege of using the Ronald McDonald House of Portland, Maine.

Primary Guest s Signature ______ Date ______

FIGURE 2.1 RMH guest referral form (prior to 2011).

GUEST REGISTRATION CARD RE#__________

Names of Guests ____

____

Relationship(s) to Patient ______________________________________

Patient's Name ____

Primary Guest's Address ____

City ______________ State ______ Zip ________ County ___________

Primary Guest's Phone (Cell) ____________ (Home) _______________

***Automobile Make/Color/State __________________________________

Would you like to receive our newsletter? Email _________________

REFERRAL INFORMATION

Hospital ______________ Dept. Of Treatment _____________________

Social Worker or Referring Individual ___________________________

Room #									
Date In									
Date Out									
# Guests									

FIGURE 2.2 RMH guest registration card (prior to 2011).

- the room's capacity
- whether or not there's a refrigerator in the room
- whether or not the family has borrowed an air bed
- the number of the garage remote door opener loaned to the family
- whether or not the room has been cleaned
- the name of a "linked" room

A room is "linked" to another room if members of the same family are staying in both rooms. For example, Figure 2.3 shows that room 152 is linked with room 254.

At 9am each morning, a new day's room schedule is generated by the House Manager or a Volunteer by manually copying the previous day's schedule onto a blank form. When a family "checks out" of a room, their name is erased from the new day's room schedule and the room becomes a candidate for cleaning by the Volunteer. While guests stay at the House for free, some choose to make a nominal contribution to help offset the cost of their stay.

Because the same family may return at a later date, the House Manager maintains a Rolodex file of all guests who have stayed there in the past. Because the process requires that confidentiality be maintained, only the House Manager and the Social Worker have access to this Rolodex file.

DATE: Thurs 11/18/10

2010 GUEST ROOM LOG

Write firmly in pencil for nightly copying of room log

125 2T/PB — Meghan Jordan	126 1Q/3T/ PB ✓ 4+ guests	151 2T/PB — Richard Sarah	152 2T/PB — Crystal (with #254) Aaron extra pillows + Sheets	214 Q ✓ 2 adults 1 child 1:30-2:00pm	215 2T ✓ Mike	218 Q/PB — Annabel
223 Q —	224 3T ✓ 3+Guests	231 2T/PB — Jennifer Christopher Tyler Remote #5 Fan #4 Airbed	232 2T —	233 3T — 3+ guests Kim Jeff	243 3T/1Q/PB — 4+ guests Glenda Timothy	244 Q — Katie Kenneth #1 Remote
245 2T ✓	250 2T/PB — Jon	251 2T/PB — Jim Penny	252 Q/PB — Andy Jeannie Airbed	253 Q/PB — Teresa Robert	254 2T/PB — Jason Brittany (with #152) Airbed	255 2T/PB — Lynne Brittany

T = Twin Q = Queen PB = Privata Bath R = Remote-Garage F = Fan AB = Air Bed

3+guests = given out to families with 3 or more overnight guests

FIGURE 2.3 RMH guest room log (prior to 2011).

From this description, we can identify three major "roles" for users of the new software: House Manager, Social Worker, and Volunteer. The Social Worker generates a referral form for a new guest family, and the House Manager enters new families into the room log when they check into the House. The Volunteer can check a family out of the house and uses the room log to determine which rooms to clean when they become vacant.

2.2.1 Requirements Gathering

Drilling down from the most general level of domain analysis, we next gather an initial set of requirements from the client in preparation for developing a design document for the new FOSS product. Requirements gathering includes meeting face-to-face with the client, learning about the domain and the application, eliciting so-called "user stories," and then developing an initial set of so-called "use cases" from those stories.

For example, consider the requirements gathered for the design of *Homeroom* for the Ronald McDonald House in Portland, ME. Figure 2.3 shows that the House has 21 rooms, and on a typical night they are almost fully booked. The House is run by a full-time staff of five—an Executive Director, a House Manager, an Office Manager, a Night Manager, and a Development Director. The House is operated with the help of dozens of volunteers. Its staff works closely with social workers at the Maine Medical Center to help identify families who need a place to stay while their children are in the hospital.

2.2.2 User Stories

At a most fundamental level, the new software can be viewed as a collection of computational actions that satisfies all the user requirements for the application. Each user requirement is a concise statement from a particular type of user (called a "user role") about a feature that the software should provide and why. That statement is expressed in plain English as a "user story" (see `https://en.wikipedia.org/wiki/User_story`).

> A "user story" is an informal, natural language description of one or more features of a software system. User stories are often written from the perspective of an end user of a system.

A good user story has four key characteristics:

1. It should be written in a client-focused style, ideally by the client themselves. If the client doesn't write the story, the developer should write it in the language and style of the application domain. Technical jargon should be avoided at all costs, so that the story can be understood by everyone on the development team.
2. It should be brief, so that it can be fully expressed in less than a minute. That is, the story should be a few sentences long and focus on a single activity. Thus, a story is just a placeholder for a more detailed conversation later on, not a full specification or requirements document.
3. It should discuss a modest-sized activity by a single user role, requiring a short time to carry out. An example of such a user story is "Capture the name and contact information from a client needing a room."
4. It should be testable. That is, when the story is implemented, it should be clear how to go about testing whether it has been properly implemented. This helps clarify acceptance criteria, especially when the story appears ambiguous. For example, "Name and contact information should be entered into the system by completing an online form."

So user stories provide a quick way of developing requirements without becoming bogged down in too much formality. The intention of the user story is to be able to respond quickly and with less overhead to rapidly changing real-world requirements. Here are a few examples of user stories for each of the principal user roles for *Homeroom*:

- House Manager
  - Determine if there are rooms that are off-line and arrange for repairs to be made.

- Check a guest out of a room and note that the room needs to be cleaned.
- View a list of pending referrals.
- Book a room on a particular future date.

- Social Worker
  - Determine whether or not there's a vacant room on a particular date.
  - Refer a family to the House Manager to stay at the House on a particular date.
  - Fill out a referral form.
- Volunteer
  - Determine if there are any rooms needing to be cleaned.
  - Record the status of a room as "clean."

Thus, user stories provide small-scale and easy-to-use pieces of information. They are written in the everyday language of the user and contain little detail, thus remaining open to interpretation. Stories thus should help the reader understand what it is the software should accomplish, rather than how it can be accomplished.

2.2.3 Use Cases

A "use case," unlike a user story, describes a user activity in detail, and it contains substantially more information. That is, it describes a series of interactions between the new system and a user for accomplishing a particular task. So user stories provide the fodder from which use cases are built. That is, we may view a use case as an elaboration and combining of user stories that becomes part of the overall software design.

Developing use cases is a highly interactive process. For a larger software project, Cockburn ([9], pp. 81–88) suggests holding a workshop in which the key users tell their stories to developers. That is, they talk about their roles and activities at a somewhat detailed level. The developers in the room do a lot of listening and note-taking. Out of these discussions, user tasks and roles emerge, and a model of these tasks and their interactions can be derived. This so-called "task model" ultimately identifies the individual use cases that need to be written.

For a CO-FOSS project, the task model and its use cases should be written so that they can be modified and expanded as new features are added to the software. Once the new software is developed and installed, a professional developer will be responsible for maintaining the integrity of the software and the data. However, this developer does not fulfill a role because he/she does not interact with the software as a user.

For each user role, the task model describes every task a user must be able to perform using the software. The following specific tasks are associated with the *Homeroom* user roles.

> The House Manager should be able to view the status of each room on the current day's schedule. A room's *status* can be "clean" (but not reserved), "dirty," "occupied," "reserved," or "off-line." An off-line room is one that cannot be occupied because it requires maintenance beyond a normal cleaning. One way to indicate status on a room display is to color-code the room display in the user interface according to its status (green = "clean," brown = "dirty," white = "occupied," yellow = "reserved," and gray = "off-line").
>
> The House Manager or Volunteer should be able to change the status of a room (e.g., from "brown" to "green"), as well as identify the number and names of the guests, and all other details related to the occupancy of the room. If the room is linked to another room, a visual symbol (e.g., a teddy bear) might be used in both rooms' displays to help viewers visualize that linkage.
>
> The software should generate the next day's schedule from the previous day's schedule, allowing the House Manager to update the status of any room(s) for any guest who is checking out on the current day. In general, the room schedule should be viewable for any past or future day, and the details for any particular room should also be viewable for that day. Here is a summary of the two different views and their interactions:
>
> **Room view** should show all the detailed information about a single room on a particular date, including its room number, capacity, status, occupants, and other details describing the room (that is, a single block in Figure 2.3). A House Manager or Social Worker should be able to determine from this description whether a particular room is available and whether it is suitable for a new family to occupy.
>
> **Daily view** should show the date and a summary of all rooms on the screen at once. Each room should appear with its room number, status, and occupants (that is, it should look like all of Figure 2.3). The daily view should be linked to the **room view** whenever the user selects a particular room. It should also be linked to any other daily view on the calendar.
>
> Additionally, the House Manager should be able to keep track of all past and current clients using an online database integrated with the system. The relevant features stored for each client should match the information that appears on the guest registration card (Figure 2.2). The client database should be searchable by client name and other key features.

Finally, the House Manager should be able to view statistical reports on room occupancy—by room, for all rooms, by client, and by room type (capacity, private bath, etc.)—for any series of days, weeks, or months on the calendar.

The Volunteer should be able to view the status of all rooms on the schedule, so that they can clean any dirty rooms in preparation for new occupancy. After cleaning a room, a Volunteer should be able to change the status of a room from "brown" to "green."

The Social Worker should be able to view the status of all rooms on the schedule, in order to tell whether there are any vacancies available on dates when the client's child is scheduled to be hospitalized. While room scheduling is done by the House Manager, this view can help facilitate the referral activity. Social workers should also be able to fill out a referral form online, and all such pending referrals should be accessible to the House Manager (but not to the Volunteers, since confidential information is present).

Unified Modeling Language

The task model can be presented in the form of a *use case diagram*, which is part of the Unified Modeling Language (UML) [4]. UML is a formalization designed to help software developers graphically express the elements of a new software system in sufficient detail that the software can actually be developed using those expressions as a guide.

UML includes a set of graphic notation techniques to create visual models of object-oriented software designs. To display visual models, UML provides 14 different types of diagrams that can capture both the static and the dynamic aspects of the software and its use. The major UML diagram types are summarized as follows:

Class diagram shows the system's classes, their attributes, and their interrelationships.

Component diagram shows how the system is split into components and their interdependencies.

Use case diagram summarizes the system's actors and use cases and their interactions.

Sequence diagram shows the sequence of events that occur during the execution of a use case.

In practice, only a few of the 14 UML diagram types are widely used, especially in a CO-FOSS development setting. Often the requirements of a new software product are so fluid that encapsulating them into a rigid diagram can make the diagram prematurely obsolete.

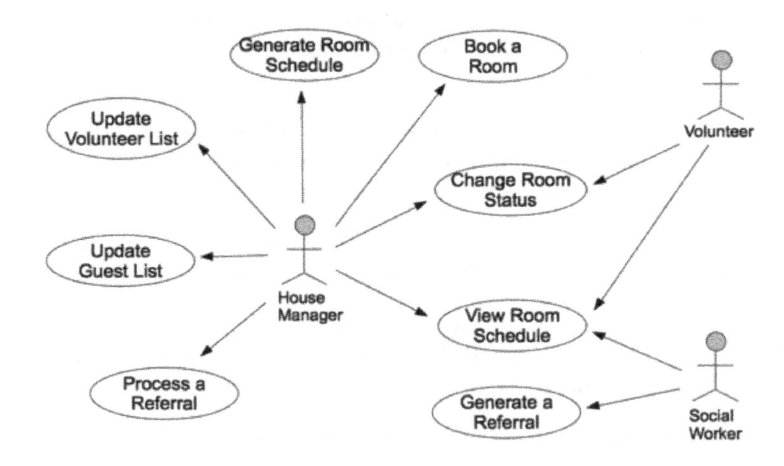

FIGURE 2.4 *Homeroom* use cases.

Nevertheless, we have found that certain UML diagram types, especially the ones identified above, provide useful tools for grounding the details of an initial design and expressing them rather precisely. For example, Figure 2.4 shows an initial use case diagram for the original *Homeroom* software.

This use case diagram delineates eight distinct activities that *Homeroom* must support in order to fulfill its requirements. The House Manager has access to seven of these, while the Volunteer and Social Worker have access to two each.

For example, all three user roles need to view the room schedule. However, only the Social Worker can generate a referral, while only the House Manager can process a referral. These distinctions are made very clear by the use case diagram. The use case diagram can thus stand as an important information resource throughout the early stages of a project's development.

Writing an Effective Use Case

A use case must convey certain information in order for it be useful in a software design. This information is presented in a stylized form, so that readers don't have to dig too deeply to find the information they need. A use case conveys the following information about a specific user-system interaction:

Use Case Name is a unique and meaningful name for the use case—a name written in the language of the domain that will be familiar to users and developers alike. For instance, in *Homeroom*, the use cases have names like "Generate Room Schedule" and "Process a Referral," which provide initial clues about the kind of interaction that it represents.

Description is a brief summary of how this use case will support the application. For instance, the following summary describes the "Process a Referral" use case:

> All the information on the referral form should be entered into the system, and an attempt should be made to reserve a room for this referral on the requested dates.

Actors identifies the roles of users authorized to perform this use case. For instance, for "Process a Referral," the lone user is the House Manager.

Preconditions describe what must be true before this use case can be started. For instance, "Process a Referral" must have all the information for a referral available for processing.

Goals or Postconditions describe the general outcome that will occur when the actor completes this use case. For instance, the "Process a Referral" has the goal of entering the guest's personal and booking information into the database, approving or disapproving the referral, and possibly reserving a room as well. A reason for disapproving a referral might be unavailability of a suitable room on the requested dates.

Related Use Cases identify any other use cases that are directly related to this use case. For example, "Process a Referral" is related to both "View Room Schedule" and "Book a Room."

Steps provide a step-by-step description of how the user will interact with the system to accomplish the goals of the use case. This is usually written as a two–column table, one column for the user's actions and the other for the system's responses. This is the most critical and intricate part of the use case to write, since it holds the key to the eventual design of the interactions in the user interface itself. In the case of "Process a Referral," the steps might look like Table 2.1.

TABLE 2.1 Process a Referral

Step	User	System
1.	Log in	Ask for Username and Password
2.	Enter Username and Password	Verify; Display the outcome and connect the user
3.	Enter a booking for the requested date.	Validate and display the booking
4.	Request room log for the requested date	Display the room log for that date
5.	If there is a room available, reserve it	Complete the booking and confirm the outcome
6.	Log out	Disconnect the user

Steps 1-2 ensure client privacy and system security. For example, only the House Manager can process a referral. Steps 3-5 describe the essence of the interaction, while Step 6 terminates the interaction.

Notice that the style of writing here reflects the language of the domain and doesn't force any technicalities of user interaction or any specific user interface design on the reader; it only indicates what must be accomplished to complete the use case. This is important; over-specifying the user-system interaction at this early stage can tie developers' hands prematurely.

2.3 SOFTWARE DESIGN

Open source software is often modeled using a client-server framework, as discussed in Chapter 1. That is because such software allows several users to simultaneously access the software's services and underlying database from web browsers on their own computers. For example, the architecture of *Homeroom* and the other open source products discussed in this chapter all use the client-server framework.

2.3.1 System and Performance Requirements

A number of initial design choices must be made for a new software project before any development begins. This section identifies those decisions in the form of performance, platform, and process requirements. While many of these may seem obvious, writing them into the initial design creates developer and user consensus at the beginning of the project.

Easy to learn and use: The user interface should be consistent with modern web-based scheduling and database applications. It should be easy to learn by persons with a variety of backgrounds and minimum technical training.

Web-based: The system should have clear and consistent Web-based user interactions, with full functionality and help screen tutorial support.

Secure and protective of privacy: Personal information in the database should be accessible on a strict need-to-know basis. The entire database should also be protected from inadvertent corruption.

Efficient: The user interface should be responsive and the entire system should use computing resources efficiently.

Reliable: The system should run correctly; all user interactions should be reversible and repeatable.

Available: The system should be available 24/7 and accessible from any Web browser connected to the Internet.

Supportive of backup and recovery: Regular system backups should be available. If the system fails or the database is corrupted, full recovery to a reasonably recent checkpoint (e.g., the previous day) should be possible.

Maintainable: The system's architecture and documentation should be designed to facilitate the correction of minor defects as well as the development of future enhancements.

Open source: The software should be freely available and adaptable by any other organization that has a similar need for room scheduling. The software may include components imported from other publicly-available open source products as well. To establish the software's identity as free and open source, an appropriate copyright notice should appear at the beginning of each module and at the bottom of each page in the software's user interface. For example, the GNU General Public License notice would look like this:

```
/* Copyright 2011 by <names of developers>. This program
 * is part of Homeroom, which is free software. It comes
 * with absolutely no warranty. You can redistribute or
 * modify it under the terms of the GNU General Public
 * License as published by the Free Software Foundation
 * (see https://www.gnu.org/licenses/).
*/
```

Server, language, and database platforms: The system should be designed and implemented using programming language, database, and server platforms that support interoperability, for example an Apache PHP/MySQL server.

Interoperability: The system should be accessible from any Windows, Mac, Linux, or mobile platform. For instance, the database and web server may be hosted on a Linux system and users may access it from any Windows, Mac, Linux, or mobile device through its Web browser.

Staging server and development tools: The system should be developed using a staging server that provides a "sandbox version" of the system that is accessible to the client for review of intermediate results throughout the development period. The software and database should be designed so that they can eventually be re-installed on the client's own server and computing environment.

Development team and process: The system should be implemented using a client-centered process.

Development timeline: The project should have a particular starting and ending date, at which time a working prototype should be completed for public demonstration. Intermediate milestones should be set so that clients and developers can exchange ideas about partial results throughout the development period.

Collaboration: Collaboration should be supported by the use of effective collaboration and code management tools, with a preference for open source tools. For example, Google Meet, GitHub, and Eclipse.

Public Availability: The software should be designed so that its source code can be placed in the public domain and be freely downloadable from a widely-used repository such as GitHub.

2.3.2 Software Architecture

Open source software products are often designed to allow different groups of developers to work independently and contribute new features in the form of *plug-ins* to the original code base. This design choice is often called "pluggable architecture." For example, Wordpress (http://wordpress.org) has a pluggable architecture.

Wordpress's developer community has several quasi-independent subgroups, each working on a separate feature, or *plug-in*, that users can add to their Wordpress Web sites. Users can choose to add any particular plug-in by simply downloading it, installing it on their Wordpress site, and activating it. For example, a particular user developing an e-commerce site may download a plug-in that implements e-commerce. Other Wordpress sites may have no interest in e-commerce, so they would not need that particular plug-in.

Many other popular software products, such as Firefox and Eclipse, also have pluggable architectures. Frameworks for adding plug-in functionality to a software product exist for many programming languages, including C++, Java, PHP, and Python. For more information, see `https://npfi.org/plug-in-computing/`.

When we consider the overall architecture of a new open source product, it is useful to model its functionality as a group of smaller, more manageable components. This "divide and conquer" approach allows us to manage the design as a coherent arrangement of individual components rather than one large clump of code. Organizing these components into a conceptually logical and manageable arrangement is thus an important challenge.

Layering, Cohesion, and Coupling

Three important principles can help guide the organization of software components: the *layering principle*, the *maximum cohesion principle*, and the *minimum coupling principle*.

An architecture that follows the *layering principle* allows developers to visualize the software as a small number of interconnected vertical layers.

> **The Layering Principle**: Each module in the code base appears in one of a small number of layers. Each module communicates mainly with modules that are in the layers that are directly above and below it.

Figure 2.5 shows a layered architecture (sometimes called "multi-tiered architecture") with five distinct layers—the user's forms, the GUI modules, the domain classes, the database modules, and the database tables. The double-headed arrows in Figure 2.5 denote two-way information flow, while the single-headed arrows denote one-way control flow. That is, $a \rightarrow b$ denotes that module b provides functional services to module a by way of a method (function) call from a to b.

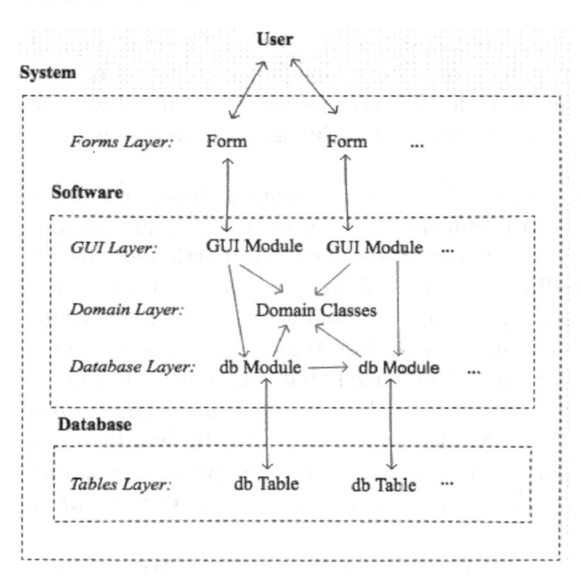

FIGURE 2.5 Layered Architecture ($\leftrightarrow$ denotes information flow and $\rightarrow$ denotes control flow).

The positioning of the arrows suggests that information flows only between adjacent layers. For example, information flows back and forth between each form and the user, as well as between each form and its corresponding GUI modules. Each GUI module manages data flow to and from a single form in the user interface. Similarly, there is a one-to-one relationship between the database modules and the database tables.

At the heart of the software lie the domain classes. They define the major types of data within the system and they implement the client's vocabulary (the so-called "namespace") that was identified during requirements gathering. This is a key distinction of customized single-client software—when the software is activated, the client will continue to use words that are familiar to his/her professional setting, rather than a vocabulary invented by developers for a mass audience. Notice also that the domain classes provide functionality to the GUI modules directly above them and the database modules below.

The domain classes should not interact directly with user forms or with database tables. Moreover, no GUI module should interact directly with a database table to obtain information. If that need exists, the corresponding database module should provide sufficient functionality to deliver that information directly to the GUI module.

The architecture of *Homeroom* exemplifies the layering principle. Modules at the GUI layer support all the interactive forms that the user sees. In turn, the GUI modules use functions provided by the domain class layer and database layer to accomplish their tasks. The database modules interact with their corresponding database tables whenever a user transaction requires that data be stored permanently. Figure 2.6 shows many of the key modules, classes, and interactions in *Homeroom*, and how they fit together to form the GUI, domain, and database layers.

The layering principle leads to two related principles of software architecture, the *maximum cohesion principle* and the *minimum coupling principle*. The former is a property of an individual module in the code base, while the latter is a property of the interconnectedness among the modules.

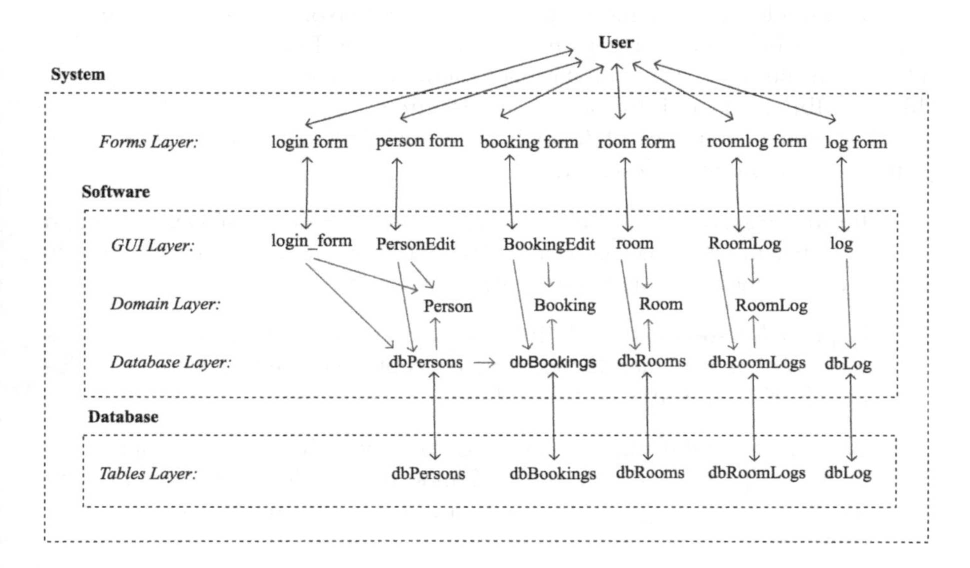

FIGURE 2.6 Layered architecture of *Homeroom*.

The Maximum Cohesion Principle: All the functions that relate to a single concept are gathered into a single module or class. A software architecture is maximally cohesive if all its modules and classes are cohesive in this way.

For example, *Homeroom* has at least two distinct concepts—persons and rooms. When designing this system, we should define two distinct classes—one for all the functionality relating to a person and another for all the functionality relating to a room. An architecture with maximum cohesion would not allow any function that manipulates an object in the Person class to creep into the Room class, or vice versa. Moreover, the Room class should encapsulate all the functionality required by other modules to manipulate any Room object.

The Minimum Coupling Principle: Two modules are *coupled* if either one sends or receives information from the other. A software system is *minimally coupled* when the number of such couplings is kept to a minimum.

Minimum coupling suggests that two modules should not be functionally connected unless they are in adjacent layers or the same layer. Moreover, no pair of modules should be coupled beyond the extent required by the use cases in the design document. Addition of superfluous couplings between pairs of modules should be avoided.

For example, the only interactions between a person and a room in *Homeroom* should be to book a person in a room via the Booking class. No other interactions between these two modules should be needed. If the code supports this, the dbPersons and dbRooms modules are said to be minimally coupled.

The main advantages of following the layering, cohesion, and coupling principles in a software design are:

Improved system readability—ideas related to a single concept are found together in the same place; loosely-related or unrelated ideas are easily ignored when reading a single module or class.

Improved system extendability—adding new features to an existing system is enhanced when the existing concepts and their interrelationships are clearly delineated in the architecture.

Support for debugging—bugs tend to appear in modules where they ought to appear; the occurrence of undesirable side effects from modifying or enhancing a piece of code in a module or class can thus be better controlled.

Support for code reuse—maximally cohesive and minimally-coupled modules and classes tend to be easier to extract and reuse in a different software system that incorporates similar concepts.

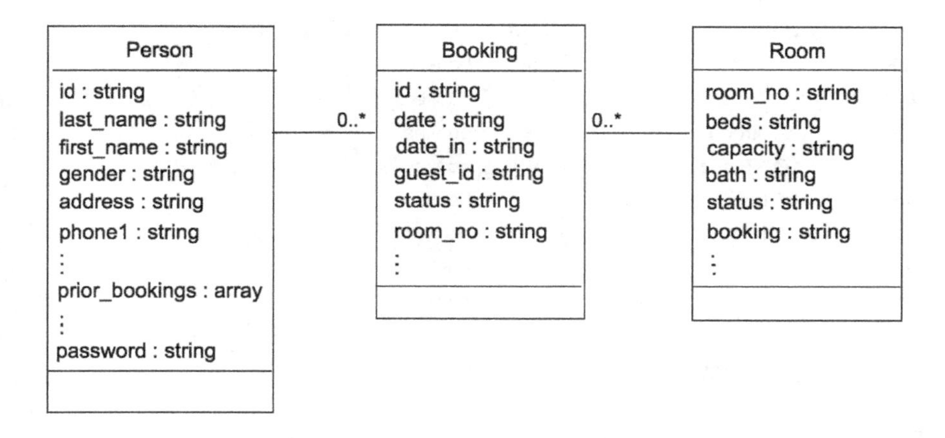

FIGURE 2.7 Some of the initial domain classes for *Homeroom*.

All of the other software in our projects—*Homebase*, *Homeplate*, and *BMAC-Warehouse*—have layered architectures just like *Homeroom*, so it has been easy for our different projects to reuse modules and classes that implement similar concepts and features rather than redesign and reprogram them from scratch. We illustrate this idea more fully in Section 2.3.4.

Domain Class Layer

One area in which UML is particularly useful is to assist with the initial domain class design for a new software system. Several open source UML drawing tools are available that can be helpful with this tedious drawing task.[1] Figure 2.7 sketches some of the initial domain classes and their key instance variables for *Homeroom*. The empty space at the bottom of each box is reserved for adding a list of the methods in that class.

Once the classes and their key instance variables and methods are sketched, the UML tool can automatically generate a new project with these classes embedded. The classes can be generated in any current object-oriented language (e.g., Java, C++, PHP, Python, or Ruby). This step establishes an initial namespace for the code base that is consistent with the one that underlies the user stories and use cases.

Database Layer

Database design involves identifying the structure of the different database tables and then designing the functionality of the modules that correspond to the different tables. In general, the system architecture diagram (Figure 2.6)

[1]One such useful tool is `ArgoUML` (see http://argouml.tigris.org/).

room_no	beds	capacity	bath	status	booking	room_notes
000			n	booked		
001			n	booked		
125	2T, strobe		n	booked		
126	1Q/3T, strobe,		y	booked		
151	2T, horn,		y	booked		
152	2T, horn,		y	booked		
214	Q		n	booked		
215	2T	2	n	booked		
218	Q	2	y	booked		
223	Q/2T, strobe	2	n	booked		

FIGURE 2.8 dbRooms table structure in *Homeroom* database.

and the domain class diagram (Figure 2.7) provide a starting point for the initial database design.

That is, the name of each table in the database corresponds to the name of its corresponding domain class with the symbols "db" prefixed and the symbol "s" suffixed. For example, in the *Homeroom* database there is a table named dbPersons corresponding to the class named Person, a table named dbRooms corresponding to the class named Room, and so forth.

Each database table has columns with names corresponding to the instance variables of its corresponding class. For example, the dbRooms table in the *Homeroom* database will have columns corresponding to each of the instance variables of the Room class (see Figure 2.7): room˙no, beds, capacity, bath, status, booking, etc. Figure 2.8 shows a snapshot of the structure for the dbRooms table.

Creation of a new Room object and insertion of that data into the dbRooms table will thus add a new row to the table. Retrieval of the data in that row will allow a Room object to be constructed. Updating a row with one or more new values corresponds to updating its corresponding object. Deleting a row from the table corresponds to deleting its corresponding object. These four functions provide a framework for implementing the functional behavior of each database module. These concepts will be carefully presented in Chapter 6, when students will be implementing the database tables for their project.

The key difference between a row in a database table and an object in its corresponding class is persistence. That is, once the user's session with the software concludes, all the active objects in that session disappear. However, all the database tables and their contents that were active during that session live on. That is, they *persist* in the same way that all the files in a person's computer remain there whether or not the person is actively using it.

Our discussions also presume that the software is using a so-called "relational database" strategy. While there are several ways of implementing relational databases, our projects all use the MySQL implementation. MySQL is widely used, embedded within the Apache stack (recall Section 1.4.2), well-documented, and easy to learn.

User Interface Layer

User interface design, on the other hand, is not particularly amenable to UML diagramming tools. Instead, a series of sketches can help to express the elements of an initial user interface and their interrelationships. Cockburn refers to these as "screen drafts" in the following way ([9], p. 181):

> Screen drafts are low-cost renditions of the way screens will look, used to explore the ways that users might interact with the system, and invent better ways for them to get their goals accomplished. Very often they are just drawings on paper.

The team may also use a simple graphical drawing tool to generate screen drafts rapidly.[2]

For *Homeroom*, several screen drafts are needed to help design the user interface. For example, each of the different views of the room schedule (room view, day view) requires a screen draft that shows the points of navigation for the client to switch from one to the other. Figure 2.9 shows a screen draft of what the room view can look like for the **House Manager** or **Volunteer** who will be using it.

In this screen draft, the simple notational convention is for existing information to be indicated by underscores, user-entered information to be indicated by empty rectangles, and links (buttons) to other views to be indicated by non-empty (named) rectangles.

2.3.3 Software Security

Security for an open source software product can be enforced in different ways. Two major kinds of security are consciously built into all of our CO-FOSS projects—*database security* and *system access security*.

Database security ensures that only that software which knows the name of the host, database, user, and password can access or modify the information in that database's tables.

Our CO-FOSS projects promote database security by defining a single function in the database layer that connects to the database with these four critical parameters. The function is called connect() and, when called, it connects the session to the database. We enforce security by encapsulating the following strict discipline in all database modules whenever they want to retrieve or modify any information in a database table:

---

[2]We have used `Evolus Pencil`, a versatile open source drawing tool that facilitates quick and easy drawings of screenshots, class diagrams, and many other types of illustrations.

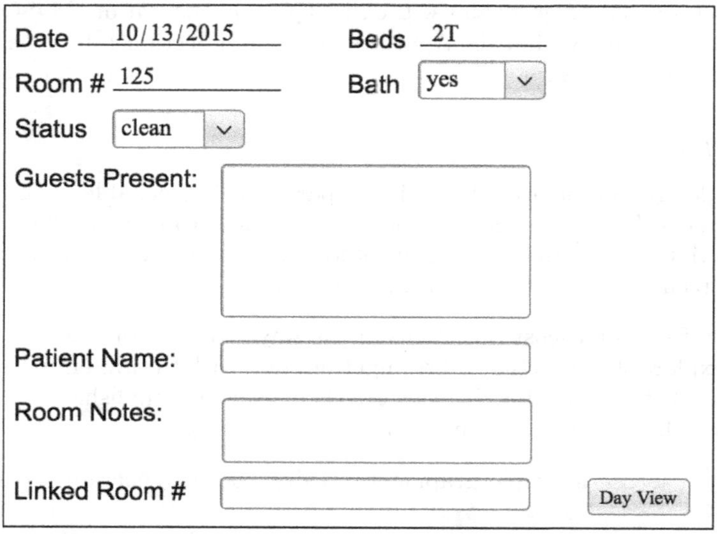

Date 10/13/2015 Beds 2T
Room # 125 Bath yes
Status clean
Guests Present:
Patient Name:
Room Notes:
Linked Room # Day View

FIGURE 2.9 Room view screen draft for *Homeroom.*

1. Connect to the database.
2. Retrieve or modify information in a table.
3. Close the connection.

Briefly, this discipline ensures that the database not remain idly connected to the user session at times when it is not needed. We will discuss and illustrate this discipline in Chapter 6 when the students are implementing the database modules for their project.

System access security ensures that only those persons who have a Username and Password can access the functionality of the system.

Our CO-FOSS projects promote system access security by requiring all users to log in before they can see the main page with the menu of functions that they can choose while using the software. Figure 2.10 shows this form as it appears in *Homebase*, *Homeroom*, *Homeplate*, and *BMAC-Warehouse.*

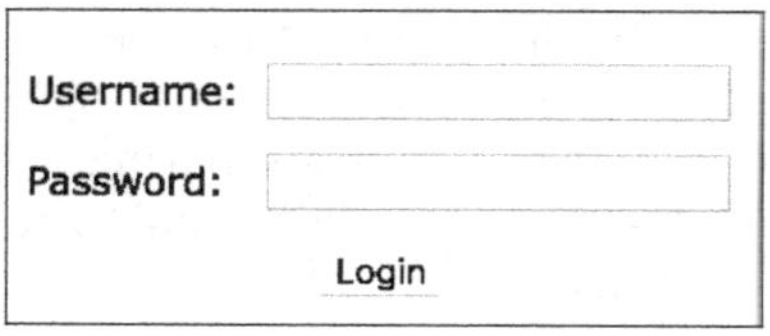

Username:
Password:
Login

FIGURE 2.10 Login Form for Restricting *Homeroom* Access.

A key element of implementing a login form is to identify the entire community of users who need to access the system and then provide them with a login Username/Password protocol that will allow any of them to login easily. Our first instinct is to look around at other software user interfaces and use the same protocol that they use. For example, logging in for access to the apps in one's Google account requires that the user has already created that account using his/her email address as a Username and setting a Password for that account as a separate prior step.

When developing software for a non-profit, this approach doesn't work. That is, the community of legitimate users for the non-profit's software cannot all be expected to have email accounts. Many of the non-profit's volunteers and clients, for example, are elderly, low-income, illiterate, or just plain not interested in having any kind of email account.

On the other hand, we can be certain that at least 99% of any non-profit's volunteers and clients have phones. Knowing that a 10-digit number uniquely identifies each person who has a phone, it makes sense to define the Username with this 10-digit number as part of it.

So in *Homeroom* we define a Username as a string that concatenates the user's first name and phone number (with no embedded blanks or special characters). For example, the Username Allen7037298111 is a legitimate Username for a person named Allen who has the phone number 703-729-8111 (please, no phone calls :-). Good usernames are ones that the user can easily and intuitively remember, and that others cannot easily remember.

In *Homeroom*, a new user's password is initially set to match their Username. When they log in, the software reminds them to change their password to something that they alone can remember. In practice, many Volunteers never do this, so the software remains vulnerable in this sense to Username/Password hijacking.

As an additional level of system security, we limit the particular pages and functionalities that a user can see depending on their Role, as discussed in Section 2.2.3. Thus, persons with House Manager access are few in number and will surely personalize their passwords. Many persons with Volunteer access are unlikely to change their default passwords, but they also have access to a much more limited level of functionality within the system.

2.3.4 Encouraging Code Reuse

A newly conceived open source software project should resolve its major design choices before coding actually begins. Most of these choices have been resolved as suggested in Section 2.3.1.

In addition, designers should look for opportunities to reduce coding time and improve system reliability by actively engaging in the practice of *code reuse* to implement a common feature. Unlike plagiarism, code reuse is an open source technique that helps developers avoid "re-inventing the wheel" by finding, downloading, and using someone else's well-worn code that already implements the same feature.

The adaptation of a pre-written FOSS component to "fit" within the new system's architecture is often easier and more reliable than writing and unit-testing the code from scratch. An important caveat of this activity is that the developer should acknowledge authorship of reused code, so that the original author receives proper credit for the work.

For example, our design of *Homeroom* requires secure login by system users in order to protect the privacy of families who are staying at the House. Managers and Volunteers must authenticate with a Username and Password before viewing or modifying any of that family or room occupancy data.

However, looking back at past projects, we recall that *Homebase* had the same need for users to authenticate with a Username and Password before accessing secure data about volunteers—their contact information and their shift schedules for volunteering at the House.

The layered architecture for *Homeroom* shown in Figure 2.6 reveals exactly where this login form should appear and what modules and classes are needed for implementing this login form. Briefly, they are:

GUI Layer: the login_form module

Domain Layer: the Person class

Database Layer: the dbPersons module

Notice in Figure 2.7 that the Person class has a field for storing the user's password (encrypted), and that the dbPersons module can update a row in the dbPersons table when that person changes his/her password. So all the machinery for managing secure logins in *Homeroom* is available for reuse by downloading and adapting their counterparts from the *Homebase* code base.

2.4 THE DESIGN DOCUMENT

With a reasonably complete set of requirements and use cases in hand, as well as the system performance requirements, sketches of the initial domain classes and screenshots, and an understanding of the importance of layering and code reuse, this section presents a framework for writing the design document itself. We have used variations of this framework for all the design documents that have guided our CO-FOSS projects since the first one in 2008.

Some of our projects implemented fresh custom software for an application that had no significant prior software support (the original versions of Homebase in 2008, Homeroom in 2011, and Homeplate in 2012). For other projects, different student teams extended an existing CO-FOSS product for the same client (Homebase and Homeroom in 2013) or redeployed it to fit similar needs for a new client (RMHP-Homebase and BMAC-Warehouse in 2015).

However, the overall structure of the design document for each of these projects remained fairly stable, regardless of whether it specified an entirely new product or it specified an extension or redeployment of an existing product. The major differences among these types of projects lie in their relative amount of code reuse and level of detail in their use case descriptions.

2.4.1 Overall Structure

The main goal of a design document is to provide guidance for the student developers and the clients throughout the project in which they are implementing a new software product. For the students, this document will be their first substantive contact with the project, so it should provide a clear picture of the user's needs, the overall software design, and where to begin. For the clients, it provides somewhat of a contract that clarifies both the features and the limitations of the product that will be developed to support their work.

Student developers will use this document for guidance on the specific intentions of its use cases and fundamental aspects of the domain, database, and GUI components. They will refine this document as they develop the software itself, especially when they present new features to the client and receive feedback at each iteration of the client-centered cycle. In this sense, the design document is an active component in the development process alongside the code.

Table 2.2 provides a template for the overall structure of the design document, along with references to earlier sections of this chapter that provide examples and clarifications.

TABLE 2.2 Overall Structure of a Design Document

Project Title, Client, Developers, Author, Date		
Section	**Heading**	**Contents**
1	Requirements	see Section 2.2.1
1.1	Domain Analysis	see Section 2.2
1.2	The Current System	see Section 2.2
1.3	Overall Goals and Features	see Section 2.2
1.4	(Current Issues to be Corrected)	see *Homeroom* design
2	Software Design	see Section 2.2
2.1	Forms and Reports	see Section 2.2
2.2	Levels of Access	see Section 2.3.3
2.3	Use Cases	see Section 2.2.3
2.4	User Interface Design	see Section 2.3.2
3	System Requirements	see Section 2.3.1
4	Software Architecture	see Section 2.3.2
4.1	Code Structure and Copyright	see Section 2.3.1
4.2	Domain Classes	see Section 2.3.2
4.3	Database Design	see Section 2.3.2
5	Project Startup Schedule	see *Homeroom* design

In this outline, note that Section 1.4 is optional, and should be included only if the new software will refine or extend an existing FOSS product for the same client. This was the case, for example, in 2013 when a student team refined and extended the 2011 version of *Homeroom.*

2.4.2 Variations

This outline should be viewed as a loose framework with plenty of room for variation depending on the starting and finishing points of the project.

For example, if the project being described begins with a fully functional code base for a very similar product, to which the developers will be adding a few new features and making minor adjustments to other features, the design document should reflect this. For example, this was the case when a different student team refitted the 2013 *Homebase* code base code that was in use at the Ronald McDonald House in Portland, ME so that it matched the volunteer calendar details of the Ronald McDonald House in Providence, RI in 2015.

In this case, the 2015 design document presented the project as a reworking of an existing code base rather than a bottom-up build of a new code base. Its Requirements section still had to show students what manual procedures the new software was going to replace, but its Software Design and Software Architecture sections used a good deal of the 2013 code base as a starting point. For a look at the details, see `https://npfi.org/homebase-design/`.

The design documents for many of our other projects illustrate more variations that may be of interest.

> The 2012 *Homeplate* `project` required two semesters to complete. Its design document identifies this partitioning of the work and describes only the first semester's piece of the project in detail. The second semester's project was actually completed by a single student as a summer internship. Its goal was to develop an Android app, whereas the first semester's goal was to develop an Apache/MySQL/PHP server side app that would communicate scheduling and activity data received via FTP from the Android and support the management of that data. If we had tried to cram both these apps into a single semester's project, our chances for success with either would have been diminished.
>
> The 2013 *Homeroom* `project` refined and extended the original 2011 version of *Homeroom.* Part of that extension would have integrated its data reporting with CiviCRM, so that the staff could view the data through the lens of a constituent/donation database system where they could manage other donor and donation data outside of *Homeroom.* We learned early in the project that this integration was not feasible in one semester, so we retreated to allow the student team to complete all the other refinements and extensions to *Homeroom.* In the initial design document, we had failed to remember the important lesson of not "overdriving our headlights" that we had learned while designing the 2012 *Homeplate* project.

The 2015 *BMAC-Warehouse* `project` developed new software, but it reused a major portion of the *Homeplate* software described earlier. In addition to reusing the login protocol described above, the student team was able to reuse *Homeplate*'s Client class two times, once for the new Donor class and once for the new Recipient class. Much of the reporting functionality from *Homeplate* was reused in *BMAC-Warehouse* as well, even though in essence they are two very different applications.

2.5 THE SANDBOX

Sometimes a new CO-FOSS project will reuse substantial parts of an existing FOSS product as a starting point. In this case, the instructor may have an opportunity to create a so-called "sandbox version" in which part or all of the existing system is launched prior to the beginning of the project. This sandbox version provides the new client and, later on, the student developers with an initial platform for design, experimentation, and testing.

Each of the four projects discussed above and illustrated throughout this book have associated sandbox databases and instructions for creating a sandbox system that can be downloaded and installed from the following sites:

`Homebase Installation and Reuse`
https://github.com/megandalster/homebasedemo2017/wiki/Installation-and-Reuse

`Homeplate Installation and Reuse`
https://github.com/megandalster/sh-homeplate/wiki/Installation-and-Reuse

`Homeroom Installation and Reuse`
https://github.com/megandalster/rmh-homeroomcivi/wiki/Installation-and-Reuse

`BMAC-Warehouse Installation and Reuse`
https://github.com/megandalster/bmac-warehouse/wiki/Installation-and-Reuse

In these and other CO-FOSS projects, the sandbox should be created using a fully anonymized database. An anonymized database can be prepared by extracting a sample of "live" data from the client's current system and then modifying all personal information offline using a spreadsheet or stylized SQL queries in a "global replace" fashion.

When modifying the database in this way, the preparer must carefully anonymize the names, addresses, phones, and emails of all persons in the database so that real people cannot be inadvertently identified by users or developers while the new CO-FOSS product is being developed.

2.6 SUMMARY

This chapter discusses the first steps needed to identify and begin developing a customized open source software product for a single client. These steps are normally found in the "requirements analysis" phase of a top-down software engineering project. Some of the tools discussed here, such as use cases and class diagrams, are fundamental in that setting as well.

However, the design of an open source development project differs significantly from the requirements analysis step in a traditional software process. That is partly because CO-FOSS development is bottom-up and fluid. New open source project requirements are written with the idea that they may change during the development process. Moreover, open source projects can (and should) freely reuse code from other open source products with similar functionalities. These two notions of agility and code reuse enable a student team to create an exciting new software product for a non-profit within the constraints of a 1- or 2-semester academic course.

The next chapter discusses the design of an academic course in which students can collaborate with the client to complete the implementation of a CO-FOSS software product using these newly-developed requirements.

2.7 MILESTONE 2

1. The instructor should find a capable client that would benefit from receiving a new FOSS product and would be willing to participate actively in its development.

2. The instructor and client should collaborate to develop an initial design document for that product, following the guidelines described in this chapter.

CHAPTER 3

Defining the Course

"The most important single aspect of software development is to be clear about what you are trying to build."
—*Bjarne Stroustrup*

Embedding an open source software project within a 1- or 2-semester course must meet two goals. As an academic requirement for graduation, the course must remain as intellectually rigorous as any other course in the curriculum. At the same time, the course must be organized in a way that students successfully complete the software itself. Organizing a course to meet these two goals simultaneously can be challenging.

In this chapter, we introduce the organization of a software development course in which the students create a real open source software product and simultaneously gain academically-creditable knowledge. The software product itself can be developed from scratch or, more likely, as an adaptation of other FOSS products that have similar components, as discussed in Chapter 2. This chapter provides a framework for defining a course that can fulfill these goals. This framework is based on our own experiences designing and teaching several such courses during the last 10 years.

3.1 SOFTWARE PROJECT ELEMENTS

A CO-FOSS project begins with the design document, which describes both the client's requirements and the overall scope and goals of the project that can fulfill those requirements. The project's organizational starting point is that document, the syllabus, and a commitment to a client-centered process, as described in Chapter 1. To complete the project, the student developers will also use a particular programming environment and set of development tools. Below is a summary of the key tools that they will master while completing the project.

3.1.1 Collaboration Tools

At the beginning of the course, the instructor, student developers, and client should establish short, rich communication paths to be used throughout the life of the project. These paths require the use of on-line tools that accommodate the fact that the developers and client are typically not within easy commuting distance – meeting face-to-face every week or two is not a normal setting. These tools include:[1]

- a private discussion forum,
- a document-sharing repository,
- a code-sharing repository,
- a synchronized face-to-face communication tool, and
- a task-management or planning tool.

The discussion forum is important because it allows team members to discuss project-related questions and suggestions in a private medium separate from their personal email accounts. An example is a private `Slack` channel, which can be established for use by the project's developers and client.

The document-sharing repository is important because it allows team members to asynchronously update the project's design document as the project proceeds. An excellent tool for document sharing is `Google Drive`. `Dropbox` also provides document sharing, and it is integrated with `Slack`.

The code-sharing repository is at a remote project hosting location where the project's code base is stored and maintained by the developers. Popular open source code-sharing repositories are `GitHub` and `SourceForge`. We discuss project hosting and code sharing more carefully in Section 3.1.3.

The face-to-face communication tool is important because it facilitates not only live discussions but also screen-sharing and text messaging between developers and clients. The details of an issue are often not resolvable without a precise example visible as a screen image or a shared URL. An excellent tool for face-to-face communication is provided by `Google Meet`, while similar tools are provided by `Skype` and `Slack`.

The task management tool is important because it lets team members interactively plan and assign tasks in advance of upcoming project milestones. It is also useful for defining user stories, use cases, and milestones. Two highly-recommended task management tools are `Trello` and `GitHub Projects`.

Since we began developing CO-FOSS projects in 2008, the landscape of freely-available collaboration tools has changed a lot. For example, we used `Basecamp` for communication in the 2008 *Homebase* project before it became proprietary. We used Google Groups in 2011 and Google Code in 2013 because

[1]All the tools discussed in this section are widely-used and provide free versions. Other tools exist in each of these categories, but most are proprietary (not free).

they had become better alternatives. In 2015, our two CO-FOSS projects used Google Hangouts (now Google Meet), Google+ Communities (discontinued), Google Docs, and Google Code (also discontinued) to facilitate communication, document sharing and code sharing.

So because collaboration tools continue to evolve, instructors should use the above recommendations as a starting point, but also remain open to choosing different collaboration tools when planning their own CO-FOSS projects.

3.1.2 Development Platform

A development platform (often called an *integrated development environment*, or IDE) is a software application that supports the development of software in a particular programming language. This support includes coding, compiling, testing, debugging, and running individual programs that comprise the code base for a software product.

Examples of IDEs include Eclipse, Microsoft Visual Studio, Vim, NetBeans, and XCode. Each of these is a multi-language IDE, but the supported language set is different in each case. Android Studio is another popular IDE, but its applications are limited to the Java-based Android platform. Similarly the PyCharm IDE is centered on the more modern Python programming language.

Another consideration in choosing an IDE for developers is cost. While some IDEs are free and open source (e.g., Eclipse, NetBeans, PyCharm, and Android Studio), others are proprietary and relatively costly (e.g., Microsoft Visual Studio). A summary of the most popular IDEs as of early 2018 can be found `here`.

Each student developer must install the chosen IDE on his/her own computer, thus gaining access to the following kinds of tools:

Source code editor that has such features as *syntax highlighting*, *syntax checking*, and *autocompletion* of application-specific terms that simplify the coding process

Source code compiler and *interpreter*

Build tools that link the source code with binaries, libraries, and other resources to make an executable application

Debugging and unit testing tools that help the developer find run-time errors and fix issues

Version control tools that help the developer synchronize his/her code contributions with the repository

Documentation tools that support writing documentation for individual source code modules and functions

For example, the `Eclipse IDE` is particularly popular. Historically, Eclipse began as an IBM Canada project, but it is now supported by the Eclipse Foundation and many software firms. Licensed under the Eclipse Public License, Eclipse is open source, cross-platform, and supportive of several programming languages and associated development tools.

Our students have used the Eclipse IDE with PHP, JavaScript, and MySQL in most of our projects since 2008. In the 2012 *Homeplate* project, we used Eclipse with Android Java to develop an Android app.

An attractive alternative to PyCharm for Python development is to configure Eclipse for Python development, using `Google's App Engine`. This combination pairs the attractiveness of Python programming with the power of Eclipse for web development.

3.1.3 Project Hosting

At the beginning of a CO-FOSS project, a project hosting service should be chosen for code sharing and supporting communication among developers and users. Such a service should provide all the communication and code development tools that the project needs. Important among these tools are version control, project Wiki, bug (issue) tracking, and project team memberships.

Prominent project hosting sites include Sourceforge, GNU Savannah, GitHub, and Bitbucket. For a more complete list, visit `https://npfi.org/project-hosting-sites/`.[2]

Started in 1999, `Sourceforge` has been a long-standing open source project hosting site. Sourceforge is a general purpose service that hosts a wide range of open source projects spanning many different programming languages and many varieties of free and open source licenses.

Begun in 2001, GNU Savannah is sponsored by the Free Software Foundation (FSF) and has two sub-domains, one that is officially part of the GNU Project and the other for non-GNU projects. GNU Savannah supports only free software projects as defined by the FSF and it bans the use of non-free formats such as Macromedia Flash.

GitHub is the most widely-used project hosting site. Founded in 2008, GitHub currently hosts over 68 million projects and has more than 24 million registered users. Unlike other hosting sites, GitHub hosts both open source and proprietary software projects. However, to create a private repository (for proprietary projects) in GitHub, registration is not free. Over the past 10 years GitHub has hosted and supported a number of prominent open source software projects, including MySQL, Python, Ruby, and Wordpress. GitHub provides the following broad services:

> *Search*: GitHub contains thousands of open source software products for a wide variety of applications.

[2]Until recently, Google Code was Google's entry in the project hosting arena, but Google discontinued it in 2016.

Community Building: Projects hosted on GitHub are often looking for developers to join in their effort. Registered users can participate in one of the existing projects by commenting on issues and submitting bug fixes.

Project Hosting: Registered users can create their own software projects. The registration process is open and accessible and provides a variety of tools and services to support project development.

Code Hosting: The project's source code can be hosted as long as it uses the Git version control system.

Issue Tracking: GitHub provides lightweight bug tracking and issue reporting services for the project. These services help manage the reporting and handling of bugs in the software as well as requests for enhancements and new features. Larger, more mature FOSS projects often use stand-alone tools for bug tracking, such as BugZilla (see Section 1.5.2 for more details.

Communication Services: Wikis, webpages, and links to other information can be added to any GitHub project.

Finally, Bitbucket is also a popular project hosting service, though its free version is restrictive on the maximum number of users (5) per project. Its repositories are all private, making Bitbucket more attractive to proprietary software projects than open source projects, whose aim is to share code widely.

All the projects discussed in this book are hosted on GitHub. Since they are open source, these projects are free to access by the public. For a list of these projects, visit `https://github.com/megandalster/`. Any of the projects listed there can be freely accessed, downloaded, exercised, modified, and reused under the terms of the GNU General Public License.

Once a CO-FOSS project is uploaded to a public hosting site, it is positioned to evolve and grow into a larger, mature project later on. The hosting service provides the project with the necessary exposure for outside developers and clients to learn about it, join the project, or download and reuse its code.

3.1.4 The Version Control System

A software project's code base usually has a large number of files that change over time as developers make additions, deletions, and modifications to it. For all but the very smallest projects, it is essential to use a *version control system* (VCS) to help manage this process.

A VCS is software that helps manage changes to a collection of source code files being made by different developers asynchronously. This software is useful even if it has only a single developer, but it is essential when there is a team of developers all sharing the same code base.

A VCS coordinates all changes to the code base, so that whenever a programmer adds a new piece of code, it doesn't simply overwrite the previous version. Rather, the VCS keeps both copies, making it possible to revert to a previous version if a current line of development needs to be abandoned.

In addition, the VCS keeps track of who contributed the code, when it was contributed, and so forth. It provides tools to compare one version of the software with another, making it easy to see what has changed, incrementally, from one version to the next over time.

A good (non-software) example of a VCS is Wikipedia. For any page in Wikipedia, it is possible to examine the history of the page, which will show exactly what changes were made, by whom they were made, and when they were made. If a user believes that a set of changes is incorrect, he or she can revert to a previous version.

In general, a VCS supports the following code management functions:

- Multiple revisions of the same code base
- Multiple concurrent developers
- Locking, synchronization, and concurrent updating of the code base
- Support for versioning history and project forking, meaning that certain parts of a project or the entire project itself can be split off (forked) into a separate project

Here are some brief descriptions of the common VCS concepts and operations:

Repository: The repository is the place where the shared code base is stored. While anyone can download and read code from a FOSS repository, only the developers with so-called "commit" privileges have authority to make changes in the repository.

Working copy: This refers to each individual developer's copy of the code base.

Check out: Any developer can retrieve a complete copy of the code base from the repository, downloading it to his or her computer, often activating the code for editing and testing within his/her own IDE such as Eclipse.

Pull/Update: The developer can download the most recent version of a code base that he/she had previously checked out. The updated version will include changes made by other programmers in the interim.

Commit/Push: A developer who has sufficient "commit privileges" with the repository can contribute new code directly. The programmer's working copy of the code is immediately merged with the existing code base, and then copied to the repository so that other developers will have access to this contribution.

Merge: Two pieces of code are merged whenever they both apply to the same file or code segment.

Trunk: A repository is organized temporally into a hierarchy, which is a tree structure. The trunk is usually the main development line for a project, and the bottom of the trunk marks the original code base. As developers contribute new code over time, the trunk grows vertically.

Branch: A branch is a copy of the software that can undergo separate and independent development. Branches are often used to package *releases* of the software or to allow experimentation. In some cases a branch can break off (or *fork*) into an entirely separate project. In other cases, branches are later merged back into the trunk.

Many different VCSs are currently in use. The oldest one is called `Concurrent Versions System (CVS)`, first released in 1990 under the GNU General Public License. But developers found that CVS had some serious restrictions. For instance, the module structure in a CVS repository could not easily be reorganized.

`Subversion (SVN)` was released in 2000 under an open source license and is supported by the Apache Foundation. Subversion addressed many of the restrictions imposed by CVS, including the one noted above.

Different version control systems take different approaches to managing a repository. Some, such as CVS and SVN, use a centralized *client-server* approach. Working copies are related to the repository as clients are related to the server – that is, in the same way that spokes of a wheel are related to the hub. Newer VCSs, such as `Git` and `Mercurial`, use a distributed approach, where there is no single centralized repository. Instead, changes are distributed in a *peer-to-peer* fashion. Git and Mercurial are among the leading VCSs in use today.

Our most recent software projects in 2015 used the Mercurial VCS and were hosted in Google Code repositories. Since the Google Code hosting service was discontinued in 2016, we migrated all our projects to GitHub and the Git version control system. Going forward, we recommend the GitHub/Git hosting/VCS combination for new open source software projects. The GitHub hosting service is complete, widely-known, and free for public repositories. The Git VCS is also free and fully functional. In the next chapter, we will provide a more detailed discussion of managing a code base using Git and GitHub.

3.1.5 Sandbox and Live Versions

While a code synchronization repository facilitates communication among developers, it is useless to the other member of the project team, the client.

To facilitate interaction among all team members, the project needs a vehicle for clients and developers to simultaneously experiment with the current version of the software so that they can see what's working and provide informed feedback on that release's strengths and weaknesses. This is particularly important when the user interface components are being developed.

The use of a *sandbox version* can support this kind of interaction. The sandbox version is hosted on a so-called *staging server*. Typically, the sandbox version has a current copy of the project's code base from its VCS repository, along with an anonymized database for clients to easily interact with the code's functional elements.

The sandbox version's functionality naturally grows with each new release. At the end of the project, the sandbox version provides a complete user interface for the software product being developed, implementing all the use cases that appear in the design document. This client-developer interaction is illustrated in Figure 3.1.

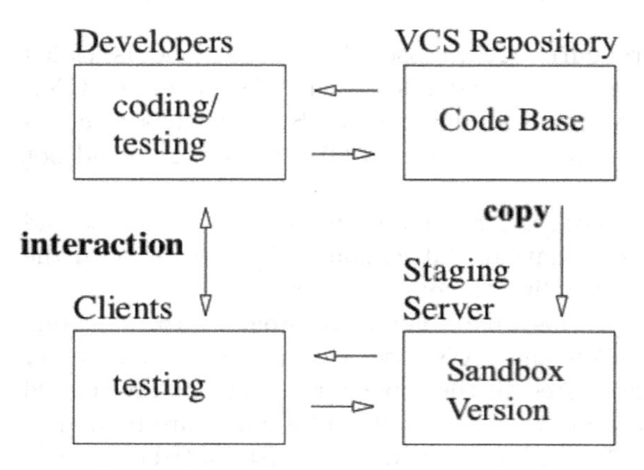

FIGURE 3.1 The sandbox version: client-developer interaction.

The sandbox version can be set up by an IT support person outside the project, by the instructor, or by another member of the development team. In any case, it is a temporary tool that will go out of business when the project reaches completion. At that point, a new *live version* of the completed software should be installed on the client's own server, using a real database replacing the anonymized database. Often, the live version is embedded within the client's Web site, in order to facilitate access by staff members and other site visitors.

As an example, a sandbox version for *Homebase* development is installed as a demo version at `https://npfi.org/homebase-demo/`. A separate instance of *Homebase* can be downloaded and installed for experimentation by any instructor or developer who follows the instructions at `https://npfi.org/homebase-install/`.

When the *Homebase* project was first completed, it was installed on the client's live server in 2008. In subsequent courses new versions were developed by different student teams, released, and installed in 2011, 2013, and 2015 on that server. The current live version of *Homebase* is in use at `http://rmhcmaine.org/homebase-login`.

3.1.6 Reading, Writing, and Documenting Code

Program reading and writing is, of course, the central activity within software development, since the program is the software. The ability to read and write program code in the language(s) of the application is a key requirement for every developer in a software project. A corollary requirement is that developers must be able to learn new language features and skills as needed to support the timely implementation of new software elements.

Many fine programming languages are in use for developing contemporary software. The example program code shown in this book is written in PHP, JavaScript, MySQL, and Android Java. Readers who are already familiar with Java or C++ should have little difficulty reading, adapting, and/or translating this code to another language, such as Python, Java, C++, or C#.

Reading code is an especially important skill. Unfortunately, the code we read in published programming textbooks is not typical of the code found in most software applications. In general, code published in textbooks tends to be more readable than code found in real applications. Code that appears in real applications may have more than one author and may contain elements that are inefficient, verbose, or otherwise difficult to read.

```
if(!array_key_exists('_submit_check', $_POST)){
    echo('<div align="left"><p>Access to <i>Homeroom</i>
        requires a Username and a Password. '  );
    echo('<ul><li>You must be a Ronald McDonald House
        <i>staff member, volunteer, or social worker</i>
        to access this system. <li> Your Username is your
        first name followed by your phone number. ' . '');
    echo(' If you do not remember your Password, please contact
        the House Manager>.</ul>');
    echo('<p><table><form method="post"><input type="hidden"
        name="_submit_check" value="true">
        <tr><td>Username:</td><td><input type="text"
            name="user" tabindex="1"></td></tr>
        <tr><td>Password:</td><td><input type="password"
            name="pass" tabindex="2"></td></tr>
        <tr><td colspan="2" align="center"><input type="submit"
            name="Login" value="Login"></td></tr>
        </table>');
}
```

FIGURE 3.2 Example code from *Homeroom*.

Figure 3.2 shows some example PHP code from the *Homeroom* project, and Figure 3.3 shows the page it produces in a Web browser. As shown,

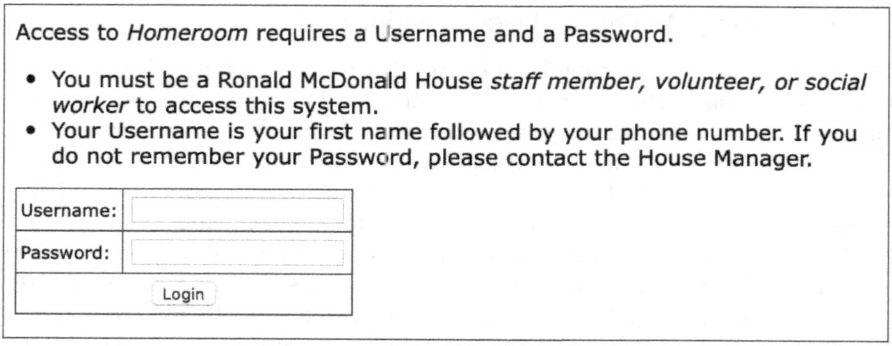

Access to *Homeroom* requires a Username and a Password.

- You must be a Ronald McDonald House *staff member, volunteer, or social worker* to access this system.
- Your Username is your first name followed by your phone number. If you do not remember your Password, please contact the House Manager.

Username:	
Password:	
	Login

FIGURE 3.3 Output of the example code in Figure 3.2.

HTML can be embedded inside the PHP script that defines the layout of a Web page. The HTML tags `<table>`, `<tr>`, and `<td>` appear inside the PHP `echo` functions to accomplish the display shown in Figure 3.3. Whenever this code is executed, the page is displayed.

As a general rule, effective software development requires that the code not only be well written but also be well documented. Software is well documented if a programmer unfamiliar with the code can read it alongside its requirements and gain a reasonable understanding of how it works. Minimally, this means that the code should contain a documentary comment at the beginning of each class and non-trivial method, as well as a comment describing the purpose of each instance variable in a class. Additionally, each complex function may contain additional documentation to help clarify its tricky parts.

When reading a code base for the first time, a new developer may find a shortage (sometimes a complete absence) of commentary documentation. If the code is well written, the reader may still be able to deduce much of its functionality from the code itself. In fact, it is a good exercise for a developer new to a code base to add commentary documentation in places where it is lacking. That is, it improves one's own understanding and it contributes to the project by improving future readers' understanding of the code.

Writing code is a craft. That is, when adding a new class, method, or module to the code base, developers use agreed-upon norms for programming style, including indentation, naming, and documentation. Each programming language has its own coding culture, so to speak, where the norms are well-established and tend to carry over from one project to the next.

In a CO-FOSS project, student developers need to learn and follow the coding standards established for their project, being careful not to superimpose their own stylistic idiosyncrasies in a way that would make their own code less understandable for the next developer who happens to read it. A good starting point in this regard is to rely upon the domain classes as the source for naming new elements. For example, if the Person class has an instance

```
/*
 * class Shift characterizes a time interval in a day
 * for scheduling volunteers
 * @version May 1, 2008, modified 9/15/08, 2/14/10, 2/5/15
 * @author Alex, Malcom, and Xun
 */
include_once(dirname(__FILE__).'/../database/dbShifts.php');
include_once(dirname(__FILE__).'/../database/dbPersons.php');
class Shift {
    private $yy_mm_dd;       // String: "yy-mm-dd".
    private $hours;          // String: e.g., '9-1' or 'night'
    private $start_time;     // Integer: e.g. 10 for 10:00am
    private $end_time;       // Integer: e.g. 13 for 1:00pm
    private $venue;          // "house", "fam", or "mealprep"
    private $vacancies;      // number of vacancies in this shift
    private $persons;        // array of ids and names
    private $removed_persons;// persons previously removed
    private $sub_call_list;// id of SCL it exists, else null
    private $day;            // string name of day "Monday"...
    private $id;             // "yy-mm-dd:hours:venue unique key
    private $notes;          // manager notes
    ...
```

FIGURE 3.4 Inserting comments into the 2015 version of the *Homebase* Shift class.

variable last_name, one should resist the temptation to create "surname" as a synonym for it when "last_name" would work just fine.

Documentation standards exist for most current programming languages, including Python, Java and PHP. These latter are supported by the JavaDoc and PHPDoc tools, respectively. They implement a standard layout for the code and its comments, and they can also automatically generate a separate set of documentation for the code once it is fully commented.

For example, consider the Shift class in the *Homebase* application. Its documentation contains a stylized comment of the form:

```
/**
 *
 */
```

at the head of each class and each non-trivial method, as well as an in-line comment alongside each of its instance variables. This is shown in Figure 3.4.

TABLE 3.1 A few PHPDoc Tags and their Meanings

Tag	Meaning
@author	name of the author of the class or module
@version	date the class or module was created or last updated
@package	name of the package to which the class or module belongs
@return	type and value returned by the function or method
@param	name and purpose of a parameter to a function or method

Notice that this stylized comment at the top contains so-called *tags*, such as `@author` and `@version`. When used, each of these tags specifies a particular aspect of the code, such as its author, the date it was created or last updated, and the value returned by a method.

A short list of the important PHPDoc tags with a brief description of their meanings is shown in Table 3.1. A more complete summary and tutorial on using PHPDoc can be found at `https://docs.phpdoc.org/`.

Java has similar tagging conventions, which are summarized at `https://en.wikipedia.org/wiki/Javadoc`. It is important for developers to learn their language's documentation tagging conventions before adding documentation to the code base.

When the PHPDoc or JAVADoc tool is integrated into the developer's IDE, it can automatically generate complete documentation for all the classes and modules in the code base. For example, Figure 3.5 shows the documentation generated by PHPDoc for an earlier version of the `Shift` class shown in Figure 3.4.

3.1.7 Unit Testing

A *test suite* is separate code that is written solely for the purpose of testing the classes and modules in the code base individually. Each class or module being tested is called a *unit*, and it has a single *unit test* in the test suite associated with it.

> *Managing and running an effective test suite is essential to software development.*

Traditional development models placed testing late in the process, after the coding had been completed. Years of experience proved that that was the wrong placement. Contemporary models place testing hand-in-hand with the coding process. In fact, sometimes it is best to design a unit test from a use case or a client's suggestions before coding its corresponding class or module. This approach is called *test-driven development* or TDD for short.

TDD is especially valuable because it keeps the client "in the game," so to speak. That is, frequent discussions between developers and clients can evoke examples of system and user interface behavior before the underlying code is written. These examples thus become the unit tests for testing the code to be sure that it captures the essence of those discussions.

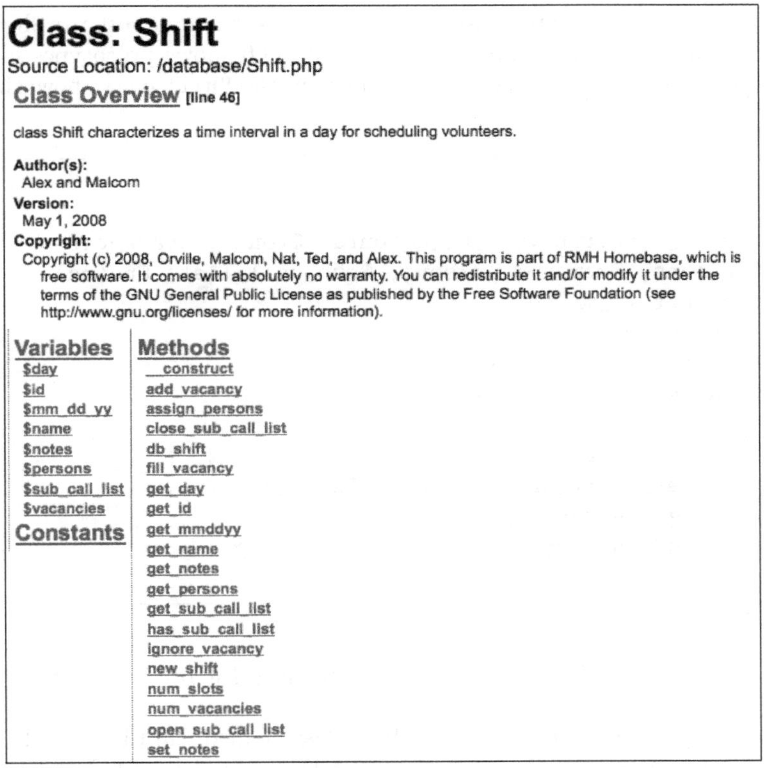

Class: Shift

Source Location: /database/Shift.php

Class Overview [line 46]

class Shift characterizes a time interval in a day for scheduling volunteers.

Author(s):
Alex and Malcom

Version:
May 1, 2008

Copyright:
Copyright (c) 2008, Orville, Malcom, Nat, Ted, and Alex. This program is part of RMH Homebase, which is free software. It comes with absolutely no warranty. You can redistribute it and/or modify it under the terms of the GNU General Public License as published by the Free Software Foundation (see http://www.gnu.org/licenses/ for more information).

Variables	Methods
$day	__construct
$id	add_vacancy
$mm_dd_yy	assign_persons
$name	close_sub_call_list
$notes	db_shift
$persons	fill_vacancy
$sub_call_list	get_day
$vacancies	get_id
Constants	get_mmddyy
	get_name
	get_notes
	get_persons
	get_sub_call_list
	has_sub_call_list
	ignore_vacancy
	new_shift
	num_slots
	num_vacancies
	open_sub_call_list
	set_notes

FIGURE 3.5 PHP documentation generated for the 2008 version of the *Homebase* Shift class.

A good IDE, such as Eclipse, supports this process of test suite development alongside coding. All contemporary programming languages have their own unit testing frameworks that prescribe how to write and run the unit tests themselves. For example, Java has the JUnit framework, Python has the PyUnit framework, and PHP has the PHPUnit framework. Visit `this site` for a more complete list of these frameworks for many languages.

No matter what language is used, the code base for a CO-FOSS project should be organized to support unit testing. That is, a separate directory of unit tests should live within the code directory itself, and each unit test should correspond to a single class or module. Keeping the test modules together with the code is good practice, even though it adds a little storage overhead to the system as a whole. This practice also facilitates unit testing of existing code whenever new functionality is added to the software at a later date.

For example, the *Homebase* test suite is found in the directory **tests** within the code base. Each unit test is paired with a particular class or database module in the system and named accordingly. That is, the unit test testShift is paired with the class Shift, and so forth.[3] Figure 3.6 shows prototypes for the constructor and several methods that appear in the **Shift** class shown in Figure 3.4.

```
class Shift {
    ...
    function __construct($yy_mm_ddhours, $venue, $vacancies,
        $persons, $removed_persons, $sub_call_list, $notes) {}
    function get_yy_mm_dd() {}
    function get_hours() {}
    function get_start_time() {}
    function get_end_time() {}
    ...
    function get_vacancies() {}
    function set_start_end_time($st, $et) {}
    function ignore_vacancy() {}
    function add_vacancy() {}
    ...
```

FIGURE 3.6 Some of the functions in the Shift class for unit testing.

A unit test for the Shift class is itself a class named ShiftTest, and it contains three principal functions – setUp, tearDown, and testShift. SetUp creates an object of class Shift by invoking the constructor, testShift tests the Shift methods, and tearDown deletes the object from the system. To test the methods, testShift itself contains a series of **assert** statements that must all be true for the Shift class to "pass" the test. Each **assert** statement typically exercises a call to a single method within the class, using both legal and illegal values for its arguments.

To ensure code coverage, the unit test should have at least one call for each non-trivial method in the module or class. For example, Figure 3.7 shows parts of the unit test for the Shift class shown in Figure 3.4.

More detailed developer guidance on designing and running unit tests is given in Chapters 5 and 6.

Unit Testing Tools

Software tools for running the unit tests display the results as a list of "successes" and "failures," as each assert statement in the unit test is true or false, respectively. For example, the results of running the ShiftTest unit test

[3]The **tests** directory has no unit tests for the user interface modules. Instead, those modules are tested by the client who can exercise all the paths through the use cases on the sandbox version. More will be said about this in Chapter 7

```
class ShiftTest extends TestCase {
    function testShift() {
        $noonshift = new Shift("08-03-28:1-5", "house", 3,
            array(), array(), "", "");
        $this->assertEquals($noonshift->get_hours(), "1-5");
        $this->assertTrue($noonshift->get_id() ==
            "08-03-28:1-5:house");
        $this->assertEquals($noonshift->get_yy_mm_dd(),
            "08-03-28");
// Test start and end time getters
        $this->assertTrue($noonshift->get_start_time()==13);
        $this->assertEquals($noonshift->get_end_time(),17);
        ...
```

FIGURE 3.7 Elements of a unit test for the Shift class.

in *Homebase* are displayed in Figure 3.8. If this figure were in color, the horizontal bar right would be green when all tests had run successfully with no failures. A blue bar would indicate that one of the `assert` statements had failed, while a red bar would indicate that a run-time error had occurred in the code itself. So this feedback provides the developer with a quick visual on the results after running a unit test.

As a final note on testing, it is important to acknowledge its fallibility. That is, a successful unit test does not guarantee that the code is free of errors. It only affirms that the code runs correctly for the test cases given. To ensure full correctness for a software artifact, the testing process would require the use of so-called *formal methods*, which is an entirely separate discipline.[4]

3.1.8 User Help

Documentation serves the needs of developers who are interested in understanding the functionality of a software system. However, it does not serve the needs of clients who want to learn how to use the system. For this purpose, separate elements are needed, which can take the form of on-line help pages that walk the user through the use cases supported by the software.

So developers can be expected to create on-line help pages that are integrated within the software itself. These pages should be written in a language and style familiar to the user, and they should use terminology that is common to the user's domain of activity.

Each help page should correspond to a single use case in the software

[4]The study of formal methods is beyond the scope of this book. For more information about formal methods in software development, please see `https://en.wikipedia.org/wiki/Formal_methods/`.

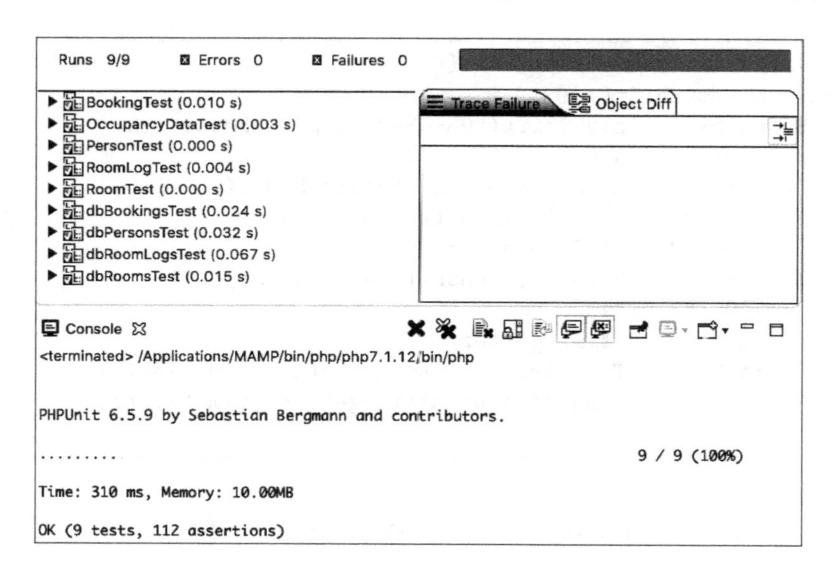

FIGURE 3.8 Results of running the TestShift unit test.

design, which is typically implemented as one or more related forms in the software itself. The word "Help" in the navigation bar should take the user to that particular help page corresponding to the form with which the user is currently working. The help page itself should describe a short sequence of steps, with illustrative screen shots as examples, that shows how to accomplish the use case that the form implements.

For example, the *Homebase* software has nine distinct use cases. One of these is called "Change a Calendar" and allows the user to find and fill a vacancy for a shift on a particular day of the week. The corresponding form for that use case is shown in Figure 3.9.

If the user needs assistance, the help page for filling a vacancy can be selected on the menu bar, and the user receives the step-by-step instructions shown in Figure 3.10. The help page should always open in a separate window from the user's current form, so that he/she can work with the form while simultaneously reading the help page.

3.2 THE COURSE

This section discusses the organization of an academic course in which a CO-FOSS project is embedded. That includes classroom scheduling, forming student teams, creating a schedule for achieving milestones, ensuring progress, laying out the syllabus, assignments, and grading.

Portland House Shift: Friday September 21, 2018 12pm to 3pm

2 slots for this shift: Add a Slot
Move this Shift

Find Volunteers To Fill Vacancies: Generate Sub Call List

Ellen Jones: Remove Person

vacancy: Assign Volunteer
Remove Vacancy

Back to Calendar

FIGURE 3.9 Form for filling a vacancy on a shift.

3.2.1 The Classroom

We view this CO-FOSS course as a lab course with mini-lectures, team presentations, and discussion sessions. In particular, the classroom scheduled for the course should be equipped with appropriate computing, networking, and other technologies that support small group meetings, videoconferencing, and computer screen projection. For our courses, we used a laboratory room that had a projector and screen, 12-15 workstations, and sufficient table spaces for teams to meet and work separately. Wireless support for students to connect their laptops to the institution's network should also be provided.

The scheduling for this course should be consistent with the institution's normal daily course schedule options. Our courses preferred two meetings per week for 1-1/2 hours per meeting. Daytime class meetings should be preferred over evening meetings, mainly to accommodate regular videoconferencing with the project's clients. For example, our classes would typically meet on Tuesdays and Thursdays from 1 to 2:30 in the afternoon, which was convenient for our clients as well. Instructor "office hours" should be organized in a way that encourages student developers to seek help at appropriate times, especially as each milestone is approaching.

Each class meeting can be organized in such a way that teams can make optimal use of that time to make progress on their projects. Some class time will require an instructor to present a "mini-lecture" on a key topic that may be new to most developers, such as database design. Other class meetings are best spent by teams working individually and the instructor moving among them to provide detailed guidance. Still other class meetings should be organized around a videoconferencing session with the client, where students present their work, clients suggest improvements to what they have seen so

Adding/Removing a Volunteer from a Shift

To begin, you must have already selected **(edit this week)** at the top left of the calendar:

Step 1: Click on a calendar shift, like this:

(Each upcoming calendar shift will turn gray whenever the mouse passes over it.)

Step 2: This will give you a shift form that looks like this:

Removing a Volunteer from a Shift: If a slot already has a volunteer, there will be a **Remove Person** button that looks like this:

Click that **Remove Person** button to remove that volunteer from that shift's slot, leaving an empty slot.

Adding a Volunteer to a Shift: If a slot doesn't have a volunteer yet (or you just removed a volunteer from that slot), there will be two buttons: **Assign Volunteer** and **Remove Vacancy**.

Click the **Assign Volunteer** button to come to a page where you can choose a new volunteer:

FIGURE 3.10 Help screen for filling a vacancy.

far, and tasks are assigned for the upcoming week. So instructors should plan to organize each of these three types of class meetings.

Finally, the network and software support for this classroom should provide a contemporary configuration of operating systems (Linux, Mac, or Windows), Internet browsing, downloading, and uploading tools, so as to support the instructor's, students', and clients' frequent needs to share documents and resources as they collaborate. Special care must be taken to ensure that clients can securely access the server that hosts the "sandbox" version of the software, especially when that server is managed by the institution's IT department.

3.2.2 Team Formation and Dynamics

A software development course with one or more CO-FOSS projects must have teams that include all the students enrolled in the course. We have taught this course several times, once with a single project and a 4-student team and once with 4 projects and 16 students (4 per project). While our preference would be for a class of 10-15 students and 2-3 projects, we recognize that many outside factors combine to influence both the size of the class and the number of projects that can be reasonably completed using the CO-FOSS approach.

Given a 10-15 student class and 2-3 projects, the instructor can assign students to teams in a way that ensures a high likelihood of success for each project. Our own preference is to balance the teams based on students' academic performance in earlier computer science and related courses, and to include consideration for racial and gender diversity. Ideally, it is good to assign one of the "best and brightest" students in the class to each team. Also, it is good to have one or more students with very good communication and interpersonal skills to each team, given the highly collaborative nature of the CO-FOSS approach.

It is also desirable to balance the teams equitably by spreading the students who have struggled in earlier computer science courses evenly among the projects. Opportunities for mentoring and pair programming in this class are significant, and they should be utilized to ensure that each student achieves better results than would have been achieved if working alone.

To assist with team formation, we also recommend the `CATME Team-Maker` tool. This tool uses team-formation principles based in engineering education research. Specifically, it accounts for diversity by avoiding isolation of women and students belonging to racial minority groups, and accommodating students' schedules and availability to meet outside of class.

Finally, we think it is not a good idea to let students self-select into teams, since that often upsets the team-balancing goals expressed above. In this sense, the instructor can and should play the role of "benevolent dictator" in the interest of maximizing each student's performance and each project's success.

With these considerations, a new CO-FOSS project can have the following team organization:

- one client,
- one lead developer (the instructor), and
- 3-5 student developers with no prior experience this sort, but with at least the data structures course or its equivalent as a prerequisite.

Assignment of tasks to individual students is mainly the job of the instructor, but also with the advice and consent the team. Task assignments will naturally vary during the life of the project. For example at the end of the project all team members may be writing user documentation, while earlier in the project they are all coders, testers, and committers.

No matter how tasks are assigned, it is important that every student gains experience learning all the skills of software development: coding, testing, writing user documentation, and communicating with the client. So special care must be taken to equitably distribute each of the coding, testing, writing, and communicating tasks among all the student developers on the team.

3.2.3 Scheduling and Milestones

The schedule for a software project must respect the constraints of individual team members. It must also be set so that the goals of the project are met at the end of the 1- or 2-semester development period. While flexibility in the accomplishment of particular tasks is important, some overall view of the main goals of the project must be apparent at all times. That is, the team members must remain confident that the project can be completed within the time frame allotted, and that each developer and the client can play an important role in reaching that outcome.

A project *milestone* is an intermediate goal that must be met by a particular calendar date. Examples of goals that can be met as milestones include completion of the project's database modules, completion of the on-line help screens, completion of user training, and so forth. Below are the final milestones that had to be completed at the end of the 2008 *Homebase* project:

Wednesday, 23 April, 2008—Complete draft of help tutorials
Thursday, 24 April, 2008—User training session I at RMH
Thursday, 1 May, 2008—User training session II at RMH
Friday, 9 May, 2008—Deliver final system to RMH

A *to-do list* is a collection of smaller units of work that occur in pursuit of a milestone. Unlike a milestone, a to-do list is an assignment of particular tasks to particular individuals on the development team.

Below is an example of developing a to-do list that was aimed at the accomplishment of the milestones listed above for the 2008 *Homebase* project.

Agenda for 4/16 Videochat
From: Alex
Date: Mon, 14 Apr 2008 at 9:44am
Category: Agendas & Meeting Notes

Here is what I see needs to be done to finish the project, along with final debugging. Feel free to add items that I have overlooked.

A. (Nat and Alex) Write and edit the remaining tutorials. Three are now completed, edited, and submitted. Here's a summary of where they stand:

1. *Logging on to the Web site—I don't think this one will be needed (the login page is pretty self-explanatory)*
2. *Persons: Searching—done*
3. *Persons: Editing—done*
4. *Persons: Adding—done, except .gif images need to be fixed*
5. *Calendar: Viewing*

6. *Calendar: Editing (includes removing persons, SubCallList editing, and adding persons to a shift)*
7. *Calendar: Managing (includes creating slots, generating weeks, and publishing)*
8. *Schedule: Viewing and editing*

B. (Ted and Nat) Organize and conduct workshop sessions at RMH for 4/24 and 5/1.

C. (Alex) Draft evaluation form for workshop participants.

D. (Orville) Write help.php.

E. (Malcom and Alex) Go through the code and add final comments for PHPDoc and future developers. Add the copyright notice at the top of each source file.

F. (Karl) Ensure that the RMH technical person prepares the RMH server with PHP and MySQL so that the system can be installed there. Add a button to the Web site so that users can access the Homebase login page.

Orville Mon, 14 Apr 2008 at 2:35pm

help.php is written and committed. Nat, take a look at editPersonHelp.inc.php in the tutorial folder. That's the format your final tutorials should follow. Then look at help.php—it just pulls the relevant tutorial (defaulting to what I call index.inc.php which is the list of all tutorials).

Note the difference in the image paths now, in the editPersonHelp.inc.php file.

To see what it will look like, log in, start editing a person, and click "help."

Also, might I suggest we always include a help footer? With links back to "help index," the URL of which would simply be help.php with no argument passed to it.

Alex Tue, 15 Apr 2008 at 7:29am

Question for Orville and Malcom: When Ted and Nat do the workshops at RMH on the 24th and 1st, they will be using the tutorials which are full of examples from our little "Jones" sandbox database. However, the live system at hfoss.bowdoin.edu has an empty database (except for admin/admin) which makes the tutorials pretty useless. Can we set up a temporary "training version" of the database with the "Jones" family live, so that the tutorials and the workshop will work together?

It seems that this "training version" of the database might be useful to RMH staff in the future. Ted, I think you need to ask Gina and others at the training sessions about this question.

Orville Tue, 15 Apr 2008 at 7:41am

I just ran testDBSchedules to populate the database for the training sessions—note that the actual shifts aren't filled yet—you need to actually start publishing some weeks.

I can show Nat or Ted how to reset the tables on-site if you guys are comfortable with unix and terminal-style OS's. If not, it's safer to not do it, because we're working on the actual server, and a mistyped command can cause a lot of damage.

Nat Fri, 18 Apr 2008 at 12:56pm

Alex, I've committed most of the calendar management help tutorials. I went with a main manage calendar tutorial which then lets you pick which specific task you are doing. This avoids redundancy since you need to get to the list of weeks page in order to edit anything.

I've completed all but the editShiftsHelp, which I've been trying to make as simple/easy to follow as possible. So feel free to critique the ones that are done; I should be committing editShiftHelp—I'm aiming for 3pm.

Notice here that Orville has taken the lead on setting up a training version of the database in preparation for the two training sessions. Meanwhile, Nat and Alex are communicating about the completion of the final help tutorials in preparation for those sessions. All of the tasks discussed in this dialogue are "to-do" items, which together lead toward achieving the help tutorial, training, and final delivery milestones.

3.2.4 Ensuring Progress

Progress on a CO-FOSS project can be measured concretely by counting such artifacts as lines of code written, tests written, use cases (user stories) implemented, domain classes completed, and database modules completed. Information provided by the VCS can assist with this bookkeeping.

To ensure progress, each weekly team meeting should be both retrospective and prospective. Looking backward, client feedback should be heard as assigned tasks are checked for completion. Looking forward, the next week's assignment should be set and to-do items should be spread among team members. In this process, it is important not to "overdrive your headlights." That is, assignments should be designed in small increments, so as to ensure success for all team members each week (see [9], pp. 51, 73).

As work progresses, conflicts may occur among team members over the boundaries among assigned tasks or the technical choices made when writing a piece of code. Sometimes team members will disagree with the task assignments initially made by the instructor and suggest changes. At other times, two or more team members want to perform the same task, or else no team member wants to perform that task. In all these cases, the team leader must resolve the issue in a way that does not compromise overall progress.

3.2.5 The Syllabus

Once a coding, testing, and development schedule for the project is established, the course can be laid out to show the process as a temporal series of milestones. Each to-do list within a milestone translates to an individual assignment on the syllabus. As each assignment is completed and each milestone is reached, the software should contain newly-written and tested code whose functionality can be reviewed by the client. Using a client-centered approach, each review session should allow the developers and client to reflect on the outcome and recalibrate goals for the next milestone.

Table 3.2 shows the schedule for the syllabus used in a 1-semester 2015 course where two 5-student teams developed two different CO-FOSS products called *RMHP-Homebase* and *BMAC-Warehouse*.[5]

Note in Table 3.2 that the Supporting Materials (readings and mini-lectures) are sequenced in a "just in time" fashion, so that students will learn each concept right before they need to use it in their CO-FOSS project. Not shown in Table 3.2 are the associated readings in the required textbook [43] for that course, which this book should replace in a future course with that type of schedule.

Several additional points should be made about the structure of this syllabus. First, each week has two 1-1/2 hour class meetings. One meeting is dedicated to introducing a new concept in either a mini-lecture or lab format. The other is dedicated to team meetings and laboratory work, where the instructor is available to help resolve design, coding, or organizational questions.

Second, client-developer meetings are scheduled at least every other week. At each meeting, clients evaluate the work most recently completed, developers listen and take notes, and the instructor uses the meeting to help refine the next week's assignment. This is one way in which client-centered development can work in the setting of a 1-semester project course.

Third, students generally come to this course with experience in data structures and class design, but little or no prior exposure to database design, GUI design, or technical writing. Thus, most of the mini-lectures pay attention to these latter subjects, filling in enough gaps so that students can complete

[5]The full syllabus, design documents, mini-lectures, and other supporting materials for these two projects can be downloaded from their GitHub repositories (see `https://github.com/megandalster/`, select one of these, and check the Project Resources on its Wiki page).

TABLE 3.2 Example Course Syllabus Schedule: Spring 2015 Semester

Week of	Milestone	Supporting Materials
Jan 19	Team formation, Intro to projects, Google project sites and tools, First Google hangouts with clients.	Collaboration Tools Mini-lecture 1
Jan 26	Setting up the development tools: IDE, code base, version control	Server, IDE, VCS Mini-lecture 2
Feb 2	Software architecture, domain classes, refactor, design	Unit Testing Mini-lecture 3
Feb 9	Domain classes: coding, testing, client review	Mini-lecture 4: Database Design
Feb 16	More domain classes: coding, testing, client review	
Feb 23	Database modules: refactor, design	Mini-lecture 5: MVC Overview
Mar 2	Database development: coding, unit testing, client review	Mini-lecture 6: GUI Forms
Mar 9	More database development, coding, unit testing, client review	Mini-lecture 7: GUI Interactions
Mar 16, 23	Spring Break (no classes)	
Mar 30	User interface development refactor, design	Mini-lecture 8: GUI Interactions
Apr 6	User interface development: coding, unit testing, client review	
Apr 13	More user interface development	Mini-lecture 9: Technical Writing
Apr 20	User documentation: coding, unit testing, client review	
Apr 27	More user documentation, client review, issue posting	Mini-lecture 10: Project Completion
May 4	Final testing and client review, issue resolution	
May 11	Final project presentations to clients	
May 15	Course wrap-up, self-evaluations, final commits, discussion	

those parts of the software effectively. These mini-lectures can be viewed in detail by downloading them from `https://npfi.org/mini-lectures/`.

Fourth, the layered architecture found in our CO-FOSS projects provides a lot of clarity about how to organize a syllabus around a new software project. That is, the students begin the course by first implementing or revising the domain classes, then implementing the database classes and modules, then implementing the GUI modules for the use cases, and finally writing the documentation and user help pages. By beginning the course with the domain classes, students can reuse their prior skills in object-oriented programming and data structures as they master the project's new application domain.

All our CO-FOSS courses since the first one in 2008 are organized the same way. The layering principle also allowed us to easily identify and reuse similar code from earlier projects.

3.2.6 Assignments and Grading

To meet each milestone, the development team has a weekly assignment, which includes design, coding, testing, and usually presenting its work to the client. Each assignment should be divided equitably among the 5 team members, and all students should commit their work to the shared code repository.

The requirements document's use cases should provide continuous oversight for the scheduling and writing of individual weekly assignments. At the beginning of the project, it is important to start with a relatively easy assignment, allowing team members to get used to working together and find early success in the first iteration of the delivery cycle ([9], p. 48). One such first assignment is to have students implement and unit-test the domain classes, as suggested above.

A different approach to a first assignment would have students develop a "walking skeleton" ([9] p. 49) so that all layers (domain, database, and user interface) of the architecture are exercised to complete a simple artifact. For example, that assignment could be to implement a login and password validation artifact, using a database table of authorized users to validate the Username and Password. This approach also brings the client face-to-face with a bit of the software's user interface early in the project.

Each later assignment should provide one or more meaningful tasks for each team member to complete more-or-less independently.[6] For example, Figure 3.11 shows Assignment 3 in the BMAC-Warehouse project in the 2015 course mentioned above. At this point, the students had already completed Assignments 1 and 2, where they set up their own IDEs and personal sandbox, and then coded and unit-tested the individual domain classes. For Assignment 3, each student used a completed module called *dbPersons.php* as a model for coding and unit-testing one of five additional database modules.

[6]Pair programming is also a feasible way to encourage students to work together to complete an assignment.

Look at the database module dbPersons.php and the associated testdbPersons.php unit test. Using these two as models, design and test new database modules dbContributions.php, dbCustomers.php, dbProducts.php, dbProviders,php, and dbShipments.php.

- *Each team member should "own" one of these 5 modules.*

- *For the module that you own, and using dbPersons.php as a model, add a copyright header, author documentation, and functions create, retrieve, getall, insert, update, and delete.*

- *Using the testdbPersons.php unit test as a model, add a new unit test called testdbContributions.php, testdbCustomers.php, testdbProducts.php, testdbProviders.php, or testdbShipments.php to correspond to the module that you own.*

- *Now unit-test your new module to be sure it is working properly. That is, each of its functions should work correctly with its associated table in the bmacwarehousedb database. A green bar at the top of the Result View window will confirm success.*

When you are satisfied with your new module and have tested it properly, "pull" the code base again, "commit" your changes, and "push" them to the repository.

FIGURE 3.11 Assignment 3 in the BMAC-Warehouse project.

It should be clear from this example that each member of the team takes responsibility for independently completing one part of an assignment. So it is incumbent on the instructor to design the assignment so that:

1. There are enough parts that each student can work on a part more-or-less independently, and

2. The whole assignment can be reasonably completed by the team by the end of the week.

A complete set of assignments used by the *BMAC-Warehouse* team in the 2015 course can be viewed and downloaded at `https://npfi.org/assignments/`.

The 2015 course used the Mercurial VCS and the Google Code repository. Since Google Code was discontinued in 2016, these two projects (and all the other projects discussed in this book) have migrated to GitHub and they now use the Git VCS. For all practical purposes, Git and Mercurial are similarly intuitive for students to learn and use productively. The subject of version control is covered in more detail in Chapter 4.

Professional documentation standards should also be followed for each new module developed, allowing each student's personal authorship of specific software components to be identified. Grades should be assigned on the basis of

1) the success of the project, and 2) the quality of each student's personal contribution to that success.

Confidential self-assessments can be used at the end of the semester to help sort out the quality and quantity of work contributed by each team member. The self-assessment questions answered by each student in the 2015 course are shown below.

1. What was the quality of your team's overall effort this semester? Include such factors as communication, workload distribution, code sharing, etc.

2. What was the quality of your personal contributions to this effort? Identify 3 specific examples of your most important contributions.

3. How did the quality of your contributions compare with that of each of your teammates?

4. Were there any assignments during the semester in which you personally could have done better? Which one(s)?

While both projects in this course were completed successfully, individual student grades in this course ranged from A to C. The self-assessments, the quality and quantity of committed code, and the instructor's own sense of how the workload was distributed among the team members were the major factors in determining these grades.

3.2.7 Alternatives: The Two-Semester Software Projects Course

Some undergraduate computing programs offer a 2-semester capstone experience as a vehicle for students to complete the requirements for a major. In this setting, it can be opportunistic to offer a 2-semester variant of the CO-FOSS course described above. The rationale here is that one semester does not provide enough time for anyone but the most motivated students to learn all they need to learn about databases, programming tools, and user interface design while at the same time developing and testing a complete software product for a real client.

Most of our CO-FOSS experience used the 1-semester course model. The one exception occurred in 2012 when a team of 3 students developed the database and administrative interface for *Homeplate* during the spring semester, and then a single student developed a related Android app called *Homeplate Mobile* during the ensuing summer term as an independent study project. The software design and toolset used for this latter task were entirely different from that used by the 3-student team. Yet the two software products had to interact with the same database in order to bring *Homeplate* to full functionality. For more information on *Homeplate* and *Homeplate Mobile*, see `https://npfi.org/projects/the-homeplate-project/`.

Other programs have implemented a 2-semester model for the software projects course in different ways. For more detailed discussions of this and other alternative HFOSS course designs, see [8] and [25].

3.3 SUMMARY

This chapter provides guidance on the details of organizing a CO-FOSS projects course, beginning with the project design document and ending with the syllabus. Key topics such as team formation, project milestones, developer tools, version control, self-assessment, and grading are covered in detail, using examples from the several successful courses that we have taught since 2008.

3.4 MILESTONE 3

1. The instructor should select the tools to be used in developing the product identified at the end of Chapter 2, including collaboration tools, a development platform, and a project hosting and version control system.

2. The instructor should identify the course in which this project will be embedded, and define the development team organization, the course's milestones, and the course syllabus.

II

Development Stage

CHAPTER 4

Project Launch

"What we have to learn to do, we learn by doing."
—*Aristotle*

During the first week of the project, a lot of organization goes on. The team is formed, roles and responsibilities are clarified, communication channels are opened, and new software tools are installed. Most importantly, the project's initial design document is read, the client representative is introduced, and the initial code base is introduced.

This chapter provides clarification for these initial steps, which we call *project launch*. At the end of project launch, the team should be ready to begin developing their CO-FOSS product in earnest.

4.1 THE TEAM

The development team for a new CO-FOSS project is formed by the instructor prior to the first day of classes. Team members will normally fill the following roles (some are optional, as noted):

> The *analyst* role is filled by a person who understands the user's domain, elicits requirements, defines use cases, evaluates risks and alternatives, sketches the initial design of the software to be developed, and schedules project milestones.
>
> The *team leader* role is filled by a person who oversees the development process. The team leader's tasks include setting weekly agendas and milestones, assigning tasks to team members, coordinating overall system architecture, teaching other team members about new techniques, leading regular team discussions, and helping resolve design issues.
>
> The *developer* role is filled by a person who writes test cases from requirements, and reads, writes, tests, debugs, and refactors program code.

Developers also refine requirements and use cases, write documentation and on-line help pages, and solicit feedback from users at each stage in the project. In short, a developer is a programmer who can also read, write, and communicate effectively with users.

The *user* role is filled by persons knowledgeable about the application's domain. A user's tasks are to review use cases, provide data for test cases, review the output of code written by developers, provide feedback on the quality of that output, and report bugs in the software after each iteration has been deployed.

The *IT technician* role is filled by persons who configure and maintain the technical environment used by the team. Their tasks are to set up and oversee the code repository, the staging server, the videoconferencing system, the team discussion site, and other tools needed by the team to develop the software effectively.

(Optional) The *observer* role can be filled by a person interested in watching the project develop and/or whose professional expertise may provide occasional high-level advice on technical or procedural details. Overall, the observer role is a fairly passive one.

In our experience with CO-FOSS-based software development classes, we formed teams in the following way.

The *analyst* role was initially filled by the instructor, who developed the design document prior to the beginning of the class. As the project evolved, some students assumed analyst responsibilities, especially as they helped update the requirements document and added new features.

The *team leader* role was also filled by the instructor. The instructor assigned tasks to the students and provided technical and other support (mini-lectures, links to tutorials, etc.) throughout the life of the project.

The *developer* role was filled by a group of 3-5 students enrolled in the class. At a minimum, they had prior programming experience and familiarity with common data structures and algorithms, such as searching and sorting. The student developers completed the design, coding, and testing of nearly all the components for the new software.

The *user* role was filled by the client representative, who met regularly with the team, reviewed the partially-developed software, and provided regular feedback to the developers on their work.

The *IT technician* role was partially filled by the institution's IT department and partially filled by the instructor. In one instance, a student developer set up and oversaw the code repository.

In one of our projects, the *observer* role was filled by a colleague in a nearby college who was interested in this way of teaching the course. In another project of which we are aware, a professional software developer participated as an observer and an occasional technical resource throughout the development period.

The above shows that the same person may play different roles and assume different tasks as different needs arise. For example, the team leader may intermittently play a developer role when introducing a new concept or technique to the student developers. In general, fluidity in role-playing and task assumption is a key element of a CO-FOSS software project, though this may be more true for a new project than for a mature project.[1]

At this point, the newly-formed team can begin using the syllabus, the initial design document, and the initial code base as a starting point to begin working together on their project.

4.1.1 Team Dynamics

The team should begin working together in an initial team meeting on the first week of the class. To illustrate, we captured the team dynamics that occurred at its first meeting in January 2008, where a 4-student team began developing the original *Homebase* volunteer scheduling system introduced in Chapter 1. This example is typical of how each of our projects got off the ground.

The 2008 *Homebase* team members and their role(s) are identified below.[2]

analyst—Alex (the instructor)

team leader—Alex

developers—Orville, Malcom, Ted, Nat (the students)

users—Gina (the client)

IT technicians—Ellis, Orville

observers—Riccardo, Truman

Throughout the semester-long project, individual roles changed in minor ways. For example, early in the project, Orville temporarily became an IT technician when he set up the project's code repository.

Prior to their first meeting with the client, the developers and the team leader took a road trip to the client's location, which was about a 25-mile drive from their college. Whenever possible, an initial site visit during project

[1]It is worthwhile to contrast this relatively simple team organization with that which characterizes mature open source software projects. These projects typically have a much larger development team and client base, as suggested in Chapter 1.

[2]The names used here are pseudonyms, so as to protect the privacy of the actual team members.

launch is invaluable; student developers get a first-hand feeling for the mission of the non-profit, as well as for the paper-based system that their software will replace. If this luxury is not possible, it may be helpful for the client to conduct a brief virtual tour of their facility and describe the activity that the new software will replace.

Here is a partial transcript for the first *Homebase* meeting along with the follow-up notes taken by the student developers:

> **Subject: Videochat, 1/30**
> **From: Alex**
> **Date: Tue, 29 Jan 2008 at 7:03pm**
> **Category: Agendas & Meeting Notes**
>
> *Here's a sketch of an agenda for tomorrow's meeting:*
>
> 1. *Discuss each of our new PHP classes.*
> 2. *Discuss PHP tutorials that will be most useful to us.*
> 3. *Determine standards for coding, testing, and documenting classes as they develop.*
> 4. *Determine what each of us can reasonably accomplish for next week.*
> 5. *Identify any relationship of the project with other volunteer scheduling systems (e.g., VMOSS).*
>
> *Feel free to add anything to this list that I've overlooked.*
>
> *"See" you at 1:00 tomorrow. I'll try to connect with Orville using my iChat and AIM id = alex. If we get stuck, my phone here is 207-729-1234.*
>
> **Orville Wed, 30 Jan 2008 at 2:07pm**
> *We had a videochat today. Present were: Alex, Malcom, Nat, Orville, Ted, Gina.*
>
> *After brief introductions, we covered our agenda and resolved the videochat bugs. Worked well overall for a first time. Please reply in the comments with the PHP class you will personally work on for next week.*
>
> **Malcom Wed, 30 Jan 2008 at 2:08pm**
> *I'm working on Calendar, and researching existing PHP Calendar classes.*
>
> **Orville Wed, 30 Jan 2008 at 2:08pm**
> *I'll break people.php into separate class files, then work on Person (which needs to be done first since the others extend it). Then I'll work on Applicant and Manager. Also, once Alex decides on a structure for our packaging, I'll implement it in the repository.*

Nat Wed, 30 Jan 2008 at 2:55pm
I am currently working on SubCallList. I'll post again and take up one (or more) of Orville's classes that extend Person.

Ted, Thu, 31 Jan 2008 at 9:05am
I am looking at the MasterScehdule class. It looks complicated – I may need some help defining its features.

As you can see, Alex initiated the discussion by posting the agenda for the meeting. The developers Orville, Malcom, Nat, and Ted had a short follow-up exchange, in which they summarized the meeting and assigned themselves to the tasks of developing the core classes Calendar, Person, SubCallList, and MasterSchedule during the upcoming week.[3] Finally, all the references to PHP in the above discussion could just as easily have been references to, say, Java or Python had the project's programming language been different.

While roles and tasks are often pre-assigned by the team leader, team members may self-select into and out of tasks as the project evolves, depending on personal preferences, skills, and other commitments during that week.

This fluidity is particularly important in the larger open source development world, where team members are not usually working full time on any one project. They usually have commitments to other work, for instance other course work (if they are students) or other work commitments (if they are professional developers).

The catalysts for effective and fluid team dynamics lie mainly with the instructor and the milestones defined in the course syllabus. As project leader, the instructor sets the tone for effective project collaboration. This tone is characterized by trust, inclusiveness, and confidence that the project will succeed and every team member's participation will be crucial to that success.

The project leader sets the schedule for each weekly or bi-weekly meeting, providing a brief agenda (like the one above) that may be refined by the other team members. Each meeting should yield both a review of the prior week's accomplishments/issues and a new set of goals for the upcoming week, accompanied by assignments of tasks to team members.

In the above example, we illustrated two distinct forms of communication among team members, *asynchronous communication* and *synchronous communication*. Each form has its own role in the project and different options for set-up, as discussed in the next two sections.

4.1.2 Asynchronous Communication

Asynchronous communication among team members takes place in an unscheduled way. It is represented as a collection of discussion threads related to

[3]The current version of these classes (at `https://npfi.org/homebase-classes/`) shows slightly different names from these original names; the original class Calendar is now named Shift and SubCallList is now named SCL. However, notice in the current version that the original 2008 authorship notices remain intact, along with the original GPL licensing statement.

the current and recent activities of their project. Two characteristics distinguish asynchronous communication in a CO-FOSS project from an ordinary email discussion: 1) its discussions are private and limited to the team members, and 2) its discussions are organized into separate threads, each thread focusing on a single topic.

Historically, the common asynchronous communication tool for FOSS projects was the *mailing list*. A mailing list can be set up as a private distribution list whose membership includes all the members of the development team. When a member sends a message to the project's list server, or so-called *listserv*, the message is distributed to all other members automatically.

In recent years, the use of mailing lists for team communication has been replaced by *discussion forums*. These provide a unique location where a project's discussion threads can be initiated and maintained by team members, and from which every new message is broadcast (via e-mail) to all members of the team who are participating in that thread. Forums thus allow users to subscribe to *really simple syndication* (RSS) feeds that notify them whenever a new post is made, but only for the list(s) that they choose.

Several modern tools that support discussion forums are available. Some of these tools are free, such as `Slack` and `Google Groups`. Others are proprietary, such as `Basecamp` and the more feature-rich versions of Slack.

Our first CO-FOSS project in 2008 used Basecamp — at that time it was freely available. Our later CO-FOSS projects used Google Groups and Google+ Communities as our discussion forum. These tools worked well for us, though they had some important differences. For example, Google+ Communities allowed a person to join using their personal email address, while Google Groups required team members to establish separate Gmail accounts before joining. Since Google+ Communities has been discontinued, we recommend the free version of Slack as an excellent choice for setting up a new CO-FOSS project discussion forum.

Finally, an important requirement for team members who participate in asynchronous communication is that they use appropriate etiquette. Many communities, such as `the Sahana project`, publish documents describing their etiquette guidelines. Generally, such guidelines remind team members to post comments that are relevant to the topic, are brief and to the point, and show respect for other team members' points of view.

Aside: Mature FOSS Projects

Mature FOSS projects with large development and user communities support different lists/forums for different aspects of the project, such as coding, testing, documentation, and management. For example, the Sahana Foundation project maintains the following `mailing lists`:

> The *Sahana-Discuss* list enables users, developers, and administrators of *Sahana* to submit queries about using, deploying, and administering the software.

The *Sahana-Agasti* list is for software developers and focuses on technical design, quality assurance, documentation, and deployment of the PHP-based *Sahana* implementation.

The *Sahana-Eden* list is for software developers and focuses on technical design, quality assurance, documentation, and deployment of the Python-based *Sahana* implementation.

The *Sahana-Localization* list addresses queries and issues related to translations of *Sahana* into different languages.

An important byproduct of asynchronous communication lists like these is that they provide an historical *archive* for the project, keeping a record of important design and policy decisions that take place throughout the life of the project.

But for a new CO-FOSS project, a single discussion forum is sufficient to support asynchronous communication throughout the life of the project.

4.1.3 Synchronous Communication

In the early days of software development, all the members of a software team would work in the same physical space throughout the life of the project. From time to time, they would meet face-to-face to discuss their progress and plan next steps. This model, however, is no longer realistic. In today's world, a project's team members are geographically dispersed, and yet the need for regular face-to-face meetings persists.

To accommodate this need, the team must synchronize their meetings using a tool that facilitates face-to-face communication. These "virtual meetings" are used to set goals, evaluate progress, and discuss other global issues as they arise. Synchronous communication tools are thus complementary to the asynchronous tools discussed in the previous section.

A classical synchronous communication tool is Internet Relay Chat (IRC), which provides real-time text communication over the Internet to support *synchronous conferencing.*[4] In a typical IRC environment, the same channel is used by all community members who have joined the session. Such a dedicated communication channel can also be used on an *ad hoc* basis to support one-on-one or small group discussions. Many different IRC clients are freely available, as listed at `http://ircreviews.org/clients/`.

Videoconferencing is a more modern tool for managing synchronized communication. It has the advantage of allowing participants to see each other, share screens and other information visually, and use voice and text messaging as well. Good tools for videoconferencing are freely available, for example `Skype` and `Google Meet`. A more comprehensive survey of videoconferencing tools for small teams can be found at `https://npfi.org/videoconferencing-tools/`.

[4]IRC is like a telephone "conference call" where each member dials into a common number and enters a secure password.

Our earlier projects (from 2008 to 2012) used Skype for videoconferencing, while our more recent ones used Google Hangouts for the same. Both tools worked well for us, though in each case our first virtual meeting had a few technical and training bugs for the team members to work out. It is important here to assume that most people on the team have never used such a tool prior to this project, but that the ones who have will step up and help others find a level of comfort. After our first session, the rest of our videoconferencing sessions in the project were smooth and free of technical issues, on both the developer's side and client's side.

In a CO-FOSS project, each scheduled videoconferencing session should include all team members (students, instructor, and client representative), have an agenda that reviews each team member's accomplishments since its last session, identify upcoming tasks and assignments to team members, and be open to discussion of any new issues that arise. However, this scheduled session is not the best place for discussing low-level detailed technical issues or bug fixes; those are better handled off-line on an individual basis, using discussion forums or other asynchronous methods.

4.1.4 Shared Documents

Maintaining shared documents—such as project designs, to-do lists, milestones, and screenshots—is vital to a collaborative software project. All team members should have sufficient access to these documents for reading, commenting, and editing. Here are the basic shared documents that arise in a CO-FOSS project:

To-Do Lists are lists of tasks to be completed in the near term, along with the names of team members who have committed to completing those tasks. Issue lists that accompany the code repository are a particularly popular form of to-do lists.

Milestones are calendar events that identify deadlines for completing major steps in the project. In a CO-FOSS project, these milestones are laid out in the class syllabus, which is an essential document for sharing.

Files and Screenshots are artifacts like design documents, client-provided scenarios, sketches of GUI forms, community outreach documents, mini-lectures, assignments, and help pages.

Project Wiki is a publicly-viewable description of the project for the team to manage and others to read. It is editable by team members and plays a useful role in briefing newcomers about the overall nature and goals of the project.

Project collaboration tools like Slack support document sharing using a linked cloud storage account, such as Google Drive. Simultaneous document editing is also supported on Google Drive whenever the shared document is compatible with the Google Docs format. This allows editing of the project's design document to be shared among all the team members.

In a Skype setting, document sharing can be accomplished using, say, Dropbox or Office 365. Also, Skype can be paired with Google Groups and Google Drive, so there are many possible combinations here. Unlike the Google tools, however, many of these other combinations are proprietary and come with a monthly fee.

Simultaneous document editing is also supported by the project's wiki. Like a version control system, the wiki has features such as history and change tracking, as well as discussion threads associated with individual pages.

4.2 THE DEVELOPMENT TOOLS

As we saw in Chapter 1, the CO-FOSS model has a series of 1- or 2-week cycles that repeat until the project is complete. Each cycle requires that the client review intermediate results and provide feedback as each element of the software is coded and tested (See Figure 1.3).

To ensure that the developers make positive progress at each repetition of this cycle, they should use effective communication techniques as described above, as well as effective development tools as described below:

1. An appropriate programming language (or languages) for coding the application,
2. An effective software platform (server, language, and database stack) for the application,
3. An integrated development environment (IDE), and
4. A version control system,
5. A unit testing framework,
6. A staging server where the client can review the software as it evolves.

The first five of these tools are used only by the developers, while the last is used by both the client and developers.

The four sections below discuss and illustrate the first four of these development tools. The last two development tools are discussed and illustrated in Chapter 5.

4.2.1 Programming Languages

The choice of programming language for a new CO-FOSS project is determined by many factors. Often, the choice is constrained by the existence and nature of an initial code base. On the other hand, if the software is being developed relatively from scratch, more choices may be available. However, the language with which the students are most familiar from prior courses (e.g., Java, C++, or Python) may or may not be the language used in the project.

So in this course, student developers should prepare to learn a new language, or at least some significantly new language features, as they dive into the project. Moreover, they should not expect the instructor to teach them the new language as if they were taking a traditional introductory computer science course. Most language learning in this course is self-directed.

Importantly, the language used for the new CO-FOSS project should be one that is well-known in the software world. The language should have strong support — efficient compilers and debuggers, an active user community, and on-line documentation, tutorials, textbooks and exemplary code snippets, functions, modules, and classes. Moreover the project's underlying initial code base can provide many clues about the basic structure of the language and its similarities with other languages that developers have already used.

The following sections identify some prominent languages that are good candidates for CO-FOSS development, along with recommendations for tutorials and on-line documentation. Our recommendations are not by any means complete, and developers should always rely on the instructor to provide more focused references for their project, including print books and eBooks as appropriate.

JavaScript

JavaScript is a dynamic, interpreted programming language that is widely-used for Web applications. It supports object-oriented, imperative and functional programming styles. Mozilla Firefox and MongoDB are two well-known open source applications developed with JavaScript, though many more applications combine JavaScript with another language in their source code. For example, several of our CO-FOSS projects such as *Homebase* combine JavaScript, PHP, HTML, and CSS in their source code.

Tutorials `https://www.w3schools.com/js/DEFAULT.asp`

Full Documentation `https://developer.mozilla.org/en-US/docs/Web/JavaScript/Reference`.

Python

Python is a dynamic general purpose programming language used for developing both enterprise and Web applications. It supports object-oriented, imperative and functional programming styles. Dropbox, Mercurial, and Django are three popular open source products developed with Python. Django is a framework for developing Web applications in Python, and so is a useful tool for developing a CO-FOSS project with Python.

Tutorials for Python can be found at `https://www.tutorialspoint.com/python/index.htm`.

Full Documentation for Python syntax and function libraries can be found at `https://docs.python.org/3/reference/index.html`.

Java

Java is a well-known object-oriented general purpose programming language used for developing server-side enterprise and Web applications. Java programs are compiled to "Java bytecode" and run on the Java Virtual Machine (JVM), which in turn runs on Windows, Linux, and MacOS systems. Two prominent software products developed with Java are Eclipse and NetBeans.

Tutorials `https://www.tutorialspoint.com/java/index.htm`.

Full Documentation `http://www.oracle.com/technetwork/java/javase/documentation/api-jsp-136079.html`.

Ruby

Ruby is a dynamic general purpose language that supports imperative, object-oriented, and functional styles. Rails is an add-on framework for Ruby that makes it simple to build Web applications, and so is a useful tool for developing a CO-FOSS project with Ruby. Popular applications that were developed using Ruby include Rails itself, GitHub, Basecamp, and Whitepages.

Tutorials `https://www.tutorialspoint.com/ruby/index.htm`. Tutorials for getting started with Rails can be found at `https://guides.rubyonrails.org/getting_started.html`.

Full Documentation `http://ruby-doc.org/`.

PHP

PHP is an open source scripting language suitable for server-side Web development. It is an interpreted language with interpreters for almost all platforms. PHP is extensively used in the development of open source projects like Drupal, Joomla, WordPress, and Moodle. As mentioned above, all of our CO-FOSS projects use PHP and JavaScript as their core development languages.

Tutorials `https://www.w3schools.com/php/` and `https://www.tutorialspoint.com/php/index.htm`.

Full Documentation `http://php.net/manual/en`.

HTML and CSS

HTML stands for "Hypertext Markup Language," and virtually all web pages are defined using an HTML specification. HTML is used to describe the content of a Web page. CSS stands for "Cascading Style Sheets," and is a language used for describing the layout and style of a Web page described by HTML.

Strictly speaking, HTML and CSS are not really programming languages because they are only declarative – they contain no procedural, functional, or object-oriented elements. We include HTML and CSS here because developers must be able to read, understand, and write HTML and CSS code whenever they are developing a web page.

Tutorials `https://www.w3schools.com/html/` and `https://www.w3schools.com/css/`.

Full Documentation `https://developer.mozilla.org/en-US/docs/Web/HTML/Reference` and `https://developer.mozilla.org/en-US/docs/Web/CSS`.

Other Languages

Other prominent languages include GO, Swift, C, C++, and C#.

Go is a relatively new programming language that made its first appearance in 2009. It is a compiled programming language created by Google. It is mainly used for the development of enterprise and Web applications.

Similarly, Swift is a general purpose programming language introduced by Apple in 2014 for developing applications for iOS, macOS, and Linux operating systems. It is derived from the earlier language Objective C.

C, C++, and C# are older languages that remain widely used for software development. For various reasons, these languages may be less well-suited for use in a CO-FOSS project than the others mentioned above.

Readers interested in further evaluating these and other programming languages for use in their CO-FOSS projects are encouraged to visit `https://opensourceforu.com/2017/03/most-popular-programming-languages/`.

4.2.2 Software Platforms

The software platform needed for developing a CO-FOSS product combines a programming language and a database, and also a server if the product is a client-server application. Once the programming language is selected, the choices for database and server are more straightforward.

The database language SQL (standing for "Structured Query Language") is a widely-used declarative language for managing relational databases. SQL has several implementations called "database management systems," which include the popular open source products `MySQL` and `PostgreSQL` and the proprietary product `Microsoft SQL Server`. For mobile devices, `SQLite` is a popular database system. Unlike MySQL and PostgreSQL, SQLite is designed for a single-user app rather than a client-server system. In general, it is hard to imagine a CO-FOSS project that does not have a significant database component.

All our projects use MySQL, since it is bundled within the Apache framework and well-connected with PHP via a so-called API (Application Programming Interface). MySQL has other APIs that connect it with the Python, C, C++, and Java platforms.

Tutorials for learning MySQL can be found at `https://www.tutorialspoint.com/mysql/index.htm`.

Full Documentation for MySQL depends on the programming language with which it is connected. For PHP, that documentation can be found at `https://dev.mysql.com/doc/apis-php/en/`.

Tutorials for PostgreSQL can be found at `http://www.postgresqltutorial.com/`.

Full Documentation for PostgreSQL depends on the programming language with which it is connected. For Python, that documentation can be found at `https://www.tutorialspoint.com/postgresql/postgresql_python.htm`.

Tutorials for SQLite can be found at `https://www.tutorialspoint.com/sqlite/sqlite_overview.htm` and `http://www.sqlitetutorial.net/`.

Full Documentation for SQLite can be found at `https://sqlite.org/docs.html`. However, more useful documentation is available for each programming language with which it is connected.

The details of the interface between the database system and the programming language are discussed in Chapter 6, where database design and programming are presented more fully.

If the software is an open source client-server application, either Apache or Nginx is usually used for the server. If the software is a single-user mobile app, then the platform is just the programming language and the database. The following sections discuss several particularly useful platforms for developing CO-FOSS applications.

The Apache/MySQL/PHP Server

The Apache/MySQL/PHP server family is a classical and widely-used combination for PHP Web application development. It fully integrates an Apache server, a MySQL database system, and a PHP language processor. Three compatible versions are available for download and installation: one for Windows (called `WAMP`), one for Macintosh (called `MAMP`), and one for the Linux (called `LAMP`) operating system.

The availability of all three allows different PHP developers to work together even though they use different operating systems. All of the CO-FOSS projects discussed in this book use the Apache/MySQL/PHP server family.

Server-Side Java

Java database and server connectivity are mitigated by JDBC (Java Database Connectivity) and JSP (Java Server Pages), respectively.

JDBC is an application programming interface that defines how a Java program may access an SQL database. For connection to a MySQL database, the `Connector/J` driver implements the JDBC interface. For connection to a PostgreSQL database, the `PostgreSQL JDBC` driver implements the JDBC interface. The `JSP technology` is part of Java itself and provides a simplified, fast way to create dynamic web-based applications that are server- and platform-independent.

Since Java runs on Windows, Linux, and Macintosh platforms, developers can work on the same project using different desktop systems. They only need to agree on their choice of a Java database system, which is typically MySQL or PostgreSQL for open source development.

Python

Developing Python applications for the Web is often done using the free open source Django platform. Django bundles all the necessary web development features by default rather than providing them as separate libraries. Tutorials for getting started with Django can be found at `https://www.djangoproject.com/start/`.

Django uses its “object-relational mapper” (ORM) to map objects to database tables. It supports all three SQL database systems – PostgreSQL, MySQL, and SQLite – and migrating a project from one to another is straightforward. For more information on using ORM with Django, see `https://www.fullstackpython.com/django-orm.html`.

Ruby

Developing Ruby applications for the Web is often done using the free open source Ruby on Rails (or just “Rails”) platform.

Like Django, Rails uses an “object-relational mapper” (ORM) to map objects to database tables, freeing the developer from learning the complex semantics of SQL syntax and queries. For more information on using ORM with Rails, see `https://guides.rubyonrails.org/active_record_basics.html`.

4.2.3 IDEs for Development

The integrated development environment (IDE) unites a server, language, and database platform within an environment where the developer can read, write, test, debug, run, and manage the code for an application independently from other developers working on different parts of the code for the same application. The IDE is a stand-alone software tool that runs on the developer’s desktop or laptop and interfaces with the project's code repository whenever the developer needs to coordinate his/her work with that of the rest of the team.

Several widely-used IDEs are discussed in the sections below. Unless otherwise noted, each one is a FOSS product, runs on Windows, Linux, and Macintosh, and is widely used by developers working on projects like the ones discussed in this book.

There are many more choices for IDE than those discussed below. However, it is a good idea to stick with a popular and well-supported IDE, and ideally one that is familiar to the instructor and/or student developers. Learning a new IDE along with all the other demands of a CO-FOSS project may add more complexity for the project than necessary.

Eclipse IDE

The `Eclipse IDE` has a *plug-in architecture*, which allows it to be extended to different programming languages and incorporate new features. It can be configured for Java, C, C++, PHP, JavaScript, or Python programming. For example, we used Eclipse configured for PHP and JavaScript with different student teams at Bowdoin and Whitman Colleges as they developed the *Homebase* (2008, 2011, 2013, 2015), *Homeroom* (2011), *Homeplate* (2012), and *BMAC-Warehouse* (2015) applications.

Figure 4.1 shows a snapshot of an Eclipse session, in which a developer's version of *Homeroom* is integrated with an Apache server, a MySQL database, and the PHP language. This developer's version of the code is also synchronized (using the Git VCS) with the *Homeroom* project hosted on GitHub.

The Eclipse user interface can show different *perspectives* or *views* of the project. In the default view in Figure 4.1, the *Homeroom* file structure is shown on the left-hand pane of the Eclipse window. It provides easy access to different PHP classes and modules in the code base. The developer can examine and work on any of these files by selecting it in the left-hand column. The middle pane displays the code itself, while the bottom pane shows the result of running the code. Not apparent here is Eclipse's use of colors to distinguish different features of the program, such as keywords, comments, and HTML snippets.

The run itself uses a built-in Apache server (called "localhost") and a MySQL database that are configured for *Homeroom*. Eclipse provides many options and tools for developers as they write, debug, and run the code for an application. For example, Eclipse *auto-completes* code. That is, when the developer types the name of a particular class of object, Eclipse will display a pop-up window that includes all the methods that can be called with that object. While many of these options are intuitive, Eclipse has good `tutorials` that provide an overview for the beginner.

A debugging tool is always built into an IDE to help the developer locate and correct run-time errors. Whenever a run crashes, the debugger shows the instruction that was executing at the time of the crash. Debuggers also provide

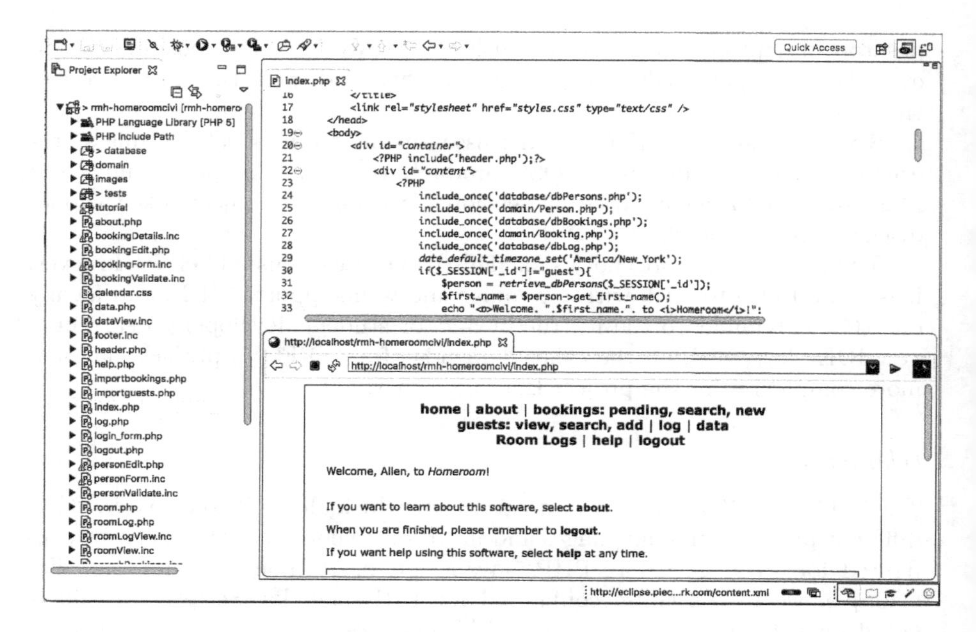

FIGURE 4.1 Developing *Homeroom* with the Eclipse IDE.

tools like running a program step-by-step, stopping at some pre-specified step (called a *breakpoint*), and tracking the values of pre-selected variables whenever a breakpoint is reached.

Python IDEs

The Eclipse IDE can be configured with the `PyDev` plugin that supports Python development. Alternatively, the very popular `PyCharm` IDE also supports Python development. Both are well-integrated with Python's `Django` web development framework.

Ruby IDEs

`RubyMine` and Aptana Studio are two popular IDEs that support Ruby development. Both are well-integrated with Ruby's `Rails` web development framework.

Java IDEs

While Eclipse can be configured well for Java server-side development, there are several others that can be considered as viable alternatives. In no particular order, these include `NetBeans`, `IntelliJ IDEA`, `BlueJ`, and `JDeveloper`.

Choosing and Installing an IDE

We have briefly presented a few of the most popular IDEs for software development. Our advice for your project, of course, is to pick one that supports your preferred programming language, and secondarily one that is familiar to your developers. Whatever IDE you choose, it will need to be downloaded and installed on each developer's desktop or laptop computer.

If the application is to be run on a server, the IDE will need to be integrated with a server, a database, and the code base. If the app is to be run on a mobile device, the IDE will still need to be integrated with a device emulator. The IDE must also take into account the version control system being used, as discussed in the next section.

4.2.4 Working with the VCS

Software systems have a large number of files that change over time as additions, deletions, and modifications are made to the code base. For all but the very smallest projects, it is essential for developers to use a *version control system* (VCS) to help manage this process.

A VCS (also known as *revision control* or *source control*) is software that helps manage the changes being made to a collection of source code files (a so-called "code base") by several developers. A VCS is useful even for a. single developer, but it is essential when there is a team or a community of developers all sharing the same code base and making changes asynchronously.

A VCS tracks all changes to the code base in a central location called a *code repository.* Whenever a developer adds or changes a piece of code to the code base, it doesn't simply overwrite the original. Rather, the VCS keeps both copies of the code, making it possible to revert to a previous version if a current line of development proves to be incorrect.

In addition, the VCS keeps track of who made each change, when it was made, and so forth. It provides tools to compare one version of the software with another, making it easy to see what has changed from one version to the next, as the code base evolves.

In general, a VCS supports the following code management functions:

- Multiple revisions of the same code base
- Multiple concurrent developers
- Locking, synchronization, and concurrency control, which protects the code base from two developers simultaneously overwriting the same file
- Support for versioning history and project forking, meaning that certain parts of a project or the entire project itself can be split off (forked) into a separate project

Some of the common VCS concepts and operations that support these functions include:

Repository: The repository is the place where the shared code base is stored. While anyone can download and read code from a FOSS respository, typically only certain developers—*the core team* or *committers*—have authority to write code into the repository.

Working copy: This refers to the individual developer's copy of the code base running inside the IDE on his/her own computer.

Check out: Individual developers can check out a working copy of the code base, downloading it to his or her computer, and linking the code to an IDE such as Eclipse.

Update: The developer can download the most recent version of a code base that was previously checked out. This updated version will reflect changes made by other developers on the team.

Commit: Developers who have sufficient write privileges with the repository can contribute new code directly. These developers' working copies of the code are immediately merged with the existing code base. The VCS figures out which pieces of code have changed in the working copy and need to be merged.

Merge: Two pieces of code are merged whenever they both apply to the same file or code segment.

Patch: A patch is a piece of code that is added to the code base. It usually describes what code is to be added or removed from the code base.

Trunk: Repositories are organized into hierarchies, which are tree structures. The trunk is usually the main development line for a project.

Branch: A branch is a copy of the software that can undergo separate and independent development. Branches are often used to package *releases* of the software or to allow experimentation. In some cases a branch can break off (or *fork*) into a separate or competing project. In many cases, branches are later merged back into the trunk.

A good (non-software) example of a VCS is Wikipedia. For any page in Wikipedia, it is possible to examine the history of the page, which will show exactly what changes were made, by whom they were made, and when they were made. If a user believes that a set of changes is incorrect, he or she can revert to a previous version. A major difference between Wikipedia and a mature FOSS code repository is that not every FOSS developer has direct write access to the code base. Even in a CO-FOSS project, only the developers and team leader can modify the code base; the client representative cannot.[5]

[5] In a mature FOSS project, only a few team members have privileges to write new code directly to the code base. These team members serve as gatekeepers who review and approve new code contributions by other developers before adding them to the code base. So in a mature project, every developer has *read access* but only a few have *write access*.

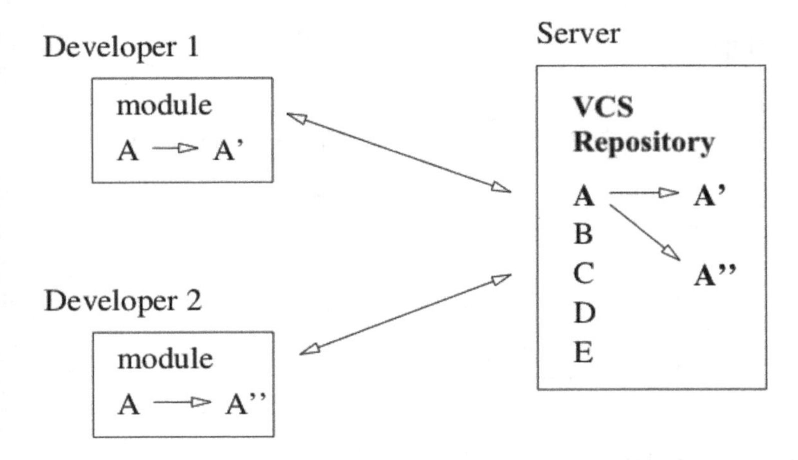

FIGURE 4.2 The code synchronization problem.

While a VCS helps manage the code base, it does not automatically resolve all possible *conflicts* that can arise during a collaborative project. To illustrate, suppose two developers decide simultaneously and independently to work on module A in a software project, whose code base has, say, five modules A, B, C, D, and E. The code base sits on a separate server, called a *code repository*, from which each developer has downloaded a working copy to their own local machine (see Figure 4.2).

Since both developers initially have a copy of module A, they begin working on it. After completing their work, these two developers typically produce two different variations, say A' and A", of the original module A. Neither variation incorporates the changes contained in the other, so no matter which variation finally replaces A in the code repository the other will be lost. This problem is illustrated in Figure 4.2.

The use of a VCS can avoid this problem. Here's how it works. Suppose Developers 1 and 2 both want to modify module A in the repository. Each of them starts working with a local copy of A that is identical with module A in the repository. That is, each one "Pulls" a copy of module A. Suppose further that Developer 1 completes his/her modifications first and *Pushes* the changed module A' to replace A in the repository.

Now when Developer 2 completes his/her modifications, A", and attempts to save them in the repository, the repository will reply that his/her modifications are based on an *out of date* version of module A. That is, a new version A' has replaced A in the repository since Developer 2 began modifying A to obtain A".

However, there is good news: Developer 2 can first *Pull* the new version A' so that it can be compared locally with A", as shown in Figure 4.3. This step will reveal either of two situations to Developer 2:

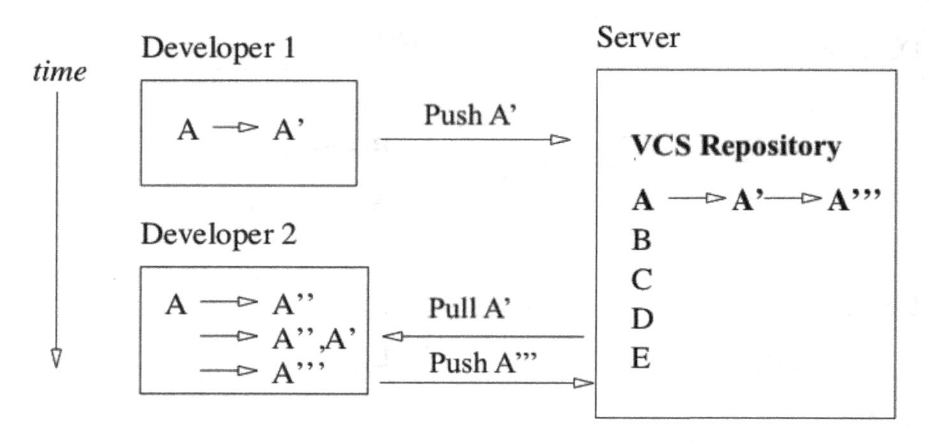

FIGURE 4.3 Resolving the problem: Copy-modify-merge.

1. The changes in A' are separate and distinct from those that appear in A".
2. The changes in A' *conflict* or overlap with those that appear in A".

Resolving situation 1 is simple. Developer 2 simply asks the synchronization repository to update his/her version A" with the changes in A', and then pushes the resulting version A"' back to the repository.

Resolving situation 2 is not as simple. In this case, Developer 2 must examine each part of A' that conflicts with a change that he/she has made to module A, and then manually combine these parts together so that the intended functionalities from both developers are properly implemented in the new version A"'. Sometimes a conversation between the two developers is needed to resolve this kind of conflict.

In some cases, it may be strategic for one developer to obtain a temporary and exclusive *lock* on a module. Until the lock is released, no other developer can make changes to that module. This strategy is useful in cases where situation 2 would create so many conflicts that their resolution would be better handled by a single developer working alone on all the modifications of module A rather than distributing the modifications among several developers working simultaneously.

In any case, the example illustrates that the use of a code synchronization repository by project developers is not a panacea. Developers must still collaborate effectively to resolve (or prevent) conflicts in the repository whenever they arise.

In our projects, a typical coding assignment shared among 4 or 5 developers would have each one working simultaneously on a separate module, so that conflicts like the one discussed above do not normally occur. This is easier to achieve in a modest CO-FOSS project with a few developers and a project leader overseeing the organization of each assignment.

Over the last several years, we have used different repositories and version control systems in our different CO-FOSS projects. Recently, we merged all these projects into a unified collection of Git repositories on GitHub (see `https://github.com/megandalster`). GitHub provides an ideal hosting environment for CO-FOSS development. Git and GitHub are normally well-integrated within each of the IDEs discussed in the foregoing section. For example, Figure 4.4 shows the menu of options provided to an Eclipse developer for pulling, committing, and pushing code to a remote Git repository.

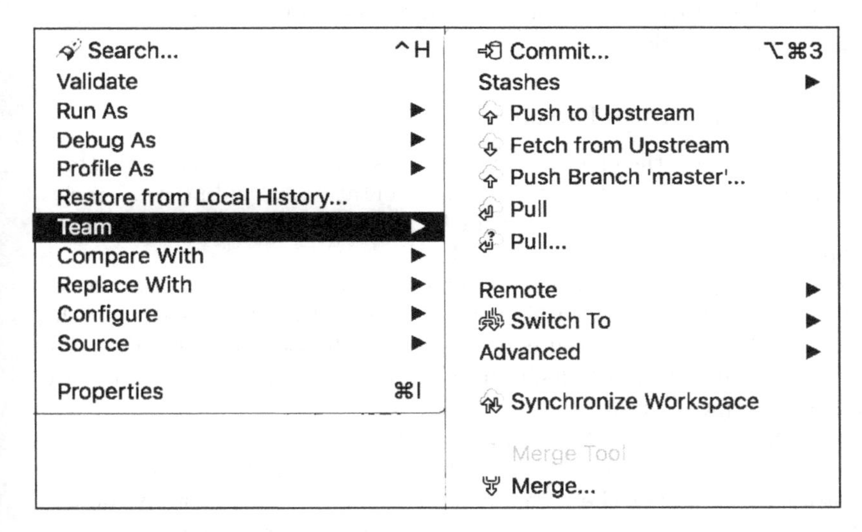

FIGURE 4.4 Git Menu Options (on right) from within an Eclipse IDE.

Here, the "Commit" option is used when the developer has finished coding and testing a module in the project – it "commits" those changes to his/her local copy of the code base, creating a new "version" along with a brief description of those changes. The "Push to Upstream" option is used to update the shared repository on the server with this new version. The "Fetch from Upstream" option is typically used by the developer before committing any local changes, so as to bring the local version into sync with the current version on the server.

Students who are new to using version control should take some time to gain a clear idea of both the concept of code sharing and the steps required whenever they begin and end a specific coding task. Excellent tutorials on getting started with Git and GitHub are available on-line, such as:

`https://guides.github.com/activities/hello-world/`

`https://www.tutorialspoint.com/git/`

Doing a simple GitHub exercise early in a CO-FOSS project will help developers become comfortable code-sharing. That comfort will reap benefits later in the project, when the code base and the coding assignments become more complex and more intertwined.

4.3 THE PRODUCT

With all the staging finally behind us, we can now begin thinking about developing the software itself. This section provides some pointers on what to look for when reading the project's initial design document, reading the initial code base, harvesting code from other projects, and identifying the open source license in the code.

4.3.1 Reading the Design Document

In a CO-FOSS project, the client's view captures the developer's initial attention. When the project has some design documentation, including user stories or use cases, developers naturally begin by reading that documentation to understand the software's goals in some detail.

In the end, developers must also understand the code base that underlies the software and serves as the basis for adding new features. Ideally, all the classes and modules in the initial code base are well documented. In the absence of such documentation (which, in practice, *may* happen), the code base becomes its own documentation.

Agile methods advocate developing the requirements for a new project simultaneously with developing the code. In this approach, the team starts by evoking stories from users to help sketch an initial design document. We do not advocate this degree of open-endedness because it creates uncertainty about the overall scope and limitations of the project. Having a design document available at the outset, the development team can immediately be clear about the outcome, and thus can more accurately predict and reach each milestone that will be needed to complete the project.

A design document typically contains a domain analysis, a requirements analysis, and a set of use cases. The domain analysis summarizes the setting in which the new software will be used, the types of users who will be accessing the software, and the ways in which the new software will influence the work conducted in that setting.

The *domain analysis* should also describe how the *current* application works, including any existing software that may be used to support it. This description should provide enough detail so that readers can understand the existing system's strengths and weaknesses.

The *requirements analysis* identifies the activities that the new software must support. Emphasis here is on the word *must*—a software product is incomplete if it doesn't fulfill all of its requirements. The requirements statement has two key parts: a collection of *use cases* and a statement about the *platform and other constraints* that will govern running the new software.

The *use cases* provide a transformative view, showing how the system should support specific kinds of interactions between types of users (called *user roles*) and the system. A use case may be written in a highly stylized way that identifies actors and roles, preconditions, results, and other details necessary for the implementation to be effective. Or it may be written as a brief 1-paragraph summary of what the use case is meant to accomplish and by whom. This latter form of a use case is often handy when upgrading an existing FOSS product rather than developing it from scratch; in this situation, the use case is already well-understood but it may need to be recast to reflect new features in the upgrade.[6]

For example, the **Process a Referral** use case discussed in Section 2.2.3 appears in the original 2011 *Homeroom* design document. At that time, the software was developed from scratch to replace an entirely manual room scheduling system. This use case describes how a user should interact with the system when entering a new referral (creating a new booking) in the system. The brief 1-paragraph version of this use case was renamed **Edit Pending and Active Bookings** in the 2013 *Homeroom* `upgrade design document`. Its brief description appears below:

> **Edit Pending and Active Bookings** The Volunteer or House Manager will continue to be able to edit all active and pending bookings as she does now, except that all bugs and issues identified in Section 1.5 of the design document for the *Homeroom* upgrade should be corrected.

Between 2011 and 2013, the "Referral" class had been renamed in *Homeroom* as the "Booking" class. This 1-paragraph version of that use case implicitly acknowledges that clients and developers can exercise the implementation of this use case in the 2011 version of *Homeroom* if they need to learn about it in more detail.

When we read a use case, we need to perform three different kinds of activities in order to understand or add new features to the code base.

1. Identify classes and modules.

2. Identify instance variables.

3. Identify methods and functions.

Each of these activities is further discussed below.

---

[6]For a detailed discussion of how to develop the domain analysis, requirements analysis, and use cases, readers may want to look back at Chapter 2.

Identify Classes and Modules

When reading the requirements for a new system, developers look for *"big picture" nouns*, which are those that can be abstracted into *classes* or *modules*, which are the major elements in the system. They also look for *"supporting" nouns* that can be abstracted as *instance variables* in a class or module.

For example, look at the **ProcessAReferral** use case in Section 2.2.3 which suggests a need for the following classes that correspond to "big picture" nouns:

Person — the House Manager and a Client who needs a room,

Booking — a referral requesting a new Booking, and

RoomLog — a record of all Rooms available for Booking and their current status.

Identify Instance Variables

To associate a class's instance variables with a use case, we look for details in the use case that characterize the state of an object in a class. Consider the RoomLog class, for example, in the *Homeroom* code base alongside the **ProcessAReferral** use case. That class contains an array of 21 `rooms` that are available in this particular Ronald McDonald House for booking.

Other less-obvious instance variables for a RoomLog are the `date` and the `status` of the RoomLog on that particular date. The idea here is that there is one RoomLog object in the database for every day, and its status may be "unpublished," "published," or "archived" depending on whether or not that RoomLog is publicly viewable or for a date prior to the current date. These variables are less apparent from reading the **ProcessAReferral** use case than is the array of rooms. Thus, identifying all of a class's instance variables is seldom achieved by examining a use case in isolation.

Thus, answering one question leads to more questions, such as:

What sort of Booking can be assigned to a room?

How do we determine whether or not a room is available to be booked?

Answers to these and other questions can only be achieved after reading all the use cases and related discussions in the design document, alongside corresponding elements in the existing code base itself.

Identify Methods and Functions

The third kind of information that comes from a use case is information that helps designers specify *actions* — methods or functions — that need to be implemented for each class and module in the new system. These actions are suggested by the *verbs* that appear in the use case.

For instance, the **ProcessAReferral** use case identifies at least the following functions: *verify* the user logging in as a House Manager (step 2), *create* a new Booking (step 3), and *add* a Booking to a room on the RoomLog (step 5).

One answer leads to more questions here, too:

- What additional supporting actions are needed to realize these new functions?
- What underlying database functionality is needed to implement these actions so that they are permanently recorded?

Answering these questions requires examining more use cases and discussions in the design document, as well as related sections of the code base itself.

When identifying a new method or function, it is important to determine what class should contain that method (if there's a choice), what to name that method, and what (if any) parameters it needs.

Consider, for example, the process of adding a Booking to a room on the RoomLog for a particular date. The method could be called `reserve_room` and implemented in the Booking class, with the `room_no` of the Room and `date` of the reservation supplied as parameters. The method appears in the Booking class as follows:

```
/*
 *  assign a room to a booking after client has confirmed.
 *  Book it immediately if the date is past
 */
function reserve_room ($room_no, $date) {
    $r = retrieve_dbRooms($room_no,$date,"");
    if ($r) {
        if ($date<date('y-m-d')) {
            $r->book_me();  // change status to "booked"
            $this->status = "active";
        }
        else {
            $r->reserve_me();  // change status to "reserved"
            $this->status = "reserved";
        }
        $this->date_in = $date;
        $this->room_no = $room_no;
        update_dbBookings($this);
        return $this;
    }
    else return false;
}
```

Notice here that additional statements change the status of the room and the booking. These details can be skipped in an initial reading of this code.

4.3.2 Reading the Code

Often the design document for an existing code base is unavailable or out of date with respect to the current version of the software. In this case, developers need to begin by reading and extracting information directly from the code base itself. If it is well-organized and legible, the code base can provide a reliable starting point from which to launch a new software project. In this case, reading the code has three key goals:

1. To understand the overall architecture and functionality of the existing software.
2. To learn the vocabulary established by the domain classes.
3. To identify the extent to which the code must be refactored before it can be modified to support the functions of the new system.

Reading the code is often accompanied by running the software itself, allowing developers to understand the relationship between various user actions and the corresponding classes and modules that support those actions.

When reading code for the first time, developers are well-advised to keep the following strategies in mind:

- Start from the top.
- Look for classes with unique keys.
- Avoid the temptation to refactor.

Each of these is discussed more fully below.

Start from the Top

Reading a code base is a disciplined and orderly activity. Starting at the top, we begin at the highest structural level and then work our way down gradually to the details of individual classes, variables, and functions.

So begin reading by examining the code's overall structure, looking for answers to the following questions:

What are the domain (core) classes, and how do they characterize the principal objects that are active when the software is running?

What are the database modules, and what tables are needed in the database? How do the database modules relate to the domain classes?

What are the graphical user interface (GUI) modules, and how do they relate to the use cases described in the design document?

What particular domain classes and database modules are needed by each GUI module? What are the key methods and functions in these classes and modules that implement individual user actions?

How does the software enforce the client's security requirements?

What other modules are in the code base, and what is their purpose?

What unit testing strategy was used for developing the software, and what is its current condition?

It is important to have the current version of the code base running during this reading, so that whenever a particular element of the code is read, its run-time role and behavior can be quickly validated. For example, Table 4.1 summarizes the directory structure of the *Homeroom* code base.

TABLE 4.1 Overall Structure of the *Homeroom* Code Base

Domain Classes	Booking OccupancyData Person Room RoomLog
Database Modules	dbBookings dbInfo dbLog dbPersons dbRooms dbRoomLogs
User Interface	about bookingDetails bookingEdit bookingForm bookingValidate data dataView log footer header index log login_form logout personEdit personForm personValidate room roomLog roomLogView searchBookings searchPeople view viewBookings viewPerson
On-Line Tutorial	addPersonHelp dataHelp editBookingHelp editPersonHelp roomLogHelp searchBookingsHelp index indexHelp login searchPersonHelp viewBookingsHelp viewCalendarHelp
Unit Tests	testBooking testdbBookings testdbPersons testdbRoomLogs testdbRooms testOccupancyData testPerson testRoom testRoomLog

Look for Classes with Unique Keys

For every class we examine, it is important to determine whether or not each member of the class has an instance variable that distinguishes it from all the other members. Such an instance variable is called a *unique key*, a *unique identifier*, or just an `id` for short.

For example, in *Homeroom*, every Person must be distinguishable from every other Person in the database. Each Person's `id` must be unique, since querying the database for that person must retrieve exactly one record, not two or three or more.

Defining the unique key for a Person can be done using various strategies. For example, the strategy used in *Homeroom* is to define each Person's `id` as the concatenation of their first name with their 10-digit primary phone number. This pretty much guarantees uniqueness, since it is unlikely that two different Persons will share the same first name and phone number.[7] This strategy also has the advantage that it is easy to remember—few people forget their first name or their phone number.

For the same reason, every Booking, every Room, and every RoomLog must have a unique id in its respective database table. More detailed discussion of unique keys and their uses in database design appears in Chapter 6.

Avoid the Temptation to Edit the Code

While reading an existing code base will suggest areas that need rewriting, we need to avoid the temptation to edit or rewrite bad code (called "refactoring") when we first encounter it. Why?

1. Refactoring must be accompanied by testing, to ensure that the refactored code doesn't introduce new bugs.

2. Once refactoring starts, it threatens to never end.

Combined, these two perils can bog a project down and divert developers from their primary code-reading goals—to understand the architecture and core vocabulary of the system. We will have plenty of opportunities to refactor later in the development process.

At the same time, we should not avoid the temptation to add or refine existing documentation. That activity has no effect on the functionality of the code, and adding good documentation can improve the readability of the code base. So whenever you come to a certain understanding of *what* a particular class, module, method, or complicated function does, take a moment to add a brief documentary comment if it is not already there.

Assuming somebody else will reread that code later in the development process, our enhancement can help others avoid having to reread the code at the same level of detail. Here's an example piece of undocumented code from *Homebase*:

```
function remove_availability ($a) {
    $index = array_search($a, $this->availability);
    if ($index !== false)
        array_splice($this->availability, $index, 1);
}
```

This little method in the Person class searches a Person's `availability` array for a certain time slot `$a` and, if present, removes it using the PHP function `array_splice`. This code is a bit technical, requiring readers to remember what is returned by the `array_search` function and also what is accomplished by `array_splice`.

[7] Unlikely, but not impossible—for example, a parent and child of the same gender may live in the same house and share the same first name and land line.

So if you think the title of the method doesn't fully reveal its purpose, you should feel free to add a comment at the top of this method that will help clarify its purpose for the next reader of this code:

```
/**
  * remove time slot $a from Person's availability, if present
  */
```

4.3.3 Reading and Writing Code

Writing good code requires using conventionally accepted coding and documentation practices. This is especially important when the software is being developed by a team. Unfortunately, much of the code that underlies actual software products does not reflect the use of good practices. Thus, some of a developer's work includes rewriting poorly written code to make it more readable and receptive to the addition of new features.

Below is a list of widely used coding standards for common program elements. These standards are illustrated here in PHP, though similar standards exist for Java, Python, and other contemporary programming languages.

Naming and spelling—Class, function, variable and constant names should be descriptive English words. Class names should be capitalized; if they consist of more than one word, each word should be capitalized—e.g., Person, RoomLog.

- Multiple-word function and variable names should separate their adjacent words by an underscore—e.g., `$primary_phone`. Alternatively, these names can be written with each non-first word capitalized—e.g., `$primaryPhone`.
- Global variable names should be written in all-caps and begin with an underscore—e.g., `_MY_GLOBAL`.
- Constant names should be written in all-caps—e.g., `MY_CONSTANT`.

Line length—A line generally contains no more than one statement or expression. A statement that cannot fit on a single line is broken into two or more lines that are indented from the first line.

Indentation—Consistent indentation should be used for control structures and function bodies. Use the same number of indentation spaces (usually four) consistently throughout the program—e.g., see Figure 4.5.

The use of coding standards such as these is sometimes met with resistance by professional developers. They argue that pressures to complete projects on time and on budget prevent them from the luxury of writing clear and well-documented code all the time. But this argument breaks down when the software being developed is subject to later revisions and extensions. This is

```
/**
 * fill a vacancy in this shift with a new person
 * @return false if this shift has no vacancy
 */
function fill_vacancy($who) {
    if ($this->vacancies > 0) {
        $this->persons[] = $who;
        $this->vacancies=$this->vacancies-1;
        return true;
    }
    return false;
}
```

FIGURE 4.5 Documentation practice using indented blocks and control structures.

especially true in the FOSS development world, where the same code typically passes through several sets of developers' eyes as it evolves.

4.3.4 Code Reuse

Open source development benefits greatly from having code from other similar projects freely available for reuse. That is, the project stands a greater chance of success if not all the code is developed from scratch. Most projects share common features with other similar projects, and so it is important at the outset to learn just what features are already fully-developed and can be "harvested" for reuse in the current project.

Harvesting and reusing good code is not plagiarism, particularly when the documentary authorship of the original code is retained when the code is reused. A careful look at some of the different code bases in our repositories at `https://github.com/megandalster` reveals just how much different student teams benefited from reusing code developed by other teams in earlier projects, going all the way back to the original *Homebase* project in 2008.

A valuable case of code reuse occurs with the `login_form` and related modules that were originally developed for *Homebase*. A careful examination of the *Homeroom*, *Homeplate*, and other later projects at our `GitHub site` reveals that those very same modules are reused in each of these projects for the exact same purpose – secure login and user authentication.

For example, `https://npfi.org/login-form/` creates the `login_form` module that appears in the *Homeroom* code base – at the top of this form, we read that it was originally coded for the *Homebase* product in 2008 by a different developer.

By reusing the original `login_form` and related code without having to reinvent it, student developers in these later projects saved significant development time and gained a head start on their own projects.

4.3.5 Licensing

Every FOSS product must be freely available to and adaptable by any other organization that has a similar need. To implement and disseminate this information for future users and developers, the particular license chosen (e.g., the GPL) should be identified on all pages of the user interface (usually as a minor footer message), as well as in the headers of all major classes and modules in the source code.

Copyright © 2011-2014 by: Felix Emiliano, Luis Rojas, David Phipps, Ruben Martinez, Alex Lucyk, Jesus Navarro, and Allen Tucker.
Homeroom was designed at Bowdoin College for the Ronald McDonald House of Portland, Maine. It is free software and comes with absolutely no warranty. You can redistribute and/or modify it under the terms of the GNU General Public License as published by the Free Software Foundation.

FIGURE 4.6 Showing the open source license notice in the user interface.

Figure 4.6 shows how this notice appears in the footer of every page of the user interface for *Homeroom*. Typically, the original authors of the software share its copyright and the users of the software acknowledge that their use will comply with the terms of its particular license.

Figure 4.7 shows how this notice appears at the top of each class or module in the *Homeroom* code base. This notice is written for developers who are debugging the code or modifying it for future use. The GPL basically requires developers to retain this notice in all future versions and distributions of the modified code.

```
/*
 * Copyright 2011 by Alex Lucyk, Jesus Navarro and Allen Tucker
 * This program is part of RMH Homeroom, which is free software.
 * It comes with absolutely no warranty.  You can redistribute
 * and/or modify it under the terms of the GNU Public License
 * as published by the Free Software Foundation
 * (see <http://www.gnu.org/licenses/).
*/
```

FIGURE 4.7 Displaying the open source license notice in the source code.

Recall from Chapter 1 that licensing a FOSS product affirms that the software is *free* as in *free to use, modify, and distribute.* This freedom provides a strong incentive for other developers to examine the code base and freely adapt and reuse its best elements in other FOSS projects.

4.4 SUMMARY

This chapter provides guidance for launching a new CO-FOSS project, which typically takes place during the first week of the course. Project launch includes setting up a lot of scaffolding that will be used throughout the project's development period, whether that is a semester or a longer period.

Project launch also includes an initial team meeting, so that clients, student developers, and others will gain an initial comfort level with the project and their respective roles in it. A first assignment for the developers should naturally come out of this meeting, so that developers can begin working in earnest toward eventual completion of the project.

Team members should review the project's initial design document in preparation for this meeting, and should feel free to ask questions that will help clarify the team's understanding of its goals and methodologies.

4.5 MILESTONE 4

1. The team leader should provide each developer and client representative with a copy of the syllabus, the project's design document, and the team's synchronous and asynchronous collaboration tools. Developers should also have access to the project's sandbox server and database.

2. Each developer should set up a working IDE for the team project, and synchronize the IDE's local server and code base with the sandbox database and code repository.

3. The entire team (leader, developers, and client) should conduct an initial meeting where people are introduced to each other and the project's goals and schedule are reviewed. A set of goals for the next meeting should be set and agreed upon.

CHAPTER 5

Domain Class Development

"A designer knows [s]he has achieved perfection
not when there is nothing left to add, but
when there is nothing left to take away."
—*Antoine de Saint-Exupery*

This chapter presents the first really productive stage in our journey through the CO-FOSS development process. By finishing this chapter, developers will have completed some new code and conducted their first client review.

Developing new code has three significant activities – coding, unit testing, and system testing. The developers work independently on each of the first two activities, and then they system-test their combined work on the sandbox version so that the client can review the results.

The first section of this chapter discusses and illustrates strategies for developing the domain classes. This includes both finding and reusing code from earlier open source projects, as well as coding the classes from scratch.

The second section focuses on unit testing and debugging principles and practices. It provides guidance for setting up and running a unit test for each domain class that was just developed. Unit testing is such a central activity in CO-FOSS development that we pay particular attention to its nuances both here and in Chapters 6 and 7.

The third section introduces the practice of merging the results of each team member's coding and testing efforts, and testing them together under the umbrella of an "integration test." This activity ensures that the combined work of individual developers has no conflicting overlaps or inconsistencies.

The fourth section suggests ways to obtain constructive feedback from a client review of the domain classes that were just developed.

5.1 CODING THE DOMAIN CLASSES

The domain classes are the central element in a CO-FOSS product that is customized to fit the needs of a single client. They capture the vocabulary used in the client's own working environment, so that the terminology reflected in the software has familiar meaning to the clients who use it.[1]

Because establishing this vocabulary is the foundation for developing the database modules in Chapter 6 and the user interface modules in Chapter 7, we strongly recommend that the software development process begins by fleshing out and testing the domain classes. This also makes sense because it draws heavily on student developers' experience with data structures and object-oriented programming in prior courses.

5.1.1 Reusing External Legacy Code

Some domain classes are so common that they reoccur in many different open source products. For example, the Person class reappears in all of our past software products, from *Homebase* to *Homeroom* to *Homepate* to *BMAC-Warehouse*. In each recurrence, the Person class carries a collection of basic instance variables that identifies an individual — their name, address, phone, and email for example. But in each occurrence, the Person class also carries instance variables that clarify the precise nature of a person's relationship with the client organization. Here, for instance, are four different occurrences of the Person class.

In *Homebase*, the Person class carries instance variables that characterize a volunteer who signs up for shifts on the calendar. So in addition to the basic instance variables mentioned above, the *Homebase* Person class carries variables like "availability" (dates and times when the person can work a shift) and "schedule" (shifts actually scheduled).

In *Homeroom*, the Person class carries instance variables that characterize a person as a guest who occupies a room for a series of days. So in addition to the basic instance variables, the *Homeroom* Person class carries variables like "patient" (name of the child receiving medical treatment) and "prior bookings" (past dates when the client stayed at the house).

In *Homeplate*, the Person class carries instance variables that characterize a volunteer who drives a food rescue truck on a scheduled basis. So in addition to the basic instance variables, the *Homeplate* Person class carries variables like "drivers license number," "availability" (dates and times when the person can drive a route), and "schedule."

[1]This simple idea is often lost in "one size fits all" software products, which force all users to assimilate an unfamiliar vocabulary before they can begin to use the software productively.

In *BMAC-Warehouse*, the Person class is used only to identify those persons authorized to log in and manage the food receipt and delivery schedule and the food inventory database. Therefore, only the basic instance variables are needed for this Person class.

Finally, all four CO-FOSS products have one or more persons in an administrative role who use all features of the software and manage the entire underlying database. For these persons, a different "manager" level of access to the database is needed. Thus, a "type" field is needed for the Person class in order to distinguish this administrative level of access from that required by volunteers and others needing more limited access to the software.

These four examples suggest that any new CO-FOSS project should first look outside its own domain for similar well-developed domain classes in other projects that can be adapted and specialized to fit the special characteristics of its own group of volunteers or beneficiaries.

For instance, in the years after *Homebase* was initially developed in 2008, a new student team developed *Homeroom* in 2011, another team developed *Homeplate* in 2012, and finally a different team developed *BMAC-Warehouse* in 2015. Each of these latter three projects reused and adapted the Person class from the original *Homebase* project to fit its own specific requirements.

How can this be done? Here is a simple 3-step process:

1. First, download the original Person class from the older project's code repository and incorporate it into the new project's code base.

2. Next, identify and retain the basic instance variables and their related functions for reuse, and remove the instance variables that are of no use to the new project (along with their related functions).

3. Finally, add new instance variables that complete the vocabulary for that class for the new application (along with their so-called "getter" functions). The design document should provide enough guidance about what specific instance variables are needed.

Here is an example of code reuse that happened when *Homeroom* was first developed in 2011. The team looked over its shoulder and downloaded the Person class from its 2008 predecessor *Homebase*, whose instance variables are shown in Figure 5.1.

From here, the developer used the *Homeroom* design document to come up with the collection of instance variables for its own Person class shown in Figure 5.2. Looking carefully at these two lists, we see that the first 11 instance variables are identical. Moreover, the remaining instance variables in the *Homeroom* list were obtained by 1) removing all instance variables from the *Homebase* list that were not needed, and then 2) adding new instance variables that were suggested by reading the *Homeroom* design document's use cases.

It is usually the case that not all instance variables for the new class can be predicted at the outset. For example, in the Person class for *Homeroom*,

```
class Person {
  private $id;       // id (unique key) = first_name + phone1
  private $last_name;  // last name - string
  private $first_name; // first name - string
  private $address;       // address - string
  private $city;          // city - string
  private $state;         // state - string
  private $zip;           // zip code - integer
  private $phone1;        // primary phone
  private $phone2;        // alternate phone
  private $email;         // email address
  private $password;    // password for secure access
/* */
  private $type; // "volunteer", "applicant", "sub", "manager"
  private $birthday;    // format: 64-03-12
  private $employer;    // current employer or school
  private $position;    // job title or "student"
  private $motivation;  // App: why interested in RMH?
  private $specialties; // App: special hobbies related to RMH
  private $convictions; // App: ever convicted of a felony?
  private $status;  // "applicant", "active", "LOA", "former"
  private $availability;// array of day:hours:venue triples
  private $schedule;    // array of scheduled shift ids
  private $notes;       // manager notes
```

FIGURE 5.1 Reusable *Homebase* Code in 2008.

the variable `$county` was added when *Homeroom* was upgraded in 2013 to refine its room utilization statistics reporting. The county of residence is now automatically computed by *Homeroom* from a person's zip code.

5.1.2 Reusing Internal Legacy Code

Many CO-FOSS projects involve upgrading an existing open source product to meet additional requirements and fix serious shortcomings in that product. This type of activity requires a more substantial effort than fixing a bug or resolving a usability issue. For example, when *Homeroom* was first implemented in 2011 it had the Booking class shown in Figure 5.3.

Homeroom was later refined in 2013 by a different team to meet new requirements, as described in the `2013` *Homeroom* `design document`. At that time, the Booking class needed additional instance variables and related functionality. Looking at Figure 5.4 and comparing it with Figure 5.3, notice the

```
class Person {
  private $id;      // id (unique key) = first_name + phone1
  private $last_name;  // last name - string
  private $first_name; // first name - string
  private $address;        // address - string
  private $city;           // city - string
  private $state;          // state - string
  private $zip;            // zip code - integer
  private $phone1;         // primary phone
  private $phone2;         // alternate phone
  private $email;          // email address
  private $password;    // password for secure access
/* */
  private $type;        // 'guest', 'socialworker', 'manager'
  private $gender;      // gender of person - string
  private $employer;    // employer of the person - string
  private $patient_name;      // array of up to 3 patients
  private $patient_birthdate; // format: 11-03-12
  private $patient_gender;    // "Male" "Female" or "Unknown"
  private $patient_relation;  // relationship, eg "mother"
  private $prior_bookings;    // array of booking ids
  private $mgr_notes;         // manager's notes
  private $county;            // county in Maine
```

FIGURE 5.2 Adapting the Code for Reuse in *Homeroom* in 2011.

new variables $date'submitted, $auto, $linked'room, and $health'questions that were added in 2013 to accommodate these new requirements.

One lesson we draw from this experience is that the original version of new software reflects a fairly naive effort to implement its use cases in a straightforward way. Subsequent improvements to that software benefit from experience with the original and a better understanding of its new requirements.

5.1.3 Coding a Domain Class from Scratch

Coding most of a new application's domain classes is not as straightforward as suggested by the foregoing examples. In practice, most domain classes for a new application are unique to that application. So the idea of customizing a new class from an existing class elsewhere is less likely than defining it from scratch, using the design document for guidance.

For example, *Homeroom* has five domain classes: Person, OccupancyData, Booking, Room, and RoomLog. All but the Person class are unique to the room scheduling application, and so they had to be designed from scratch.

```
class Booking {
  private $id;          // identifier: current_date + guest_id
  private $date_in;     // check-in date: "yy-mm-dd"
  private $guest_id;    // id of guest, eg "John2077291234"
  private $status;      // "active" or "closed"
  private $room_no;     // id of the room
  private $patient;     // patient name
  private $occupants;   // names of people staying in the room
  private $date_out;    // check-out date "yy-mm-dd"
  private $referred_by;// id of referring social worker
  private $hospital;    // patient's hospital
  private $department; // patient's treatment department
```

FIGURE 5.3 Original Booking Class for *Homeroom* in 2011.

The design document can provide a sketch of some instance variables for each class to the extent that the lead developer understands them during the requirements-gathering stage (see Chapter 2). For example, Figure 5.3 shows the initial instance variables for the Booking class that was designed from scratch in the original 2011 *Homeroom* design. The Room and RoomLog classes are simpler, in the sense that they have fewer instance variables. For more details on these classes, visit `https://github.com/megandalster/rmh-homeroomcivi`.

5.1.4 Adding Functionality: Constructor and Getters

In general, not all the functions in a new class can be predicted at the time it is first coded. However, it is safe to create an initial level of functionality for the class by coding:

A constructor for a new object in the class, and

A series of so-called "getter" functions, one for each instance variable.

As readers may recall from earlier experience, a "getter" function is a simple function that returns the current value of its associated instance variable. In general, getters provide security for the class because they obviate the need to declare public instance variables. For example, a sketch of the instance variables, constructor, and getters for the original version of the Room class in *Homeroom* are shown in Figure 5.5.

Defining new classes, constructors, and getters should be familiar territory for the development team, and thus it provides a comfortable starting point for the project as a whole. Moreover, when the number of classes to be implemented is nearly the same as the number of developers on the team, assigning

```
class Booking {
  private $id;          // identifier: current_date + guest_id
  private $date_submitted;// date booking was submitted
  private $date_in;     // check-in date: "yy-mm-dd"
  private $guest_id;    // id of guest, eg "John2077291234"
  private $status;      // "pending", "active", or "closed"
  private $room_no;     // id of the room
  private $patient;     // patient name
  private $occupants;   // names of people staying in the room
  private $auto;        // automobile make:model:color:state
  private $linked_room;// family may need 2 rooms
  private $date_out;    // check-out date "yy-mm-dd"
  private $referred_by;// id of referring social worker
  private $hospital;    // patient's hospital
  private $department; // patient's treatment department
  private $health_questions; // 11 questions for the family
```

FIGURE 5.4 Revised Booking class for *Homeroom* in 2013.

one class to each developer provides a fairly equitable division of work. Thus, each developer will "own" a different and essential part of the development process. Moreover, this distribution of ownership will provide a reasonable basis for division of work when the database and user interface modules are developed later in the project.

5.2 SOFTWARE TESTING

Maintaining a systematic and effective testing strategy for new code is a key part of CO-FOSS development. Traditional approaches would place testing sometime after all the coding is done. That is the wrong placement.

Contemporary approaches to software testing integrate testing throughout the coding process; whenever a new class or unit of code is written, it is immediately tested. A *most* aggressive approach places testing hand-in-hand with requirements gathering and before any coding is done. This approach is called *test-driven development* or TDD.

Contemporary approaches to testing are especially valuable in CO-FOSS development because they provide a vehicle that keeps the client in the game. That is, discussions between developers and clients can evoke concrete examples of system and user interface behavior, both before and throughout the period when the code is being written.

Contemporary approaches also encourage thoughtful and effective coding. Initially, the new code in TDD may be only a trivial prototype, since it is designed to respond specifically to satisfy the test at hand. However, once an

```
class Room {
  private $room_no;       // room number, like "125" or "233"
  private $beds;          // bed configuration: 2T, Q, K, etc.
  private $capacity;      // maximum number of persons
  private $bath;          // "y" if there's a private bath
  private $status;        // "clean","dirty","booked",
                          // "reserved", or "off-line"
  private $booking;       // the current booking id for this room
  private $room_notes;    // room-specific notes
// constructor
  function __construct ($room_no, $beds, $capacity, $bath,
         $status, $booking, $room_notes) {
    $this->room_no = $room_no;
         ...
  }
// getters
  function get_room_no () { return $this->room_no; }
  function get_beds () {...
  function get_capacity () {...
  function get_bath(){...
  function get_status () {...
  function get_booking_id() {...
  function get_room_notes () {...
}
```

FIGURE 5.5 Room class constructor and getters for *Homeroom*.

iteration or two has taken place, the code (and the tests) begin to mature and become more robust.

A good integrated development environment, such as Eclipse, supports continuous testing alongside coding. Testing frameworks, like PHPUnit (for PHP) and JUnit (for Java), are normally integrated into the IDE for easy access. The examples below illustrate the use of a testing framework in an actual CO-FOSS development setting.

The architecture of the code base should also support continuous testing, in the sense that a separate directory of test cases should coexist among the code directories themselves. Keeping the tests together with the code is good practice, even though it adds a bit of storage overhead to the system as a whole. This practice also facilitates systematic testing of new code whenever new functionality is added at a future date.

As a final note on testing, we need to acknowledge its fallibility. That is, even the most systematic and thorough testing strategies do not guarantee that the code is free of errors — it only ensures that the code being tested did not fail for the particular tests that were run. Only an approach to testing called *formal methods* would guarantee that the code is free of errors. This characteristic is called "correctness." However, for a variety of reasons, the use of formal methods is not common in most software development settings. This book does not include the study of formal methods. For a more careful overview of the use of formal methods in software development, readers may want to visit `https://en.wikipedia.org/wiki/Formal_methods/`.

5.2.1 Test Case Design

All that said, how do we systematically test our software while we are developing it? As a start, we need to create a collection of individual tests that cover all the classes and other modules in the system. This collection is called a *test suite.*

When the software being developed is an extension of an existing system, the existing code base may already contain a test suite. If it does not, our first step is to develop a test suite for the existing system—maybe not a simple task in itself. This will allow us to develop our new code in a setting that is anchored in a reliable initial test suite.

What should a test suite look like? This is a complicated question in general. Our discussion identifies a basic approach to test suite design, but it should not be considered a complete treatment of software testing. For a more thorough treatment, see `https://en.wikipedia.org/wiki/Software_testing`.

Here is a straightforward test suite definition and organization scheme that has been effective in our CO-FOSS projects:

> A *test suite* is a collection of "unit tests." The suite should have one unit test for every module or class in the code base and one unit test for every use case.
>
> Each unit test should contain a group of calls that exercise all of the module's functions (and constructors, in the case of a class). Each such call is written as an *assertion*, or Boolean function that delivers "true" exactly when that call is successful, and "false" otherwise.
>
> The test suite should aim for *100% code coverage and 100% use case coverage.* That is, every line of code in every class or module should be exercised by at least one assertion among the unit tests in the test suite. Every use case step should also be covered in the same way.

In fact, aiming for 100% code coverage can be a weak testing strategy. For example, there may be boundary conditions not mentioned in the use case that are not covered by the code, but nevertheless should be tested.

A layered software architecture and development strategy is especially hospitable to building a test suite of this sort, since the domain classes are developed first, then the database modules, and finally the user interface modules that satisfy the use cases. At each of these three stages, its unit tests can be developed and run alongside the coding of their respective classes and modules.

Designing a unit test for a domain class or database module requires writing at least one assertion for every function and constructor in that class or module. Designing a unit test for a user interface module is more difficult, since each user interface module contains a mix of PHP, JavaScript, and HTML code that supports a particular user form.

Testing an entire use case should be done only after the separate testing of all the underlying domain classes, database modules, and user interface forms has been completed. To test a use case, we begin by identifying as "units" those user interface elements—forms and associated modules—that combine to implement that use case. All together, use case testing provides a basis for "acceptance testing" the entire system in the eyes of the user.

Because testing is such a complex subject, we spread its treatment over three separate chapters. In this chapter, we illustrate unit testing for just the domain classes. In Chapter 6, we discuss unit testing of the database modules, a process that has its own peculiarities. Finally in Chapter 7 we discuss testing the user interface.

In our CO-FOSS projects, student developers designed test suites with unit tests for all the domain classes and database modules. However, unit tests for the user interface modules were not developed, and for two principal reasons:

1. The user interface can be actively and effectively tested by the client using the sandbox server.

2. Limited time is available for developing complete unit tests for the user interface modules, given the strict end-of-semester deadline for completing the entire project.

So our projects relied on critical client review and active issue posting during the development of the user interface modules. More detailed discussion of these ideas appears in Chapter 7.

5.2.2 Unit Testing Frameworks

Software frameworks for unit testing provide a convenient means for running all the unit tests in a test suite at once, relieving the developer from the need to run each unit test individually. In PHP, for example, a popular open source framework for running a suite of unit tests is called *PHPUnit*. The `PHPUnit website` provides downloads for PHPUnit, along with documentation and tutorials for installing it, designing unit tests, and running a test suite.

To illustrate, the test suite for *Homeroom*'s domain classes and database modules is kept in the `tests` directory of the code base, as summarized in Figure 5.6. Whenever PHPUnit is run, it exercises all these unit tests and

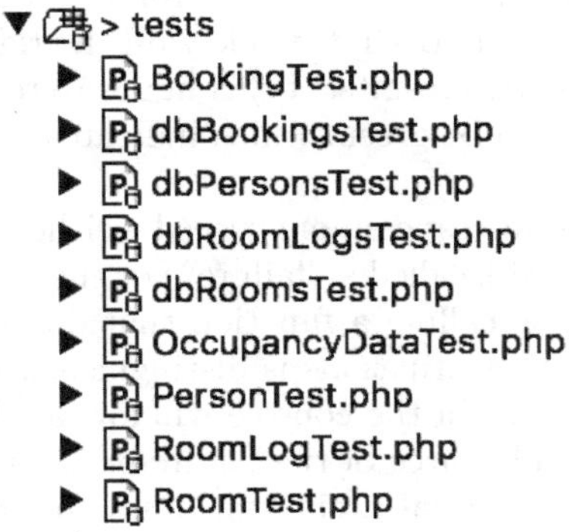

- tests
  - BookingTest.php
  - dbBookingsTest.php
  - dbPersonsTest.php
  - dbRoomLogsTest.php
  - dbRoomsTest.php
  - OccupancyDataTest.php
  - PersonTest.php
  - RoomLogTest.php
  - RoomTest.php

FIGURE 5.6 Test Suite in the *Homeroom* `tests` Directory.

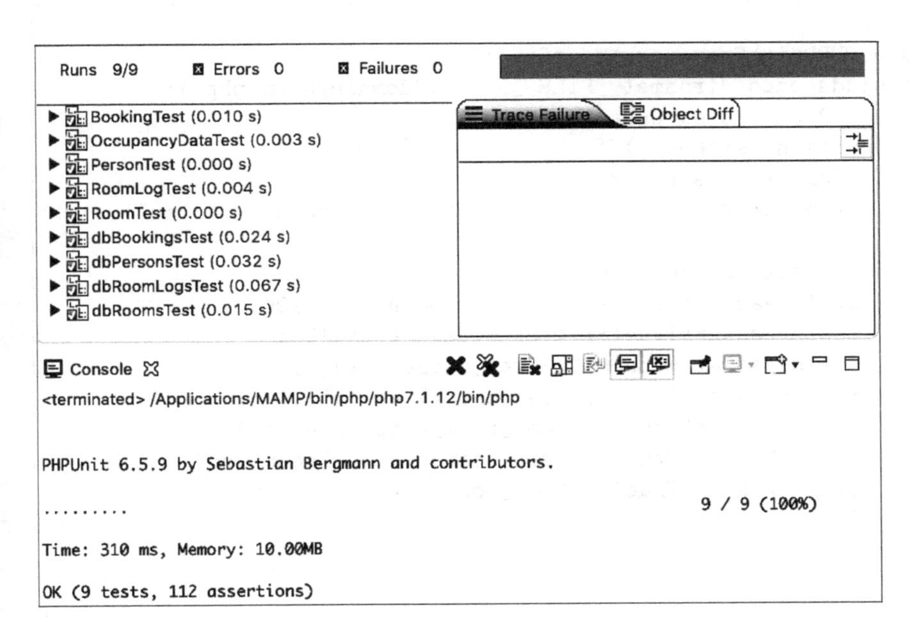

FIGURE 5.7 Results of running a Test Suite.

reports their results. The results of running all the *Homeroom* unit tests in its `tests` directory are displayed in Figure 5.7.

The console window at the bottom provides details and error messages when an individual test does not succeed. The PHPUnit window on the right lists the tests that were run and their respective results. If this figure were in color, the horizontal bar at the top right would be green, which indicates that all tests ran successfully.

There are two different ways in which a unit test cannot succeed – either an assertion in the unit test is not confirmed (that's called a "failure") or the code itself contains a run-time PHP error, such as calling a function that doesn't exist (that's called an "error"). Either of these situations is distinguished by a maroon bar at the top, along with a message in the console window at the bottom that further pinpoints the nature and source of the failure or error.

To illustrate further, let's look at an actual unit test and the result of running it. Figure 5.8 shows a (minimal) unit test for the Room class shown in Figure 5.5. The assertions in this unit test are executed in order whenever PHPUnit is run. The two most prominent kinds of assertions are `assertTrue(boolean)` and `assertEquals(expression1, expression2)`. The assertTrue statement succeeds only if its Boolean is true, and the assertEquals statement succeeds only if its two expressions are equal.

```
<?php
use PHPUnit\Framework\TestCase;
include_once(dirname(__FILE__).'/../domain/Room.php');
class RoomTest extends TestCase {
  function testRoom() {
  // Construct a new room
    $r = new Room("126", "2T", "3", "y", "reserved", null, "");

    // Test each of its getters
    $this->assertEquals($r->get_room_no(), "126");
    $this->assertTrue($r->get_beds() == "2T");
    $this->assertEquals($r->get_capacity(),3);
    $this->assertTrue($r->get_bath() == "y");
    $this->assertTrue($r->get_room_notes() == "");
    $this->assertTrue($r->get_status() == "reserved");
    $this->assertTrue($r->get_booking_id() == null);
    }
}

?>
```

FIGURE 5.8 A Unit Test for the Room Class in *Homeroom*.

A unit test is typically written by constructing an object in the class being tested, and then adding assertions that test each of its getters to be sure that the values of its respective instance variables have been properly assigned by the constructor. For example, `$this->assertTrue($r->get_bath() =="y");` literally asserts that the question "Does the `$bath` instance variable for the object `$r` have the value 'y'?" should produce a true result.

To illustrate a failure in a unit test, let's change the constructor in the RoomTest unit test so that the room number for room `$r` is 125 rather than 126; the unit testing run will report a failure, as shown in Figure 5.9. Note here that the console window on the left reports the details, showing what the assertion expected and what the actual value of the room number was that caused the failure.

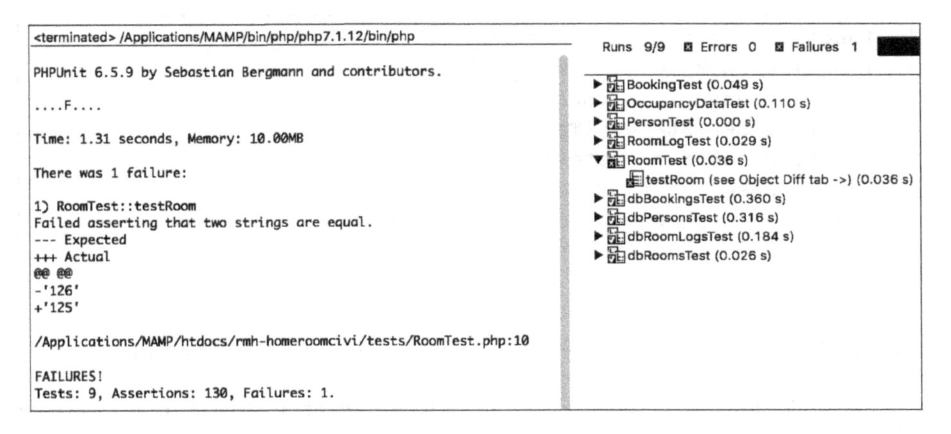

FIGURE 5.9 Reporting a Unit Test Failure.

A similar level of detail would have been reported if a PHP run-time error had occurred in any of the code being exercised by the unit test. Especially valuable in this case is the traceback to the line number in the code where the error actually occurred.

For a complete summary of the different assert statements that can be used with PHPUnit, visit `the PHPUnit reference manual`. In addition, we note that there are many tutorials on the Web that introduce the practice of test case design and unit testing for PHP (and other languages) more gently.

For other popular programming languages, their respective unit testing frameworks are similarly well-supported and documented. For example, a popular framework for Java is `JUnit`, while a popular framework for Python is `unittest`, which is part of the Python standard library. For a complete list of testing frameworks available for various languages, please visit `the list of unit testing frameworks on Wikipedia`.

5.2.3 Unit Testing the *Homeroom* Domain Classes

The *Homeroom* code base has 5 domain classes, 7 database modules, 32 user interface modules and 8 use cases. In this section, we discuss the development of unit tests for the 5 *Homeroom* domain classes.

At the time the domain classes are being developed, the test suite should receive one unit test for each domain class. We have already seen the beginning of a unit test for the Room class in Figure 5.8, but it only tests the class's constructor and getters. During the course of *Homeroom* development, new functions (called "setters") are added to the Room class that modify the values of various instance variables. The profiles for those functions are listed in Figure 5.10.

```
function reserve_me ($booking_id)
function book_me ($booking_id)
function unbook_me ($booking_id)
function set_status ($new_status)
function set_room_notes($notes)
function set_beds ($beds)
function set_capacity($newCapacity)
function set_bath($newBath)
```

FIGURE 5.10 Setter Functions for the Room Class in *Homeroom*.

As suggested by their names, some of these setter functions simply change the value of an instance variable. However, others are more complex, like the `book_me`, `reserve_me`, and `unbook_me` functions. Each one of these not only resets a room's status but also updates its entry in the database. To complete the unit testing of the Room class, therefore, we need to combine it with the unit testing of its associated database module, dbRooms.php.

Finally, a complete unit test of the Room class must include calls to functions that cover all the likely variations that will occur when they are utilized. For example, consider the `set_status` function. To make sure that it properly sets a room's status, we need to ensure that a room's status can have one of only five valid values: "clean," "dirty," "reserved," "booked," and "off-line."

Thus, the `set_status` function should be designed to screen for these five values, and the RoomTest unit test should assert that such screening actually works by calling `set_status` with both valid and invalid arguments. So we would add at least the following assertions to the RoomTest unit test shown in Figure 5.8:

```
$this->assertEquals(($r->set_status("clean"))->get_status(),
    "clean");
$this->assertFalse($r->set_status("invalid status");
```

The first assertion sets the status to "clean" and then immediately gets the status to confirm that particular value was assigned. The second assertion checks to be sure that only a valid status will be assigned to a room. Concurrently, the `set_status` function should contain the code that actually does the work of screening for a valid call, as shown below.

```
function set_status ($new_status) {
  if ($new_status=="clean" || $new_status=="dirty"
    || $new_status=="reserved" || $new_status=="booked"
    || $new_status=="off_line") {
      $this->status = $new_status;
      return $this;
  }
  else return false;
}
```

When designing a unit test for the Booking class in *Homeroom*, we need to write assertions that test the constructor and every getter and setter in the class. A partial unit test for the Booking class is shown in Figure 5.11, where the constructor is first called, and then the getters and some of the setters are exercised. The full unit test is shown in the *Homeroom* `code base.`

A similar approach to test case design applies to each of the other domain classes. Before moving forward to developing the database modules, the team must have confidence that all the domain classes are solid, in the sense that they all give the "green light" when their constructors, getters, and setters are exercised.

5.2.4 Unit Testing the *Homebase* Domain Classes

Here is an illustration of the unit testing process for *Homebase*, which was first developed in 2008 and then upgraded in 2011, 2013, and 2015 in three separate CO-FOSS projects. In each project, the domain classes were developed first, then the database modules, and finally the user interface modules.

The 2015 code base for *Homebase* has 7 domain classes, 8 database modules, and 30 user interface modules. The examples below discuss the development and unit testing of the *Homebase* domain classes, with an emphasis on the upgrading of unit tests from one version of *Homebase* to the next.

Adding a new instance variable, adding/modifying a function, or adding an entirely new class or module to the code base requires adding assertions to their respective unit tests to ensure that the new variable, function, class, or module behaves as intended. This must be done at all levels of the architecture, beginning with the domain classes.

For example, in 2015 a suite of unit tests for *Homebase* was inherited from the 2013 code base and used as a starting point for adding and testing new

```
class BookingTest extends TestCase {
  function testBooking() {
    $today = date('y-m-d');
    $b = new Booking($today,"","Meghan2075551234","pending",
      "",array("Tiny"),array("Meghan:mother", "Jean:father"),
      "", "", "", "Mil2073631234","Maine Med", "SCU",
      "00000000000","$10 per night", "","","","","new");

    $this->assertTrue($b->get_id()==$today."Meghan2075551234");
    $this->assertEquals($b->get_date_submitted(),$today);
    $this->assertTrue($b->get_date_in()=="");
    ...
    $occ = $b->get_occupants();
    $this->assertTrue(in_array("Jean:father",$occ));
    $this->assertEquals($b->get_linked_room(),"");
    ...
    $b->add_occupant("Jordan","brother","","");
    $this->assertEquals(sizeof($b->get_occupants()), 3);
    $b->remove_occupant("Jean");
    $this->assertEquals(sizeof($b->get_occupants()), 2);
  }
}
```

FIGURE 5.11 Partial unit test for the Booking Class in *Homeroom.*

features. This original test suite had 5 unit tests and corresponding domain classes, as shown below:

```
PersonTest.php
RMHdateTest.php
SCLTest.php
ShiftTest.php
WeekTest.php
```

As students added the new features described in `https://npfi.org/homebase-design/`, some of these unit tests were modified to cover the new functionality. Moreover, two additional unit tests were added to cover the two new domain classes that were added. Thus, the new test suite had these 7 domain class unit tests (Asterisks indicate modifications and additions):

```
ApplicantScreeningTest.php*
MasterScheduleEntryTest.php*
PersonTest.php*
RMHdateTest.php
SCLTest.php
ShiftTest.php*
WeekTest.php
```

To illustrate how a unit test changes as a result of adding new functionality to an existing class, consider the elements of the ShiftTest unit test for the 2013 version of *Homebase* (Figure 5.12), and compare it with the ShiftTest unit test for the 2015 version of *Homebase* (Figure 5.13).

```
class ShiftTest extends TestCase {
  function testShiftModule() {
    $noonshift = new Shift("03-28-08-12-15",3, null, "", "");
    $this->assertEquals($noonshift->get_name(), "12-15");
    $this->assertTrue($noonshift->get_id() == "03-28-08-12-15");

// Test new function for resetting shift's start/end time
    $this->assertTrue($noonshift->set_start_end_time(15,17));
    $this->assertTrue($noonshift->get_name() == "15-17");

// Be sure that invalid times are caught.
    $this->assertFalse($noonshift->set_start_end_time(13,12));
    $this->assertTrue($noonshift->get_id() == "03-28-08-15-17");
    $this->assertTrue($noonshift->get_name() == "15-17");
    $this->assertTrue($noonshift->num_vacancies() == 3);
    $this->assertTrue($noonshift->get_day() == "Fri");
    $this->assertFalse($noonshift->has_sub_call_list());
  }
}
```

FIGURE 5.12 The 2013 unit test for the Shift class.

Comparing these two unit tests, we notice that the 2015 version has several new assertions, and also that its first few assertions are slightly different from the 2013 version. These differences are mainly to accommodate the addition of new instance variables and a new constructor. However, they also acknowledge the reformatting of the date field from the form `mm-dd-yy` to the form `yy-mm-dd` so that it supports the natural sorting of shift ids into chronological order. Also, new assertions were added to test new functions `add_vacancy`, `ignore_vacancy`, and `set_notes` for the Shift class.

```
class ShiftTest extends TestCase {
  function testShift() {
    $noonshift = new Shift("08-03-28:1-5", "house",3,
        array(),array(),"","");
    $this->assertEquals($noonshift->get_hours(), "1-5");
    $this->assertTrue($noonshift->get_id() ==
        "08-03-28:1-5:house");

// Test new function for resetting shift's start/end time
    $this->assertTrue($noonshift->get_start_time()==13);
    $this->assertEquals($noonshift->get_end_time(),17);

// Be sure that invalid times are caught.
    $this->assertFalse($noonshift->set_start_end_time(13,12));
    $this->assertTrue($noonshift->get_hours() == "1-5");
    $this->assertTrue($noonshift->num_vacancies() == 3);
    $this->assertTrue($noonshift->get_day() == "Fri");
    $this->assertFalse($noonshift->has_sub_call_list());
    $persons = array();
    $persons[] = "alex1234567890+alex+jones";
    $noonshift->assign_persons($persons);
    $noonshift->ignore_vacancy();
    persons[] = "malcom1234567890+malcom+jones";
    $noonshift->assign_persons($persons);
    $noonshift->ignore_vacancy();
    $persons[] = "nat1234567890+nat+jones";
    $noonshift->assign_persons($persons);
    $noonshift->ignore_vacancy();
    $this->assertTrue($noonshift->num_vacancies() == 0);
    $noonshift->add_vacancy();
    $this->assertTrue($noonshift->num_slots() == 4);
  }
}
```

FIGURE 5.13 The 2015 unit test for the Shift class.

```
class ApplicantScreening {
  private $type; // Unique identifier for this template
      // reflects type of position for which it will be used.
      // Eg, "volunteer2" or "guestchef1"
  private $creator; //  id of who created it
  private $steps; // array describing the individual steps
  private $status; // "unpublished" or "published"

// constructor
  function __construct($t, $c, $s, $st) { ...
// getters
  function get_type() {...
  function get_creator() {...
  function get_steps() {...
  function get_status() {...
// setters
  function set_type($new_type) {...
  function set_creator($new_creator) {...
  function set_steps($new_steps) {...
  function set_status($new_status) {...
}
```

FIGURE 5.14 New ApplicantScreening Class Added to *Homebase* in 2015.

It should be clear that these new assertions provide only minimal tests for the new functions. Adding even more assertions to gain better code coverage for a class is always a good way to start a new CO-FOSS project from an existing code base and test suite.

To illustrate how a new unit test is added alongside adding an entirely new class, consider the new unit test for the ApplicantScreening class shown in Figure 5.14. Applicant screenings were added to *Homebase* in 2015 so that the manager could keep track of the progress of each new applicant for a volunteer position by checking off a series of steps (like “application reviewed,” “interview completed,” “background check completed,” etc.) to be completed before accepting the applicant as a regular volunteer. The corresponding unit test for this new class is shown in Figure 5.15.

5.2.5 Code Synchronization and Integration Testing

Unit testing the domain classes individually should pave the way for testing the interactions among them and between them and other modules in the code base. These interactions occur when one class has functions that depend on another class or module, and testing these interactions is called “integration

```
class ApplicantScreeningTest extends TestCase {
  function testApplicantScreeningModule() {
    $myApps = new ApplicantScreening("volunteer",
      "Gabrielle1111234567", "Background_Check,Interview",
      "unpublished");

    $this->assertTrue($myApps->get_type()== "volunteer");
    $this->assertTrue($myApps->get_creator()==
        "Gabrielle1111234567");
    $this->assertEquals($myApps->get_steps(),
        array("Background_Check","Interview"));
    $this->assertTrue($myApps->get_status()=="unpublished");
  }
}
```

FIGURE 5.15 New ApplicantScreening Unit Test added in 2015.

testing." In general, integration testing occurs after all the individual units are tested independently. It purpose is to test the interactions among related classes and modules to ensure that they behave as intended.

For example, the *Homebase* classes `Week`, `RMHdate`, and `Shift` are interdependent in the sense that a week on the calendar is a collection of days, and a day is a collection of shifts. We see this interdependence explicitly in the headers of the `Week`, `RMHDate`, and `Shift` classes shown in Figure 5.16.

A key kind of interaction is the one that occurs between a domain class and its associated database module. In this example, we will eventually need to test the interactions between the Shift.php class and its associated dbShifts.php module, as suggested in Figure 5.16. But this can happen only after we have unit-tested each of these separately. We will return to discussing that activity in Chapter 6.

So before integration testing of two interacting classes or modules can take place, the following three activities must be completed by the developers who "own" the respective classes and modules:

1. Each individual class and module must be unit-tested separately,

2. Each of those tested classes, modules, and their respective unit tests must be "pushed" and "merged" into the project's code base using the VCS, and

3. The merged result must then be "pulled" into each developer's own copy of the code base.

Once these steps are completed, the task of integration testing can be distributed among the developers, who are now sharing the same code base.

```
include_once('RMHdate.php');
class Week {
    private $id;    // id of Monday, yy-mm-dd;venue
    private $dates; // array of 7 RMHdates, beginning Monday
    ...
}
    ...
include_once(dirname(__FILE__).'/Shift.php');
class RMHdate {
    private $id;        // "yy-mm-dd:venue" form of this date
    private $shifts;    // array of Shifts
    ...
}
    ...
include_once(dirname(__FILE__).'/../database/dbShifts.php');
include_once(dirname(__FILE__).'/../database/dbPersons.php');
class Shift {
    private $yy_mm_dd;  // String: "yy-mm-dd".
    private $hours;     // String: '9-1','1-5','5-9' or 'night'
    ...
}
```

FIGURE 5.16 Interdependencies among Classes for Integration Testing.

In the above example, for instance, one developer may expand the Week class's unit test to exercise its interactions with the RMHdate class. Similarly, another developer may expand the RMHdate class's unit test to exercise its interactions with the Shift class. Below is an assertion, for example, that tests some of the interactions between the RMHdate and Shift classes:

```
$my_shifts = array();
$my_shifts[] = new Shift("10-02-28:9-1", "house", 1,
    array(), array(), null ,"");
$my_shifts[] = new Shift("10-02-28:1-5", "house", 1,
    array(), array(), null ,"");
$my_date = new RMHdate("10-02-28","house",$my_shifts,"");
$my_shifts = $my_date->get_shifts();
foreach ($my_shifts as $value)
  $this->assertTrue($value instanceof Shift);
```

Prior to the addition of this code, the unit test for the RMHdate class could have been run independently for its other functionality by simply constructing a date that had no shifts and testing its other getters and setters.

5.3 DEBUGGING AND REFACTORING

Throughout the development and useful life of any software product, we can expect that errors in the code will occur. In the case of a CO-FOSS product, the client provides the first line of defense in locating software errors. This is inevitable, since the client exercises the code far more rigorously than any test suite can. Clients thus provide the most reliable and thorough source of feedback to developers, especially for the existence of errors, or *bugs*, in the software itself.

> Technically, the term *bug* refers to a *defect* or *flaw* in the code base that contributes to the occurrence of a *failure* when the software is running. A *failure* is any unacceptable behavior in the software that can be observed by the user.

In this book we use the term *bug* rather comprehensively, in the sense that it will simultaneously refer to both a failure and the underlying defect in the code that causes it.

A most severe kind of bug is one where the software totally "crashes," or goes into a permanent dysfunctional state from which the user cannot recover. One example of a crash is the occurrence of a so-called "white screen of death" (WSOD, for short), where the user suddenly ends up staring at a blank screen. Little evidence of the source of the bug is apparent when this happens. A debugging strategy for this case is to gather all the information that was available just before the WSOD happened (by restarting, repeating the steps, and gathering information along the way).

When finding and correcting a bug, the developer may add a new test to the test suite that addresses that particular error. Severe bugs may even cause developers to refactor parts of the code base. Briefly, the idea of "refactoring" suggests that parts of the code can be rewritten more simply and clearly without losing any of its functionality. Thus, the interplay among unit and integration testing, debugging, and refactoring is intimate.

In this section, we discuss the process of debugging in more detail, and then discuss the process of refactoring the code in a way that makes it cleaner and more receptive to future changes and additions.

5.3.1 Debugging

Some valuable tools are available for reporting, tracking, finding, and correcting software bugs. Every IDE contains debugging support. Simultaneously, every version control repository provides issue-tracking support for reporting and updating the status of each bug in a CO-FOSS project until it is resolved.

For example, the Eclipse PHP IDE provides debugging support through PHP's XDebug extension or the Zend Debugger, along with its built-in Debug Perspective. Many tutorials on debugging for different programming languages are available on-line, so we will not dwell on this subject here. Concurrently, the GitHub code repository provides issue-tracking support for every project. Figure 5.17 shows a list of 6 issues for the *Homeplate* project on GitHub that were open as of this writing. These issues will be fixed in a future *Homeplate* project or by the developer who provides ongoing support.

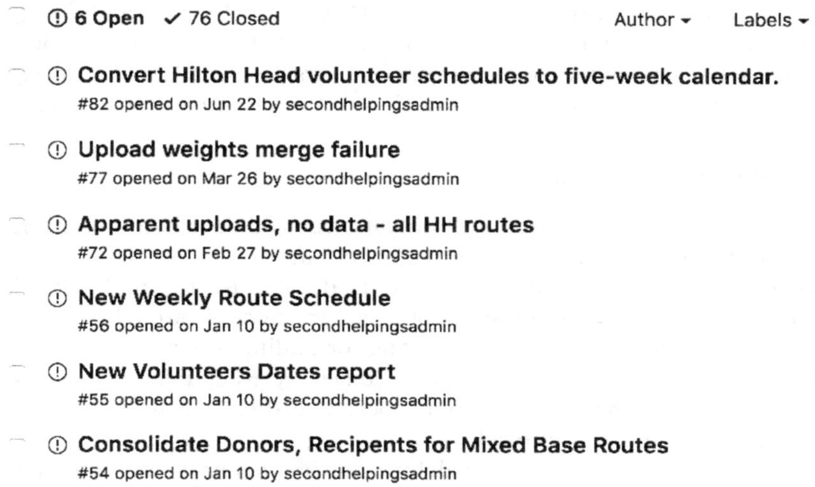

6 Open ✓ 76 Closed Author ▾ Labels ▾

Convert Hilton Head volunteer schedules to five-week calendar.
#82 opened on Jun 22 by secondhelpingsadmin

Upload weights merge failure
#77 opened on Mar 26 by secondhelpingsadmin

Apparent uploads, no data - all HH routes
#72 opened on Feb 27 by secondhelpingsadmin

New Weekly Route Schedule
#56 opened on Jan 10 by secondhelpingsadmin

New Volunteers Dates report
#55 opened on Jan 10 by secondhelpingsadmin

Consolidate Donors, Recipents for Mixed Base Routes
#54 opened on Jan 10 by secondhelpingsadmin

FIGURE 5.17 A Recent GitHub Issue List for the *Homeplate* Project.

However, tools alone are insufficient for effectively locating and removing a bug from a code base. Whether it is a new CO-FOSS project or a mature FOSS project, effective debugging relies on the full participation of users and developers to help keep the code base up to date and reasonably free of errors. Mature projects like Linux, Apache, PhP, Python, and MySQL maintain a close connection between developers and users, since the developers *are* the users. Mozilla provides a good example of how a user community participates directly and closely with developers. Developers thus rely on the active participation of users to help them test new releases and verify bug corrections in the code base.

Beyond tool use and collaboration with users, developers need two additional skills for effectively diagnosing and removing a bug from the code base: 1) a healthy understanding of the software architecture, and 2) an ability to traverse and analyze the code base to find the source of the bug. If the software does not have a coherent architecture, finding a bug in its code base can be equivalent to finding a needle in a haystack.

So an important strategy for debugging is to use the software architecture to narrow the search to a particular class or module where the bug is most likely to occur. For instance, if the bug seems to be "cosmetic" in nature, with no apparent impact on the permanent data, it is likely to be traceable to a user interface module. On the other hand, if the bug seems to have an impact on the system's data, the developer may need to follow the bug all the way down from the user interface to a related module at the database level. In either case, the developer should understand the code base at different levels, including the interdependencies among different modules and levels.

Examples in the following sections, as well as Chapters 6 and 7, illustrate the effective interplay between tools, users, software architecture, and developer skills in identifying and removing bugs from a code base. These examples represent actual errors that were discovered by users and corrected by developers after the *Homebase* code had been deployed and put into productive use.

5.3.2 Identifying Bad Smells

Before adding new features to existing software, the code needs to be examined for quality and receptivity to adding such features. Many times, such an examination reveals bugs that need to be fixed, or at least organizational characteristics that can be improved, before the new features are added.

The code we inherit for testing, debugging, or adding a new feature may reflect some so-called "bad smells" (a term coined by Kent Beck in [17], p. 75). A bad smell is a segment of code that doesn't read clearly and concisely, and hence probably can be improved.

The code may have been developed by a novice, a person not familiar with standards for good programming practice, or someone more interested in "making the program work" than "making the program readable." In any case, the original author of the smelly code may not be nearby. So it falls upon the current developers to remove bad smells from the code so that they and future developers will have an easier time understanding and working with it.

To illustrate this idea, consider the PHP program text in Figure 5.18, which contains a series of duplicate copies of a very technical bit of code.

```
$first_name = trim(str_replace('\'','\\\'',htmlentities($_POST['first_name'])));
$last_name = trim(str_replace('\'','\\\'',htmlentities($_POST['last_name'])));
$address = trim(str_replace('\'','\\\'',htmlentities($_POST['address'])));
$city = trim(str_replace('\'','\\\'',htmlentities($_POST['city'])));
$state = trim(str_replace('\'','\\\'',htmlentities($_POST['state'])));
$zip = trim(str_replace('\'','\\\'',htmlentities($_POST['zip'])));
$phone1 = trim(str_replace(' ','',str_replace('\'','\\\'',htmlentities($_POST['phone1']))));
$phone2 = trim(str_replace(' ','',str_replace('\'','\\\'',htmlentities($_POST['phone2']))));
$private_notes = trim(str_replace('\'','\\\'',htmlentities($_POST['private_notes'])));
$public_notes = trim(str_replace('\'','\\\'',htmlentities($_POST['public_notes'])));
$my_notes = trim(str_replace('\'','\\\'',htmlentities($_POST['my_notes'])));
```

FIGURE 5.18 Example bad smell—duplicate code.

Removal of this bad smell can make the code more readable. Figure 5.19 shows how creating a new function with a clear name can eliminate the bad smell of replicating obscure code that appears in Figure 5.18.

```
function sanitize_post ($s) {
    return trim(str_replace('\'','\\\'',htmlentities($_POST[$s])));
}
    ...
    $first_name = sanitize_post('first_name');
    $last_name = sanitize_post('last_name');
    $address = sanitize_post('address');
    $city = sanitize_post('city');
    $state = sanitize_post('state');
    $zip = sanitize_post('zip');
    $phone1 = sanitize_post('phone1');
    $phone2 = sanitize_post('phone2');
    $private_notes = sanitize_post('private_notes');
    $public_notes = sanitize_post('public_notes');
    $my_notes = sanitize_post('my_notes');
```

FIGURE 5.19 Example bad smell removal.

Several different types of bad smells can occur in programs. Here is a list of common ones that we have found useful in our projects (a more comprehensive list can be found in Fowler [17]):

Poorly Named Variable, function, or Class—the name does not clearly represent its purpose in the code. For example, the name `$sch` is a poor name for a variable instead of `$schedule`.

Duplicate Code—the same sequence of expressions or instructions appears in several places. For example, see Figure 5.18.

Long function—a function contains a lot of code that accomplishes several sub-tasks. For example, writing a function `AvMaxMin($list)` is a poor way to compute the average, maximum, and minimum value in a list, compared with writing three separate functions.

Large Class—a class has an unusually large and diverse collection of instance variables or functions. Often this signals a lack of *cohesion* for a class, in the sense that it is trying to represent more than one type of object at one time. For example, the class `AlarmClock` might have some features of an alarm and other features of a clock. A better design would define two classes, `Alarm` and `Clock`, and specify that a `Clock` may have an `Alarm` as a component.

Too Many/Too Few Comments—too many comments can hide bad code (i.e., they can be "used as a deodorant," so to speak), while too few can make code difficult to read. As a minimal rule of thumb, we recommend commenting each instance variable in a class, each column in a database table, and each non-trivial function.

Lazy Class—a class that is seldom used and its functions are seldom called. This usually suggests replacing those calls by calls to functions in other classes, and then removing the class from the code base altogether.

Temporary Fields—instance variables in a class that are set only in certain circumstances. All instance variables in a class should be set whenever an object in that class is created.

Aside: Using Software Metrics

Another way to evaluate the quality of a code base is to calculate its so-called "software metrics." A metric is a quantification of some aspect of the syntactic structure of a program, which aims to expose potential weaknesses (bugs, or risks of run-time failures) in that structure.

Several different metrics can be used to measure the quality of a program. When the metrics of a code base are not all within a normal range, this may indicate poor code quality. Here is a short list of metrics that are used to measure the quality of Java programs:

Count Metrics—the number of packages, classes, instance variables and functions for each class, and parameters and lines of code for each function. Good design tries to minimize these numbers in accordance with the nature of the application and its individual classes.

Clarity Metric—"McCabe cyclometric complexity" counts the number of distinct paths through each function, considering all its if, for, while, do, case, and catch statements. A good design tries to keep this number under 10 for every function in the code base.

Cohesion Metric—lack of cohesion of functions (LCOM) in a class with m functions means that the ratio of the average number of functions accessing each instance variable to $m - 1$ is near 1. That is, a class is completely cohesive if all its functions access all its instance variables, which means that $LCOM = 0$. So the more cohesive the class, the closer LCOM is to 0, which is another indicator of good design.

Coupling Metrics—indicate how strongly the packages in a code base are interdependent. Good design tries to minimize coupling.

Metrics plugins are available for most popular IDEs and programming languages, including for example `Eclipse`, `NetBeans Java`, and `PHP`. When the plugin is enabled, it recalculates the above measures whenever the code base is changed.

Bad smells and metrics tend to be complementary, in the sense that each metric corresponds to a particular bad smell that can be detected by reading the code. The main advantage of using metrics over looking for bad smells is that metrics can be quickly applied to a very large code base, highlighting areas that need a closer look. That is, reading all the text of a million-line code base, without the aid of metrics, would be a very tedious exercise. For a smaller code base, like the one typically occurring in a CO-FOSS project, metrics are not as useful as identifying bad smells by just reading the code.

5.3.3 Refactoring

Refactoring is the process of improving the structure source code without changing its behavior. As suggested above, refactoring can improve code readability and reduce its complexity, making it more maintainable and accessible to adding new features. However, refactoring is not usually done at the same time as adding new features. Moreover, any refactoring activity should be followed by running the complete set of unit and integration tests, to gain confidence that the code's overall behavior was not unintentionally changed.

Refactoring a domain class involves an examination to see if it has any redundant or useless functions or instance variables, or if it has any other types of bad smells listed above.

> A function is *redundant* if the work that it does can be accomplished by one or more other functions. An instance variable is redundant if the information it contains also appears in other instance variables.
>
> A function is *useless* if it is not called from anywhere in the code base. An easy way to examine a function for uselessness is to invoke the IDE's search tool to find all calls to that function from anywhere else in the code base.

In Eclipse, this search tool is activated by highlighting the function name and then selecting **Search->Text->Workspace** to obtain a list of all references to that function name throughout the code base. For example, Figure 5.20 shows the resulting list of all instances of the getter function named `get_address` that appears in the *Homebase* Person class. Each instance is identified by showing the class or module where it appears, along with its corresponding line number. Selecting that instance puts the developer back to the line number in the code, thus facilitating easy editing.

If such a search reports that there are no such calls, that function is useless and can be safely deleted from the code base. More important, this search technique can often be used to cleanly repair all calls to a function whenever its own profile or internal definition needs to be modified.

These considerations notwithstanding, let's take a look at some other functions in the *Homebase* Person class to see if there are any useless ones hanging around. This is critical when changing the profile of a function. Using the Search feature described above, we gain the following insights:

1. Each of the "getters" defined for the Person class is called at least once, so none of these can be removed.

2. On the other hand, the lone "setter" function `set_availability` in the Person class is not called, so it can be safely removed.

The second item above raises an obvious question. Why are there no useful setter functions in the Person class? The answer to this question is tied up in

```
database
  dbPersons.php
    38: $person->get_address() . '","' .
domain
  Person.php
    127: function get_address() {
personForm.inc
  79: <p>Address<span style="font-size:x-small; ...
reportsCompute.php (3 matches)
  150: $person->get_address(), $person->get_city(), ...
  159: "</td><td>".$person->get_address() ."</td><td>"...
  471: $person->get_address(), $person->get_city(), ...
```

FIGURE 5.20 Searching the code base for all references to the `get_address` function.

the overall design of *Homebase*. In short, whenever a Person changes any of his/her personal information on a form, an entirely new object is created using the Person constructor and that person's entry in the database is updated by removing the old entry and inserting a new entry in its place. We will clarify this question further in Chapters 6 and 7, where database and user interface design is covered in more detail.

However, for every *new* domain class we add to the code base, we should assume that all the "getter" functions are necessary. Later opportunities for refactoring each domain class occur when new features are implemented in the database and user interface modules. With *Homebase*, for instance, the new class ApplicantScreening should not require any refactoring at this early point in the process.

Nevertheless, adding a new class can suggest reexamination and possible refactoring of other related classes, especially those after which the new class has been modeled. For example, in *Homebase*, the newer MasterScheduleEntry class was modeled from the existing Shift class. The main difference is that a master schedule entry represents a shift that reoccurs on a regular basis throughout the year (weekly, bi-weekly, or monthly), while a shift is associated with a single date. Thus, when designing the MasterScheduleEntry class, its instance variables `hours` and `persons` reflect the same information as their counterparts in the Shift class, while its variables `day` and `week_no` are fundamentally different from the Shift variable `yy_mm_dd`.

What other types of refactoring can be done? We look to the list of bad smells for guidance. For example, one kind of refactoring is to eliminate instances of duplicate code, which usually involves defining a new function with one copy of that code and replacing each one of the duplicate copies by a function call. The bigger the size of the duplicated code, the worse it smells!

The type of refactoring shown in Figures 5.18 and 5.19 is called "extracting a function." Many other types of refactoring are possible (for a complete list, see Fowler [17]), as listed below.

Renaming a variable, function, or class to add clarity for the reader and consistency with the user's domain and design document.

Extracting a function to eliminate duplicate code, and then replacing each duplicate appearance by a call to that function.

Extracting functions to reduce the size of a long function. That is, divide the long function into smaller segments, extracting each segment as a separate function and then replacing the original segments by a series of calls.

Reorganizing a class to improve its logical cohesion. If two or more kinds of objects are encapsulated by the class, break it into separate classes.

Simplifying conditionals Sometimes a collection of conditionals can become so deeply nested that the underlying logic becomes impossible to understand. Such a nest can be simplified by rethinking and disentangling the logic.

Removing Useless Code When reviewing the code base alongside the design document, look for code that has no purpose in relation to the existing requirements and use cases. This may take the form of a function, a variable, or an entire class or module that is never referenced. All such code should be removed.

Removing Layering Violations When layer, cohesion, and coupling principles are violated, refactoring should be done to remove the violations. See Chapters 6 and 7 for more discussion of layering.

Merging Similar Functions/Modules During different stages in development, a developer may not be aware that a function/module already exists that (nearly) fulfills a certain requirement, and thus may reimplement that function/module. When this redundancy is discovered later, it becomes a candidate for refactoring.

Separating Model, View, and Controller Code In a graphical user interface (GUI) module, confusion can occur if the model, view, and controller code are unclearly intertwined. Separating the code for these three clarify it and make it more robust. We shall discuss this activity in more detail in Chapter 7.

Note that some refactoring activities may combine two or more of these techniques in a single step. For example, when combining two similar functions into one, we may also eliminate or add a parameter. Overall, This list reveals several types of refactoring that can be useful for a new CO-FOSS project that inherits an existing code base.

As a final refactoring example, we reexamine the 2015 *Homebase* code base. There, the Shift class has 24 functions and 1 constructor. Utilizing the search tool provided in our IDE, we find that the function `duration` is useless, since it is not called from anywhere else in the code base, and can be safely eliminated.

We also find that the functions `get_hours` and `get_time_of_day` in the Shift class are mutually redundant, since they both return the same result when called. To remove one of these, we need to change each of its calls to call the other one instead, and then run our unit tests and integration tests to gain confidence that this refactoring did not change the software's run time behavior. Notice also that `get_date` and `get_yy_mm_dd` in the Shift class are mutually redundant, and one of these can be removed using the same systematic replacement technique suggested above.

Why were these useless functions and pairs of redundant functions introduced into the Shift class in the first place? The answer probably lies in the fact that the Shift class was developed incrementally by different teams of developers in four separate projects over several years. The beginning of a new CO-FOSS project, when the domain classes are being developed, is a good time to clean house and remove bad smells from the initial code base. When preparing to add new features, we want the code to be as clean and lean as possible at the outset.

Note finally that much of our refactoring tends to *reduce* the size of the code base rather than enlarge it. Effective software, like effective writing, has little tolerance for redundancy and superfluity. Remember the words of Saint-Exupery at the beginning of the chapter!

5.4 CLIENT REVIEW AND ISSUE TRACKING

Now that the domain classes have been coded and tested, we can now consider the process of client review and issue identification for this first iteration in the CO-FOSS cycle. What can the client say about domain class development when presumably the only user interface element that exists for the software is (maybe) a secure login? And how can the client post issues for the developers when bugs are found in the software? We address these questions below.

5.4.1 Client Review

To prepare for client review, the code base should be integration-tested by the developers and then pushed to the sandbox server for clients to exercise. An initial database should also be uploaded onto the database server, including a few entries in the table corresponding to the Person class. In our applications, this table is called dbPersons and it is seeded with rows corresponding to clients who can securely log in, along with rows for other persons who play different roles in the application. In *Homebase*, for example, these rows represent people who are house volunteers, while in *Homeroom* they represent people who stay overnight at the House. Only a handful of such entries is needed at this stage, since we are not really testing the database modules yet.

So what should the client be testing with the software at this early stage? They can be testing two important aspects – the well-functioning of the secure login protocol and the completeness of the domain classes with regard to the requirements document.

To test the login protocol, each client should be able to log into the software, and should also test that entering an incorrect password prevents them from logging in. Once logged in, the client should see a simple interface that allows them to view the instance variables for each domain class and review the details to be sure that they capture all the needs expressed in the requirements document and the use cases.

In a CO-FOSS project that is adding new features to an existing product, client review may be more extensive, since much of the original user interface may still be accessible after login. With that, the client will be able to test that the original features have not been compromised by the adding of new instance variables to the domain classes. However, the removal of unwanted instance variables from the original domain classes may raise new errors, since the corresponding database and user interface modules have not yet been updated to incorporate these changes. Thus, a certain level of circumspection is needed during client review at this stage in the project.

5.4.2 Issue Tracking

In Figure 5.17, we saw an example of an issue list that had been posted for the *Homeplate* software project on GitHub. So who posts these issues, how are they managed (or "tracked"), and what is the relationship between each issue and its resolution within the CO-FOSS development cycle?

In general, a new issue occurs whenever a developer or a client discovers a new bug in the current version of the software. For a mature FOSS project with a large developer and client base, there is an elaborate process for tracking issues from their inception to their resolution. For example, some mature projects use the Bugzilla protocol outlined at the end of Chapter 1.

For a CO-FOSS project, the issue tracking process is simpler, since the code base is relatively modest and so is the number of developers and clients. Here is a simple issue-tracking protocol that we have used informally in our recent CO-FOSS projects. We use the GitHub issue-tracking terminology just to clarify the discussion.

1. An issue can be opened by a developer or a client during the Review step of any CO-FOSS cycle.

2. The issue is posted as "open" with a definitive title, a brief description, and an example.

3. Each open issue is reviewed by the project leader, who determines whether it needs immediate correction, overlaps (and can be combined) with another open issue, can be deferred until later in the development process, or is not relevant to the project.

4. The project leader assigns issues that need immediate correction to the developers in an equitable fashion. The issues are thus "owned" by the developers.

5. Once an issue is corrected by its owner, it is reviewed by the person who posted it and, if approved, the issue is then "closed."

For example, Figure 5.17 shows 6 open issues and 76 other issues that were closed after the initial development of *Homeplate* had concluded. Sometimes during the process of addressing an open issue, a text-messaging exchange can occur between its owner and the person who posted it. That discussion often involves eliciting more information in order to clarify the nature and extent of the issue. To view a few recent examples, navigate to `the` *Homeplate* `project page` at `https://github.com/megandalster/sh-homeplate/` and hit the Issues tab.

When *Homeplate* was initially developed in 2012, we did not use issue tracking. Instead, we used threaded discussions on Google Groups — not only to record and track issues, but also to assign new milestones, record meeting notes, and post all other conversations and documents that we needed along the way. At that time, the need for a separate issue tracking tool was less critical that it is now.

5.5 SUMMARY

This chapter has addressed three significant elements in the process of domain class development – coding, unit testing, and system testing. We presented and illustrated strategies for finding and reusing code from other open source projects, as well as coding the classes from scratch.

Coherent unit testing and debugging techniques are key to effective software development. The techniques introduced here will be revisited and expanded in Chapters 6 and 7. Integration testing is part of this process, and it helps ensure that the combined work of individual developers has no conflicting overlaps or inconsistencies.

Finally, this chapter has introduced the idea of obtaining client feedback from a non-developer's review of the software being developed. This idea will also be expanded in Chapters 6 and 7.

5.6 MILESTONE 5

1. Complete the initial development of all the domain classes for your CO-FOSS project. Each class should have a set of instance variables, a constructor, a complete set of getters, and a handful of setters for functions that will apparently be needed by the class.

2. For each class whose code is being adapted from another open source project, perform all refactoring that is needed in order to remove unnecessary and redundant functionality, as well as all other "bad smells."

3. Complete the unit testing and integration testing of all domain classes.

4. Conduct a client review of the domain classes and the secure login protocol for the project, identifying and posting any issues that arise out of this review.

5. Evaluate these new issues and assign ownership to individual developers for any issues that need to be resolved during the upcoming CO-FOSS cycle.

CHAPTER 6

Database Development

"Errors using inadequate data are much less than those using no data at all."
—*Charles Babbage*

This chapter focuses on the database aspects of CO-FOSS development. It assumes that readers have little or no familiarity with databases, in comparison with what they know about programming and data structures. However, in most real software projects, the role of database development is central.

To provide a minimal introduction to databases, the first section of this chapter introduces database principles using examples from our own CO-FOSS projects. These principles include the ideas of SQL, tables, queries, normalization, unique keys, and concurrency control.

Because this introduction is minimal, readers may seek supplementary readings that treat all these concepts more carefully, such as the following:

Relational databases: `https://www.tutorialspoint.com/sql/sql-rdbms-concepts.htm`

SQL tutorial: `https://www.tutorialspoint.com/sql/`

For an even more complete treatment of databases, readers are encouraged to take a full database course and/or read any one of a number of fine textbooks on the subject.

The second section introduces techniques for working with a database from within a CO-FOSS project. This discussion includes database and table creation, and searching, insertion, updating, and deletion of rows from a table, using SQL examples from our own projects such as *Homebase* and *Homeroom*. The section also discusses database security, including encrypting user passwords and locating vulnerabilities.

The third section discusses the important process of database testing, debugging, and refactoring. Unit and integration testing of the database modules has its own peculiarities, so that designing effective unit tests for database modules must be done carefully. Testing database modules can also have ripple effects on the testing and refactoring of the domain classes as well.

The fourth section provides our perspective on the process of client review and issue tracking for the database component of a CO-FOSS project. An example illustrates how finding a new issue can often affect the code in both the database layer and the domain layer, thus raising the bar for integration testing to a higher level.

6.1 DATABASE PRINCIPLES

A *database* is a collection of data organized in a particular way for efficient storage and retrieval of information. Databases can be simple, as in a collection of flat files, or more complex, as in a relational database, a distributed database, or a hierarchical database.

Databases reside at the heart of most software applications. A database differs from a domain class or object primarily because its data remains in permanent storage, separate from the memory where the program runs. Often this storage is on a different computer from the software or the user that accesses it. When this storage is part of a "server," as discussed in Chapter 1, efficient access from the program requires Internet connectivity and a client-server software architecture.

The data in a database is distinctive because its own lifetime exceeds that of the program or programs that access it. This characteristic is known as *persistence*. In practice, one program may create a database, while others may access its data and still others may modify its data. At the end of the day, the data in a database lives on, or "persists," while the different client programs that access it may have come and gone.

There are different models for organizing a database. A database model that is quite popular is called the *relational model*. Relational databases are convenient because their data can be stored as a collection of two-dimensional tables, they are reliable, they can be accessed efficiently, and their implementations are well understood throughout the software development community.

The Structured Query Language (SQL for short) is a nearly-universal database modeling language. As noted in Chapter 4, SQL has three principal open source implementations – MySQLi, PostgreSQL, and SQLite. SQL is also firmly based in relational algebra, a field of mathematics.

For these reasons, SQL databases are particularly robust, reliable, and straightforward to set up and maintain, Moreover, SQL has a lot of tutorial material available to support learning how to design and integrate SQL databases with software applications. We shall use SQL for our examples throughout this chapter.[1]

[1]SQL dates back to 1970, when IBM researchers created a simple non-procedural language called Structured English Query Language. or SEQUEL [10]. In the late 1980s ANSI and ISO published a standardized version of SEQUEL called SQL (for Structured Query Language). MySQLi, PostgreSQL, and SQLite are all implemented using the SQL standard.

Because the following presentation is also PHP-specific, we recommend one of the following supplementary materials so that readers can match these topics with the language-specific aspects of their own CO-FOSS project:

MySQLi: `https://www.mysqltutorial.org`

PostgreSQL: `https://www.tutorialspoint.com/postgresql/index.htm`

PHP PDO: `https://phpdelusions.net/pdo`

Java database connector: `https://www.tutorialspoint.com/jdbc/`

Python Django ORM: `https://www.fullstackpython.com/django-orm.html`

Ruby on Rails ORM: `https://www.tutorialspoint.com/ruby-on-rails/rails-active-records.htm`

6.1.1 Relations and Tables

A database is a collection of tables. Each table has a distinct name, a fixed number of columns (called *attributes*), and a varying number of rows. The *Homebase* database, for example, has the following eight tables:

dbApplicantScreenings

dbDates

dbLog

dbMasterSchedule

dbPersons

dbSCL

dbShifts

dbWeeks

As these names suggest, every *Homebase* table except the dbLog table corresponds to a particular domain class.

All the permanent data for *Homebase* is stored in these eight tables. So any change to a volunteer's schedule, contact information, or current status requires that one or more of these tables be modified. Database modification is a discipline whose programming principles and practices are the focal point of this chapter.

Each *attribute* (column) in a table has a unique name and a data type. In a sense, a table attribute is like a variable. A single row of a table designates a particular instantiation of the values for these attributes.

id	shifts	mgr_notes
18-08-06:portland	18-08-06:9-12:portland*18-08-06:3-6:portland*18-08-06:6-9:portland*18-08-06:12-3:portland	
18-08-07:portland	18-08-07:9-12:portland*18-08-07:12-3:portland*18-08-07:3-6:portland*18-08-07:6-9:portland	
18-08-08:portland	18-08-08:9-12:portland*18-08-08:12-3:portland*18-08-08:3-6:portland*18-08-08:6-9:portland	
18-08-09:portland	18-08-09:9-12:portland*18-08-09:12-3:portland*18-08-09:3-6:portland*18-08-09:6-9:portland	
18-08-10:portland	18-08-10:9-12:portland*18-08-10:12-3:portland*18-08-10:3-6:portland*18-08-10:6-9:portland*18-08-10:night:portland	
18-08-11:portland	18-08-11:10-1:portland*18-08-11:1-4:portland*18-08-11:night:portland	

FIGURE 6.1 A few rows in the dbDates table.

To illustrate these ideas, let's look at the three columns (attributes) of the dbDates table that are shown in Figure 6.1. These columns are named id, shifts, and mgr_notes. Each row identifies a single object from the RMH-date class. The attributes in the dbDates table have corresponding instance variables in the RMHDate class.

Note also that each row of the dbDates table has a unique `id`, designating a distinct date and venue on the *Homebase* calendar. Each row in the table identifies a collection of shifts for that venue, designated by an asterisk-separated list of Shift id's.

For example, the date with id `18-08-06:portland` has four shifts with ids `18-08-06:9-12:portland`, `18-08-06:3-6:portland`, `18-08-06:6-9:portland`, and `18-08-06:12-3:portland`. The abbreviation `18-08-06:9-12:portland` stands for the 9-12 shift on August 6, 2018 at the Portland venue.

Table Naming Conventions

The above discussion suggests an important idea that helps us understand the design of the database modules in the code base. This idea is that each of the domain classes in a software architecture has a corresponding database table. In such cases, the class's instance variables correspond with attributes in its related database table. This allows a row in the table to be interpreted as an object in its corresponding class. This relationship suggests an important naming convention for developers:

> *Use attribute (column) names that unify a database table with its corresponding class's instance variables.*

For example, the following database table and attribute naming convention is used in the design of *Homebase* and all the other CO-FOSS products our students have developed:

1. If a class has name C, its corresponding database table is named dbCs. While class C represents a type of object, its corresponding database table dbCs contains several occurrences of such objects, one per row.

2. Each attribute A in table dbCs corresponds to one or more instance variables V in the definition of class C. (Some instance variables in class C may be concatenated together and appear as a single attribute in table dbCs.)

3. Each database table dbCs has a corresponding module dbCs.php (assuming the programming language is PHP) in the code base whose responsibility is provide all the needed functionality for adding, deleting, and changing individual rows in that table.

To illustrate this convention, let's look at the instance variables in the *Homebase* Shift class, as shown in Figure 6.2.

```
private $yy_mm_dd;        // "yy-mm-dd".
private $hours;           // '9-1', '1-5', '5-9' or 'night'
private $start_time;      // e.g. 10 means 10:00am
private $end_time;        // e.g. 13 means 1:00pm
private $venue;           // "portland" or "bangor"
private $vacancies;       // number of vacancies
private $persons;         // array of persons filling slots
private $removed_persons; // persons previously removed
private $sub_call_list; // "y" if sub call list exists
private $day;             // string name of day "Monday"...
private $id;              // "yy-mm-dd:hours:venue"
private $notes;           // manager notes
```

FIGURE 6.2 *Homebase* Shift class instance variables.

The *Homebase* database has a table called dbShifts that corresponds to this Shift class. A single row in this table represents a single shift. The attribute names shown in Figure 6.3 correspond to the instance variables shown in Figure 6.2, though there are fewer attributes than instance variables.

To gain more insight into this correspondence, we can examine a few entries in the dbShifts table directly. For example, Figure 6.4 shows the entries in the dbShifts table for August 6-8, 2018 at the Portland venue.

We can learn a lot by examining this small sample of data. Notice that each Shift's `$id` in Figure 6.4 is different from all the others, since it has a unique date:time:venue value. Notice also that the number of vacancies in a shift is explicit, and the list of Persons scheduled for that shift appears as a string of ids and names concatenated by asterisks (*).

Name	Type
id	char(25)
start_time	int(11)
end_time	int(11)
venue	text
vacancies	int(11)
persons	text
removed_persons	text
sub_call_list	text
notes	text

FIGURE 6.3 Attribute names and types in the dbShifts table.

id	start_time	end_time	venue	vacancies	persons
18-08-06:12-3:portland	12	15	portland	1	Cheryl7038089589+Cheryl+Jones
18-08-06:3-6:portland	15	18	portland	0	Robin7037510984+Robin+Jones*Clai
18-08-06:6-9:portland	18	21	portland	1	Nonie7037812392+Nonie+Jones
18-08-06:9-12:portland	9	12	portland	0	Jane7038293469+Jane+Jones*Cathy
18-08-07:12-3:portland	12	15	portland	1	Cindy7035631089+Cindy+Jones
18-08-07:3-6:portland	15	18	portland	0	Becky7037725009*Betsy7038464935
18-08-07:6-9:portland	18	21	portland	0	Kara7035953232+Kara+Jones*Daniel
18-08-07:9-12:portland	9	12	portland	0	Jane7038859127*Stacey7032333522
18-08-08:12-3:portland	12	15	portland	1	John7032476256+John+Jones

FIGURE 6.4 The entries in the dbShifts table for August 6, 7, and 8, 2018 in Portland.

Finally, note that not all database tables need to have a corresponding domain class. For example, *Homebase* has the additional database table dbLog, which stores a series of entries that log individual calendar changes, when they occur, and who made them. Individual calendar changes can be made by any volunteer, for example to remove themselves from a shift or to fill a vacancy.

6.1.2 Queries

In SQL, the manipulation of data in a database is accomplished by executing a so-called *query*. Oddly, the term *query* is a misnomer. That is, a query not only can retrieve information from a table, but it also can insert, update, or delete rows from a table, and can even create or drop an entire table. Here is a brief summary of the different kinds of SQL queries and their uses.

`CREATE TABLE tablename (attribute type, attribute type, ..., PRIMARY KEY attribute)` creates a new table in the database with one column for each attribute-type pair listed. The `PRIMARY KEY` clause identifies that attribute which distinguishes one row from all other rows in the table.

`DROP TABLE IF EXISTS tablename` removes a table from the database.

`SELECT * FROM tablename WHERE relation` returns either a non-empty array of rows or false, depending on whether or not any rows in the table satisfy the given relation.

`INSERT INTO tablename VALUES (value, value, ...)` inserts a new row into the table associating each of the values with the columns defined in the `CREATE` command for that table, from left to right.

`DELETE FROM tablename WHERE relation` removes all rows from the table that satisfy the given relation.

`UPDATE tablename SET attribute = value, attribute = value, ... WHERE relation` changes the attributes' values in the table for all rows that satisfy the given relation.

Execution of an SQL query relies on the presence of a "current database." That is, a database connection must be established prior to executing any of these queries on a particular table.

When used in a query, a `relation` denotes any Boolean-valued expression that defines the criteria by which a row is included in the result of the query. The relation uses common relational operators, AND, OR, constants, and the names of table attributes as arguments. Table 6.1 summarizes the structure of the common relations found in SQL queries, along with their meanings.

TABLE 6.1 Relations in SQL Queries

Relation	Meaning
attribute = v	the attribute's value is v
attribute <> v	the attribute's value is not v
attribute IS NOT NULL	the attribute's value is not NULL
attribute LIKE pattern	the attribute's value matches the pattern
expression AND expression	both expressions are true
expression OR expression	either one or both expressions are true
NOT (expression)	expression is false

Many examples of writing SQL queries and relations are given in later sections.

6.1.3 Normalization

In database design, *normalization* is a strategy for designing tables so that they support general-purpose querying and ensure data integrity.[2] A major goal of normalization is to allow tables to be queried using a standardized language, like SQL, that is grounded in mathematical logic. A relational database table is said to be "normalized" if it satisfies all of the following criteria:

[2]E. F. Codd, inventor of the relational model, introduced the idea of `database normalization` in 1970.

1. The rows can be rearranged without changing the meaning of the table (i.e., there's no implicit ordering or functional interdependency among the rows).

2. The columns (attributes) can be rearranged without changing the meaning of the table (i.e., there's no implicit ordering or functional interdependency among the columns).

3. No two rows of a table are identical. This is often accomplished by defining one column whose values are mutually unique. This column is known as the table's *primary key.*

4. No row has any hidden components, such as an object id or a timestamp.

5. Every entry in the table has exactly one value of the appropriate type.

6. No attribute in the table is redundant with (i.e., appears as an explicit substring of) the *primary key.*

There are disadvantages to working with unnormalized tables. Querying an unnormalized table can create more complexity than is needed, and unanticipated side effects (known as "anomalies") may also occur. For example, when a table does not satisfy criterion 3 above, it permits the creation of two or more identical (redundant) rows. Updating an entry in one of those rows, without doing the same for its identical twin, will introduce an inconsistency into the table.

Another type of anomaly occurs when information cannot be recorded in the table at all. This situation arises when trying to insert a row for which only some of the entries are known (violating criterion 5 above). One way to circumvent this anomaly is to agree that absence of information in a row will always be recorded as the `null` value, or equivalently the empty string.

Thus, normalization is good database design practice. The tables in *Homebase* satisfy the first four criteria listed above. For example, the `id` attribute serves as the primary key in the dbShfits table (see Figure 6.4), since all its values are mutually unique. The importance of primary keys will become clear in later sections, when we discuss queries for retrieving and updating a single row in a table without impacting any of the other rows.

But the *Homebase* tables are not fully normalized because they do not satisfy criteria 5 and 6 above. For example, the dbDates table shown in Figure 6.1 violates criterion 5, since its `shifts` attribute contains several shift `ids`, not just one. Also, the dbShifts table shown in Figure 6.4 violates criterion 6, since a Shift's venue attribute appears redundantly as parts of the Shift's `id` attribute.

The main advantage to designing a database in this way is that it has fewer tables than its fully-normalized counterpart. Fewer tables can improve querying efficiency, which we briefly illustrate in the next section.

6.1.4 Keys

A table's *primary key* is a column whose values are mutually unique. For example, the `id` column in the dbDates table shown in Figure 6.1 serves as its primary key, since no two `ids` in that column are identical. Similarly, the `id` column in the dbShfits table in Figure 6.4 is a primary key for that table. Our CO-FOSS databases generally organize their tables so that each one has a primary key. The reasons for this will become clear as we discuss the various table updating functions in Section 6.2.3.

A database table can also have a *foreign key*, which is a column whose values are the primary keys in another related table. The purpose of the foreign key is to provide a link between corresponding rows of those two tables. In *Homebase*, for example, every date has zero or more shifts. In the dbDates table of Figure 6.1 we read that the date 18-08-06:portland has four shifts with these primary keys:

```
18-08-06:12-3:portland
18-08-06:3-6:portland
18-08-06:6-9:portland
18-08-06:9-12:portland
```

If we were to redesign the *Homebase* dbShifts table to have a foreign key that linked each shift to a single row in the dbDates table, we would add the attribute `date_id` whose value is the `id` of the date on which the shift occurs. This change would allow us to remove the `shifts` attribute from the dbDates table altogether and have a dbShifts table whose first four rows would appear as in Table 6.2.

TABLE 6.2 Redesigning the dbShifts table to improve normalization

id	date_id	...
18-08-06:12-3:portland	18-08-06:portland	...
18-08-06:3-6:portland	18-08-06:portland	...
18-08-06:6-9:portland	18-08-06:portland	...
18-08-06:9-12:portland	18-08-06:portland	...

This change would also bring the dbDates table into conformity with normalization criterion 5 discussed in the previous section. However, this change also has a downside. That is, every time we want to query the database to retrieve a single date along with the ids of all its associated shifts, we would need to query two tables rather than one.

For this reason, *Homebase*, does not use foreign keys. Each row in the dbDates table has a `shifts` column containing the `ids` of *all* the shifts that are scheduled for that date and venue, as illustrated in Figure 6.1. This variation allows us to retrieve the `ids` of all the shifts for a single date by querying a single table, dbDates. However, this design choice also creates violation of normalization criterion 5.

This brief example should make it clear that the challenge of database design with and without full normalization is a complex one. For more details, interested readers should consult any of the many excellent textbooks and `Web resources` on database design.[3]

6.1.5 Concurrency Control

In a client-server application, it is possible for two or more clients to be accessing the same database, even the same table, at the same time. Thus, the database system must define and enforce the actions that will occur whenever two or more clients try to update the same entry in a table at the same instant in time. This event is called a "collision."

What happens when a collision occurs? More importantly, how can we be confident about the integrity of the result whenever a collision occurs. The answers are wrapped up in how a database system implements "concurrency control;" that is, how it manages collisions whenever they occur.

Both MySQLi and PostgreSQL are used in client-server applications, but they implement concurrency control in different ways. SQLite is mainly used in single-user apps, as on mobile devices. But when collisions do occur, SQLite implements concurrency control in yet another different way. These three different implementations of concurrency control are outlined in the following three paragraphs

MySQLi can use either of two different "storage engines" for accessing a database, which are called MyISAM and InnoDB. With MyISAM, MySQLi provides *table-level locking* to ensure database integrity when several clients (sessions) are accessing the same table at the same time. This means that whenever a particular table in a database is queried (via a `CREATE`, `SELECT`, `INSERT`, `DELETE`, or `UPDATE`) by a particular session, that table is unavailable to all other sessions that are currently querying it. The other sessions' queries are placed in a queue until the lock is released on the first session's query. With the InnoDB engine, MySQLi uses *row-level locking* to prevent conflicts, which operates in the same way for an individual row (rather than for an entire table).

PostgreSQL also provides row-level and table-level locking, but only as an option for situations in which its native concurrency control mechanism, called MVCC, is not needed. The MVCC strategy is that each query in a session receives a snapshot of the data as it existed some time ago, rather than instantaneously. The goal of MVCC is to improve database performance, especially in applications where conflicts are infrequent or non-existent. A more careful discussion of MVCC can be found `here`.

[3] In a CO-FOSS project, the instructor should normally scope out the database design in advance, especially with regard to the definition of primary keys, inter-table linkages, and normalization in general. Most student developers in this course may not have sufficient database experience to do that design as part of the project itself.

When an SQLite database serves several clients simultaneously, it provides database-level locking for each write. Thus all other write requests must be queued, encouraging each client to do its database work quickly and move on, allowing its lock to last for only a few dozen milliseconds. However, many applications require more inclusive concurrency control strategies, as the ones found in MySQLi or PostgreSQL, so SQLite may not be the best choice for these applications.

In most cases, table-level locking is a satisfactory vehicle for concurrency control, but there are some drawbacks:

> Table-level locking enables many sessions to r`SELECT` (read from) a table at the same time, but if a session wants to `INSERT`, `DELETE`, or `UPDATE` (write to) a table, it must gain exclusive access to the entire table. So for the duration of that write, all other sessions that want to access the table must wait until that write is complete.
>
> Table-level locking can cause problems when a session is waiting because the disk is full and free space needs to become available before the session can proceed. In this event, all sessions that want to access the table are put in a waiting state until more disk space is made available.
>
> Table-level locking is also disadvantageous when a session issues a `SELECT` that takes a long time to run. If another session wants to issue an `UPDATE` on the same table, it must wait until the `SELECT` is finished.

Here are two database access strategies that can help reduce contention among sessions when table-level or row-level locking is involved.

1. Perform only one or two `SELECT`, `INSERT`, `DELETE`, or `UPDATE` queries during any single connection to the database; disconnect from the database at all times when no queries are being executed.

2. Try to ensure that each `SELECT`, `INSERT`, `DELETE`, or `UPDATE` query runs fast so that it locks a table or row for a minimum amount of time.

Of course, using either or both of these strategies can come at a price, either in coding complexity or in system performance.

6.2 DATABASE ACCESS

In this section, we present basic strategies for accessing a database and its tables, with particular attention to organizing the interactions between a CO-FOSS application and its underlying database. Our medium for presentation is the PHP/MySQLi extension, since that one provides the most detailed examples from our various projects, especially *Homebase* and *Homeroom.*

6.2.1 Connecting the Program to the Database

In general, popular database systems are well-integrated with popular programming languages by way of so-called database extensions. This means that one can embed database queries directly within the code whenever a database action is needed. Table 6.3 lists some popular programming languages with which MySQL, PostgreSQL, and SQLite are connected in this way, along with links to key references for each type of connection.

TABLE 6.3 Programming Language Database Extensions for SQL

Language	MySQL	PostgreSQL	SQLite
Java	MySQL Connector/J	PL/Java	SQSLte Java
PHP	MYSQLi	PostgreSQL	SQLite
Python	MySQLdb	PL/Python	PySQLite
Ruby	ruby-mysql	ruby pg	sqlite3

To illustrate, let's look at the PHP/MySQLi extension. Here are the key commands that enable a PHP program to work with an SQL database using that extension.

`$con = mysqli_connect($host, $user, $password)` connects the program with the database server `$host`, submitting `$user` and `$password` as access credentials. No other database actions can be performed until a connection has been established. The command returns a connection `$con` to the server if successful; otherwise it returns false.

`mysqli_select_db($con, $database)` selects a particular `$database` on the server with connection `$con`. Returns true or false if this selection succeeds or fails.

`mysqli_error($con)` returns the particular MySQLi error message that accompanies the most recent query with the selected database.

`mysqli_close($con)` disconnects the program from the server and selected database.

`mysqli_query($con,$query)` A `$query` can either create or drop a table from the selected database, or else it can select, insert, update, or delete rows from a table. We illustrate queries in more detail below.

Since establishing a connection with a database is such a frequent activity, it is useful to write a separate PHP function that accomplishes that. For example, Figure 6.5 shows the function used to connect *Homebase* to its database `homebasedb`.

Establishing a connection to a database is thus a two-step process. First, a connection to the server, or host, must be made. Second, the database must be selected from among all the databases residing on that server. Either one

```
function connect() {
    $host = "localhost";
    $database = "homebasedb";
    $user = "homebasedb";
    $pass = "homebasedb";

    $con = mysqli_connect($host,$user,$pass,$database);
    if (!$con) {
        echo "not connected to server";
        return mysqli_error($con);
    }
    $selected = mysqli_select_db($con,$database);
    if (!$selected) {
        echo "database not selected";
        return mysqli_error($con);
    }
    else return $con;
}
```

FIGURE 6.5 Connecting to the *Homebase* database.

of these may fail, in which case the `mysqli_error()` function delivers information about what went wrong. Otherwise, the function in Figure 6.5 returns the connection `$con`.

The function shown in Figure 6.5 is used in the sandbox versions of all the CO-FOSS projects *Homebase*, *Homeroom*, *Homeplate*, and *BMAC-Warehouse* that are discussed in this book. Though these particular user and password choices are not very secure, they do provide transparency to facilitate experimentation by developers in other CO-FOSS projects.

6.2.2 Table Creation and Dropping

Table creation and dropping are one-time events. Table creation serves only to establish the table in the database and identify its name, its attribute (column) names, and their associated types. Table creation can be done using a template like the one shown in Figure 6.6 for a PHP/MYSQLi table.

To use this template, the programmer should name the table "tablename" and "list of attributes and their types" by an actual table name and list of attributes and types.

The `$con = connect()` command opens a connection to the database by calling the function defined in Figure 6.5. Once a connection is established, the second command ensures that any earlier version of the table is dropped

```
function setup_tablename() {
   $con = connect();
   mysqli_query($con, "DROP TABLE IF EXISTS tablename");
   $result=mysqli_query($con, "CREATE TABLE tablename" .
       "(list of attributes and their types)");
   if (!$result)
       echo mysqli_error($con);
   mysqli_close($con);
}
```

FIGURE 6.6 Template for MySQLi table creation.

from the database. The `IF EXISTS` clause acts like a conditional statement—if the table doesn't exist, no attempt will be made to drop it.

The third command in this sequence creates the table in the database with a unique `tablename`. Following this is a list of arguments that define the names and types of the attributes (columns) that define the table. The next three lines check to be sure the table was created successfully by interrogating the `$result` variable. If not, a MySQLi error is reported. In either case, the last line closes the connection to the database.

The types of values that can be stored in a MySQLi table are many and varied. A summary of the most common ones is given in Table 6.4. When an attribute has the `NOT NULL` phrase appended, this means that the `NULL` value is not permitted in any row for that attribute. The list of attributes can also contain the expression `PRIMARY KEY (attribute)`, which means the given attribute serves as the primary key for the table.

TABLE 6.4 Common Attribute Types in MySQLi Tables

Type	Meaning
INT	an integer in the range $-2^{31} \ldots 2^{31-1}$
TIMESTAMP	a timestamp in the form yyyymmddhhmmss
CHAR (m)	a fixed-length string of m characters $(0 < m < 255)$
VARCHAR (m)	a string of up to m characters $(0 < m < 255)$
TEXT	a string of up to 65,535 characters

For example, Figure 6.7 shows a function that can create the dbDates table in the *Homebase* database using the template shown in Figure 6.6. This definition of the dbDates table requires that no row's id attribute be `NULL` and that the id attribute also serves as the table's primary key (guaranteeing that every row has a unique id).

Because table setup and dropping are one-time events, they are usually done off-line rather than from within the running application itself. One way to set up all the tables off-line is to *import* the entire database from a file

```
function setup_dbDates() {
   $con=connect();
   mysqli_query($con, "DROP TABLE IF EXISTS dbDates");
   $result=mysqli_query($con,
       "CREATE TABLE dbDates (id CHAR(20) NOT NULL, " .
       "shifts TEXT, mgr_notes TEXT, " .
       "PRIMARY KEY (id))");
   if(!$result)
       echo mysqli_error($con) ;
   mysqli_close($con) ;
}
```

FIGURE 6.7 Creating the dbDates table in the *Homebase* database.

that contains a backup copy of the database. This activity can be done by using a database administration tool, such as `phpMyAdmin` which is used with MySQLi databases. For example, the phpMyAdmin interface for our work is shown in Figure 6.8.

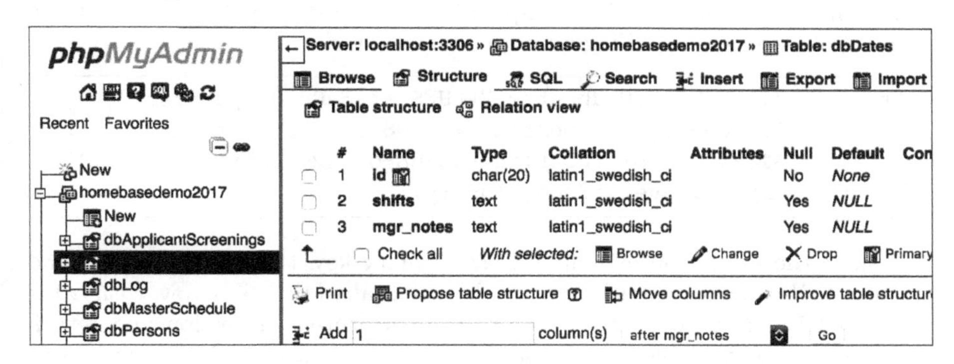

FIGURE 6.8 The phpMyAdmin tool for managing a MySQLi database.

This illustration shows the tabs used to import an entire database or a single table, as well as to drop the database or table. These actions should always be done with care, and with the availability of a backup copy should things go wrong.

6.2.3 CRUD Functions

Three important types of actions are used when maintaining a table in a MySQLi database.

1. The table must first be set up in the database.

2. Individual rows of the table can be created (inserted), retrieved, updated (changed), and deleted.
3. The table can be dropped from the database, in which case all its data is lost.

As discussed in the previous section, the first and third of these actions are usually done off-line since they are one-time events. The second of these actions occurs frequently throughout the running of a CO-FOSS application.

Correspondingly, every table in a database should be supported by a corresponding database module whose functions allow developers to *C*reate, *R*etrieve, *U*pdate, and *D*elete an individual row in that table. Together, these functions are (unflatteringly) called the `CRUD functions`. We discuss and illustrate each CRUD function in the following sections.

Create: Inserting Rows into a Table

The query `INSERT INTO tablename VALUES (value, value, ...)` inserts a new row into the table assigning each of the values to a column in that table, from left to right. It returns true or false, respectively, depending on whether the insertion was successful. For example, the following query inserts a row into the dbDates table corresponding to August 5, 2018 for the Portland venue, which has four shifts and no manager notes.

```
$query = "INSERT INTO dbDates VALUES ('18-08-05:portland', " .
        "'18-08-05:9-12:portland*18-08-05:3-6:portland".
        "*18-08-05:6-9:portland*18-08-05:12-3:portland', '')";
$result = mysqli_query($con, $query);
```

Since the id attribute is the table's primary key, this query will fail and return a "Duplicate entry" error if the table already has a row with `id = "18-08-05:portland"`. Otherwise, it will perform the insertion and return true. Thus, the variable $result can be used to check that the insertion was made successfully.

Retrieving Rows from a Table

The query `SELECT * FROM tablename WHERE relation` searches the table `tablename` and returns a so-called "resource" that contains all rows that satisfy `relation`. The latter is an expression describing attributes and values that must be true for every row returned by the search.

Omission of the `WHERE` clause causes this query to return the entire table as the resource. For example, if we want to retrieve all rows in the dbDates table, we would just say:

```
$result = mysqli_query($con, "SELECT * FROM dbDates");
```

Normalized tables are particularly suitable for general purpose searching. In particular, when searching for an entry in a table with a particular value as its primary key, we are guaranteed to receive either a single row or nothing as a result, but never more than one row.

So if we want to return a particular date from the dbDates table whose id is "18-08-05:portland" (there can be no more than one, since the id is the table's `PRIMARY KEY`), we can use the following pair of statements to issue a MySQLi query and then parse the result into an array.[4]

```
$result = mysqli_query($con,
    "SELECT * FROM dbDates WHERE id = '18-08-05:portland'");
if ($result)
    $result_row = mysqli_fetch_assoc($result);
```

The resource returned by the query in the first statement will be either that row in the dbDates table that has this particular primary key, or else `null` (if there is no such row). To separate a single row into its individual fields, we can use the function that appears in the third line. This function returns the first row of the result as an associative array, in the order in which the columns of the table are defined, with indexes identifying the attributes' names.

If we think that multiple rows will be returned by a query, we can use a loop to process each row of the result individually.

```
$result = mysqli_query($con,
    "SELECT * FROM dbDates WHERE mgr_notes<>''");
while ($result_row = mysqli_fetch_array($result, MYSQLI_ASSOC))
    // process $result_row
```

Here, each call to the function `mysqli_fetch_array` returns a single row from `$result` until there are no more left. At that point, null is returned, which causes the while loop to terminate. Thus, this loop will process all rows in the dbDates table that have a non-blank `mgr_notes` attribute. The second argument, `MYSQLI_ASSOC`, designates that the result be returned as an associative array, as described above.

Pattern matching is often useful for selecting rows from a table that satisfy a particular criterion. For example, suppose we want to find all the shifts where Jane is scheduled to volunteer. Then we could search for all shifts in the dbShifts table whose persons attribute contains the string "Jane." Here is what that query looks like.

---

[4] A note about quotes: a MySQLi query is treated as a character string by the PHP/MySQLi extension. Thus, every call to `mysqli_query` must pass a quoted string as an argument. This becomes tricky when the query itself has a character string embedded inside it, as in the current example. To keep one level of quotes distinct from the other, we can use a double quote character (") to encapsulate the query itself and a single quote character (') to encapsulate any string that is embedded inside the query.

```
$query =  "SELECT * FROM dbShifts WHERE persons LIKE '%Jane%'";
$result = mysqli_query($con, $query);
```

For example, if the dbShifts table has only the shifts listed in Figure 6.4, this query will return a $result of two shifts, one with $id = 18-08-06:portland and the other with $id = 18-08-07:portland.

We note finally that the SELECT query has the useful clause `ORDER BY attribute`. When this clause appears, it will return the resource with its rows sorted into ascending order according to the given attribute. For example, the following query returns a list of all shifts in the dbShifts table where Jane is volunteering, arranged into ascending order by id (that is, by date and venue).

```
$query = "SELECT * FROM dbShifts WHERE persons LIKE '%Jane%'
    ORDER BY id";
$result = mysqli_query($con, $query);
```

Update: Altering Rows in a Table

The query `UPDATE tablename SET attribute=value,attribute=value, ... WHERE relation` assigns new values to attributes in the table for all rows that satisfy the given relation. For example, suppose we want to alter the shifts attribute for the August 6, 2018 row in the dbDates table by changing the 6-9 shift to the night shift. We can make this change using either of two approaches:

1. Change only the shifts attribute in the row corresponding to August 6, 2018 using the UPDATE query, or

2. Completely remove and re-insert that row in the dbDates table using DELETE and INSERT queries.

Here is an UPDATE query that will make this change using the first approach:

```
$query = "UPDATE dbDates " .
  "SET shifts='18-08-06:9-12:portland*18-08-06:12-3:portland ".
      "*18-08-06:3-6:portland*18-08-06:night:portland' ".
  "WHERE id = '18-08-06:portland'";
$result = mysqli_query($con, $query);
```

The UPDATE will do nothing if there is no row in the table with id = "18-08-06:portland." Moreover, at most a single row in the table will be affected, since id is the primary key.

Delete: Removing Rows from a Table

The query `DELETE FROM tablename WHERE relation` removes all rows from the table that satisfy the given relation. For example, the following query removes the row in the dbDates table corresponding to August 6, 2018 in Portland (if it is there).

```
$query = "DELETE FROM dbDates WHERE id = '18-08-06:portland'";
$result = mysqli_query($con, $query);
```

Since id is the `PRIMARY KEY` for dbDates, there can be no more than one row with id = "18-08-06:portland."

With the DELETE query in hand, we can now see how to update a row in a table by issuing a DELETE followed by an INSERT. Here it is for the foregoing example:

```
$query = "DELETE FROM dbDates WHERE id = '18-08-06:portland'";
$result = mysqli_query($con, $query);
$query = "INSERT INTO dbDates VALUES ('18-08-06:portland', " .
        "'18-08-06:9-12:portland*18-08-06:3-6:portland".
        "*18-08-06:6-9:portland*18-08-06:night:portland', '')";
$result = mysqli_query($con, $query);
```

6.2.4 Database Security

A *secure* database is one that accomplishes all of the following:

1. It prevents unauthorized or accidental disclosure, alteration, or destruction of data.

2. It prevents unauthorized or accidental access to data considered confidential to the organization or individual who owns the data.

3. It ensures *data integrity*, which means that the data stored in the database is always valid and accurate.

To help ensure database security, access to all tables must be properly controlled. This can be done by granting privileges that limit logged-in users to see only those portions of the data to which they should have access. For data that need to be very secure, implementation of an encryption strategy for individual table attributes or individual users may also be required.

In general, definition and enforcement of a security policy for a database require careful consideration for the types of users and the different levels of access to the database that are permitted by the software itself. For example, some types of users may be granted access to parts of the database, while others should have access to all of it. Moreover, some users should have read-only access to the data while others should have read-write access.

An additional element of database security is the importance of protecting the database against malicious attacks. Since many of these attacks originate at the user interface level, we shall discuss and illustrate these security issues in Chapter 7.

A database system comes with a built-in permission model, which allows the owner to specify which users have access and what level of access each user has. In the case of MySQLi, permissions are controlled via the GRANT and REVOKE commands. Four levels of access can be controlled:

1. Server level: privileges that apply to all databases on the server
2. Database level: privileges that apply to all tables in a particular database on the server
3. Table level: privileges that apply to all columns of a particular table in the database
4. Column level: privileges that apply to an individual column of a table in the database.

The types of privileges that can be granted/revoked at each of these levels are directly aligned with the individual types of MySQLi queries that were discussed earlier—Create, Drop, Select, Alter (Insert), Delete, and Update.

The database administrator may assign privileges to individual users who connect to the database, depending on their roles. Alternatively, the administrator may simply assign all privileges to a single user. In turn, that user inherits the responsibility of limiting database access privileges to individual users of the software that interacts with the database.

In our CO-FOSS projects, this latter strategy is used. For example, whenever a *Homebase* database connection is made during an individual user's session, the connection grants all the privileges listed above to that user's session. The session, in turn, knows the particular access level that the user has (a visitor has access level 1, a volunteer has level 2, and a manager has level 3), and the software regulates their database access accordingly. A discussion of how this strategy is implemented appears in Chapter 7.

Our CO-FOSS projects take an additional step to ensure database security. That is, they encrypt all user passwords that are stored in the dbPersons table. Thus, if an uninvited visitor happens to stumble upon the dbPersons table, they are not able to read a user password and write it on the back of a napkin. Each encrypted password is stored in the database, and that is used to compare with the encrypted version of the user-entered password whenever a new user logs in.

For example, look at the dbPersons table in *Homebase*. There, the field `password` is encrypted using md5 encryption. So when we look at a person's password in the database, we will see a long hexadecimal code that is the encrypted value of the actual password. Whenever a person logs in and provides a password, the `login_form` script encrypts the typed password and compares it with the password entry in the database. So the actual password is always kept private between the person and the login_form script. Below is the excerpt from that script that does the checking:

```
1    $db_pass = md5($_POST['password']);
2    $db_id = $_POST['userid'];
3    $person = retrieve_person($db_id);
4    if ($person) { //avoids null results
5        if ($person->get_password() == $db_pass) {
6            $_SESSION['logged_in'] = 1;    // log the person in
```

Here, the notation $_POST identifies the text that the user just entered (userid or password), and $person is the person retrieved from the database with that userid. The first line encrypts what the user entered as a password, and line 5 tests to see if that matches the encrypted password stored in the database for the person with the same userid. We shall discuss the idea of $_POST more carefully in Chapter 7, which focuses on the details of user interface development.

6.2.5 Database Integrity

To ensure database integrity, each of our projects defines a separate module for each separate database table. In each database module, every CRUD function is implemented for its associated table. For example, in *Homebase* the dbDates table has an associated module dbDates.php, which contains all the functions that create, retrieve, update, and delete information in the dbDates table.

This design strategy follows from the layering principle, since it ensures that all queries in the software that affect the database originate from one of the database modules. For example, all queries to the *Homebase* dbDates table should be performed only by appropriate functions within the dbDates.php module. If this module is complete, developers should be able to resist the temptation to query the dbDates table from anywhere else in the code base.

This design strategy can also ensure concurrency control by requiring each CRUD function within a database module to be implemented using the following strict protocol.

1. Connect to the database.
2. Perform the create, retrieve, update, or delete query.
3. Close the connection to the database.
4. Return a result that reflects the outcome of that query.

This protocol ensures that conflicts among different clients accessing the same table simultaneously, as discussed in Section 6.1.5, will be minimized. When they do occur, such conflicts will be properly handled via table-locking.

When implementing a CRUD function, it would be naive to assume that whenever, say, an INSERT query is executed a new row will always be inserted into the table. For example, if a row with the same primary key is already there, the INSERT will fail.

Also, if the arguments passed in an INSERT query are in any way invalid (e.g., there are too many arguments or one of them is not of the correct type), the INSERT will also fail.

Therefore, our implementation strategy for each CRUD function is to encapsulate the function inside a wrapper that both reflects the above protocol and handles unexpected outcomes appropriately.

An example of this strategy appears in Figure 6.9, which implements the deletion of a row from the dbDates table in the *Homebase* database.

```
1  function delete_dbDates($d) {
2      if (!$d instanceof RMHdate)
3          die("Invalid argument for delete_dbDates function");
4      $con=connect();
5      $query = "DELETE FROM dbDates
6          WHERE id=\"" . $d->get_id() . "\"";
7      $result = mysqli_query($con,$query);
8      if (!$result) {
9          echo ("unable to delete from dbDates: ".
10             $d->get_id().mysqli_error($con));
11         mysqli_close($con);
12         return false;
13     }
14     mysqli_close($con);
15     $shifts = $d->get_shifts();
16     foreach ($shifts as $key => $value) {
17         $s = $d->get_shift($key);
18         delete_dbShifts($s);
19     }
20     return true;
21 }
```

FIGURE 6.9 Deleting a date from the dbDates table.

This is an interesting example because it illustrates some of the abnormal event-checking that must be done to ensure that the caller knows exactly what happened after the DELETE query was issued. Notably, the code on lines 8-12 that begins `if (!$result)...` reports a failure to make the deletion, closes the connection to the database, and returns `false`.

This example also illustrates a vulnerability that occurs if the database happens to crash just after a row is deleted from the dbDates table and just before its corresponding shifts are deleted from the dbShifts table.[5]

To illustrate the generality of this implementation strategy, Figure 6.10 shows the essence of the `retrieve_dbPersons` function implemented in the dbPersons module of *Homeroom*. Notice the structural similarity between this code and that shown in Figure 6.9.

In particular, lines 9-16 of Figure 6.10 guarantee that retrieving a person from the database returns exactly one person, or else it returns false. Reliance on the database normalization principle of unique primary keys is critical to *Homebase* as well as all of our other CO-FOSS applications.

```
function retrieve_dbPersons ($id) {
    $con=connect();
    $query = "SELECT * FROM dbPersons WHERE id = '".$id."'";
    $result = mysqli_query($con,$query);
    if (mysqli_num_rows($result) !== 1){
        mysqli_close($con);
        return false;
    }
    $result_row = mysqli_fetch_assoc($result);
    $thePerson = new Person($result_row['last_name'],
        $result_row['first_name'], $result_row['gender'],
        $result_row['employer'], $result_row['address'],
        ... );
    ...
    mysqli_close($con);
    return $thePerson;
}
```

FIGURE 6.10 Retrieving a person from the dbPersons table in *Homeroom*.

More complete and illustrative examples of this implementation strategy for database modules and functions appear within the database modules in the *Homebase* and *Homeroom* code bases themselves. These can be directly examined at `https://github.com/megandalster/`.

Finally, to enforce this protocol, all actions involving database tables that are required by non-database modules in the software system, especially the user interface modules, should accomplish those actions by calling a function defined inside that table's database module, rather than writing separate MySQLi queries from the outside. If no such function exists, then the proper

[5] A more thorough treatment of this issue might use the concept of *rollback* that would help the database gracefully recover from such an event.

development strategy is to add and test that function within its respective database module, rather than to implement that function elsewhere.

6.2.6 Adding a Database Abstraction Layer

After a CO-FOSS project has been completed using a particular programming language and database extension, as shown in Figure 6.3, the future cost for converting the code to a different extension can be severe. For example, suppose we had implemented *Homebase* using the PHP/MySQLi extension and we later determined that the PostgreSQL extension would provide better performance. To convert the project to PostgreSQL would require us to rewrite and retest all the database modules discussed above.

However, a new software project might want to retain flexibility in choosing a database extension. For example, it might need to use the MySQLi extension for a client-server installation and a SQLite extension for a mobile installation. Or else, it might need to retain the option to easily switch between database extensions to improve its performance or its database security in the future.

To retain database flexibility, a project may pair its chosen programming language with a "database abstraction layer" rather than picking one of the specific database extensions shown in Figure 6.3. The abstraction layer provides an API that allows the database coding to be done using its own SQL query protocol that maps into any one of the MySQLi, PostgreSQL, or SQLite extensions.

In PHP, this database abstraction layer is called "PHP Data Objects," or `PDO` for short. In Java, it is called "Java Database Connectivity," or `JDBC`. In Django and Rails, it is called "object relational mapping" `(ORM)`.

By developing the new software with PDO (or JDBC or ORM), the developers retain the freedom to switch between different SQL extensions without significantly rewriting the code base itself. The PDO statements are interpreted by a so-called "driver," which is associated with one of the three extensions, such as MySQLi. After the software is completed, later developers can redeploy it with a different SQL extension, say PostgreSQL, by just selecting a different driver rather than doing any recoding.

Beyond providing this flexibility, a database abstraction layer has another advantage. That is, it can enhance database security by preventing certain types of attacks on the database known as "SQL injection." Since this and other vulnerabilities usually originate at the user interface level, we shall discuss them more carefully in Chapter 7.

For a more careful discussion of the tradeoffs between an extension like MySQLi and an abstraction layer like PDO, readers should visit `https://websitebeaver.com/php-pdo-vs-mysqli`. For a more thorough tutorial introduction to PDO, readers should visit `https://npfi.org/pdo-tutorial/`.

For uniformity, most of the examples in this book use the PHP/MySQLi extension. Readers can visit our *BMAC-Warehouse* code base to see an example of the PDO abstraction layer.

6.3 DATABASE TESTING

As the database modules are developed, their corresponding unit tests should be developed alongside them. This process is similar to that which was introduced in Chapter 5 for testing, debugging, and refactoring the domain classes.

However, testing and debugging the database modules has its own peculiarities, which stem from the fact that data within the database tables are persistent. The goal of each unit test for a database module must not only include a test of that module's CRUD functions, but it must also include provisions for restoring its table to the same state that it had been in before the test began. This leads to the following 3-step process for unit-testing each database module:

1. Setup — create objects and insert new rows in the table, testing the C part of the CRUD functions.
2. Test — replace and update those rows in the table, testing the R and U parts of the CRUD functions.
3. Teardown — delete those rows from the table, testing the D part of the CRUD functions.

Following this 3-step process ensures that, if successful, running a unit test on a database table will leave that table in the same state that it had before the test began. The following sections illustrate these ideas, using examples from the *Homebase* and *Homeroom* database modules and their corresponding unit tests.

6.3.1 Testing the dbShifts.php Module

Assuming that the Shift.php class has already been unit tested and the dbShifts.php module is being coded, we can design a unit test for the dbShifts.php module, which will have implementations of the following four major CRUD functions:

C `insert_dbShifts` inserts a new shift (row) into the dbShifts table. This is the Setup step

R `select_dbShifts` selects a unique shift (row) from the dbShifts table.

U `update_dbShifts` replaces an existing shift in the dbShifts table by another with the same key.

D `delete_dbShifts` removes a shift (row) from the dbShifts table. This is the Teardown step

A unit test for dbShifts.php is shown in Figure 6.11.

It should be clear from this example that we cannot sensibly unit test the dbShifts.php module until after the Shift.php class itself has been unit

```
use PHPUnit\Framework\TestCase;
include_once(dirname(__FILE__).'/../database/dbShifts.php');
include_once(dirname(__FILE__).'/../database/dbDates.php');
class dbShiftsTest extends TestCase {
  function testdbShifts() {

    // Setup step -- test insertion into the dbShifts table
    $s1=new Shift("08-02-25:1-5","portland", 3, array(),
        array(), "", "");
    $s2=new Shift("08-02-25:9-1","portland", 3, array(),
        array(), "", "");
    $this->assertTrue(insert_dbShifts($s1));
    $this->assertTrue(insert_dbShifts($s2));

    // Test step -- test retrieval and update for the table
    $this->assertEquals(select_dbShifts($s2->get_id())->
        get_vacancies(), 3);
    $s2=new Shift("08-02-25:9-1","portland",2, array(),
        array(), "", "");
    $this->assertTrue(update_dbShifts($s2));  // Update

    // Teardown step -- test delete from the table
    $this->assertTrue(delete_dbShifts($s1));  // Delete
    $this->assertTrue(delete_dbShifts($s2));
  }
}
```

FIGURE 6.11 A unit test for the dbShifts module.

tested. That is because testdbShifts.php needs to use some of the Shift class's functions (constructor, `get_id`, and `get_vacancies`) for the purpose of testing the dbShifts module's functions themselves. Knowing *a priori* that those functions in the Shift class are reliable helps us pin down errors and failures while testing the dbShifts.php module.

Note also that the unit test for dbShifts leaves the dbShifts.php table unchanged when it completes its three steps. That is, the Setup step adds two rows to the dbShifts table at the beginning of the test, and the Teardown step deletes those same two rows at the end of the test. In between, the retrieve and update functions of the dbShifts.php module are tested by accessing only these two rows.

What if testdbShifts.php does not complete all 3 of its steps? For instance, suppose it runs into a coding error during the Test step. In this event, PHPUnit would terminate and the dbShifts table would be left with two superfluous rows, since the Teardown step would not have been reached. To restore the dbShifts table to its original form in this event, the developer must clean up the table manually by deleting those two rows using the phpMyAdmin interface shown in Figure 6.8. Failure to do this will compromise the integrity of the dbShifts table as well as future unit tests on that table.

6.3.2 Testing the dbPersons.php Module

In *Homeroom*, dbPersons.php is the database module corresponding to the Person class, and its corresponding database table is dbPersons. To provide a frame of reference, Figure 6.12 summarizes the instance variables for the Person class.[6]

```
class Person {
    private $id;             // primary key = first_name + phone1
    private $last_name;      // last name
    private $first_name;     // first name
    private $address;        // address
    private $city;           // city
    private $state;          // state
    private $zip;            // zip code
    private $phone1;         // primary phone
    private $phone2;         // alternate phone
    private $email;          // email address
    private $password;       // password:default = $id
    private $type;           // 'guest','socialworker','manager'
    private $gender;         //gender of person
    private $employer;            //employer of the person
    private $patient_name;        // array of up to 3 patients
    private $patient_birthdate;   // format: yy-mm-dd
    private $patient_gender;      // "Male" "Female" or "Unknown"
    private $patient_relation;    // relationship to the patient
    private $prior_bookings;      // array of booking id's
    private $mgr_notes;           // manager's notes
    private $county;              // county in Maine
```

FIGURE 6.12 Instance variables for the Person class in *Homeroom*.

The unit test for the dbPersons.php module is shown in Figure 6.13. It has a Setup step, a Test step, and a Teardown step. In the Setup step, an individual person is constructed and added to the dbPersons database table, thus testing the C part of CRUD. In the Test step, this person's entry in the dbPersons

[6] A complete definition of the Person class can be viewed at the *Homeroom* code repository.

table is retrieved and updated, thus testing the R and U parts of CRUD. In the Teardown step, this person's entry is deleted from the database table, thus testing the D part of CRUD and leaving the dbPersons table unchanged from its original state before this test was run.

```
class dbPersonsTest extends TestCase {
    function testdbPersonsModule() {
    // Setup step -- test the insert function
        $person1 = new Person("Smith", "John", "male", "",
             "123 College Street","Ashburn", "VA", "20147",
             7035551234, "", "email@bowdoin.edu", "guest",
             "", "Jane Smith", "98-01-01", "Female", "", "");
        $this->assertTrue(insert_dbPersons($person1));
    // Test step -- test the retrieve and update functions
        $this->assertEquals(retrieve_dbPersons(
            $person1->get_id())->get_id(), "John7035551234");
        $this->assertEquals(retrieve_dbPersons(
            $person1->get_id())->getith_patient_name(0),
            "Jane Smith");
        $this->assertEquals(retrieve_dbPersons(
            $person1->get_id())->get_patient_birthdate(),
            "98-01-01");
        $this->assertEquals(retrieve_dbPersons(
            $person1->get_id())->get_patient_gender(),
            "Female");
        $this->assertTrue(retrieve_dbPersons(
            $person1->get_id())->check_type("guest"));
        $person1->set_address("5 Maine Street");
        $this->assertTrue(update_dbPersons($person1));
        $this->assertEquals(retrieve_dbPersons(
            $person1->get_id())->get_address (),
            "5 Maine Street");
    // Teardown step -- test the delete function
        $this->assertTrue(delete_dbPersons($person1->get_id()));
    }
}
```

FIGURE 6.13 A unit test for the dbPersons module.

As shown, this unit test is far from complete. Missing, for example, are important tests of the interactions between a person and a booking, which take place whenever a new booking is created for a person and added to the dbBookings table. This is an example of the need for integration testing, and we return to it in Section 6.3.5.

6.3.3 Testing the dbBookings.php Module

A Booking in Homeroom captures the details of a single client's stay at the house while their child is undergoing surgery at a nearby medical facility. Details of a booking include the id of the client, the name of the patient, the date of the booking, the room number where the client will be staying, and so on. To provide a frame of reference, Figure 6.14 summarizes the instance variables for the Booking class.[7]

```
class Booking {
    private $id;      // primary key:  current_date + guest_id
    private $date_submitted; //  "yy-mm-dd"
    private $date_in;      // check-in date:  "yy-mm-dd"
    private $guest_id;     // e.g.,  "John2077291234"
    private $status;       // "pending", "active", "closed"
    private $room_no;        // id of the room, if assigned
    private $patient;        // up to 3 patients
    private $occupants;      // up to 6 people per room
        // E.g., array("John:father:Male:Present")
    private $auto;           // make:model:color:state
    private $linked_room;    // optional second room
    private $date_out;       // check-out date, "yy-mm-dd"
    private $referred_by;    // person requesting this booking
    private $hospital;       // hospital where patient staying
    private $department;     // department of treatment
    private $health_questions; // 11 questions (0/1 = no/yes)
```

FIGURE 6.14 Instance variables for the Booking class in *Homeroom*.

Accordingly, a unit test for the dbBookings.php module tests the correctness of inserting, updating, retrieving, and deleting a single booking from the dbBookings table. Portions of that unit test are shown in Figure 6.15. For a complete listing, please see the *Homeroom* `code repository`.

The dbBookings.php module has many additional functions that are not tested in this illustration. A more thorough unit test should include at least one assertion that tests each function in that module. Even then, the unit test for dbBookings will not be perfect. But it will provide a level of confidence in the integrity of this module before it is pressed into service by the user interface modules that will be developed later in the project.

[7] A complete definition of the Booking class can be viewed at the *Homeroom* `code repository`.

```
class dbBookingsTest extends TestCase {
    function testdbBookings() {
    // Setup -- create a booking and test insert
        $today = date('y-m-d');
        $b = new Booking($today,"","Meghan2075551234","pending",
            "",array("Tiny"),array("Jean:father", "Teeny:sib"),
            "", "", "", "Millie2073631234","Maine Med", "SCU",
            "00000000000","$10 per night", "","","","","new");
        $this->assertTrue(insert_dbBookings($b));
    // Test -- test the retrieve and update functions
        $this->assertEquals(retrieve_dbBookings($b->get_id()),
            $b);
        $this->assertEquals(($b->book_room("126",$today))->
            get_date_in(),$today);
        $bretrieved = retrieve_dbBookings($b->get_id());
        $this->assertEquals($bretrieved->get_room_no(), "126");
        $this->assertEquals($bretrieved->check_out($today,""),
            $bretrieved);
        $bretrieved2 = retrieve_dbBookings($b->get_id());
        $this->assertEquals($bretrieved2->get_status(),
            "closed");
    // Teardown -- test the delete function
        $this->assertTrue(delete_dbBookings($b->get_id()));
        $this->assertFalse(retrieve_dbBookings($b->get_id()));
    }
}
```

FIGURE 6.15 Portions of a unit test for the dbBookings.php module.

6.3.4 Testing the dbRooms.php Module

The dbRooms.php module in *Homeroom* is much simpler than the dbBookings.php or dbPersons.php module, due in part to the fact that a room has only a handful of instance variables. To provide a frame of reference, Figure 6.16 summarizes the instance variables for the Room class.[8]

A unit test for the dbRooms.php module is shown in Figure 6.17. Note here that the functions reserve˙me and book˙me are tested for reserving and booking the room for a particular person whose id is `Alison2076942604`, but no testing of the interaction between the room and that person's database record is tested. That test is part of integration testing, which is discussed more fully in Section 6.3.5.

[8] A complete definition of the Room class can be viewed at the *Homeroom* `code repository`.

```
class Room {
    private $room_no;   // room number in the house, like "125"
    private $beds;      // bed configuration: 2T, 1Q, etc.
    private $capacity; // maximum number of persons
    private $bath;      // "y" or "n" if there's a private bath
    private $status;    // "clean","dirty","booked","reserved"
    private $booking;   // the current booking id for this room
    private $room_notes; // room-specific notes
```

FIGURE 6.16 Instance variables for the Room class in *Homeroom*.

6.3.5 Integration Testing: Persons, Bookings, and Rooms

After completing the individual unit tests for the domain classes and database modules, we should look for opportunities to test the interactions among them. Of particular interest here are the interactions among modules, both across these two layers, and within the database layer.

By unit testing the database modules individually, we have already done a fair bit of integration testing. For example, by testing the dbShifts.php module we have also tested its interdependence with the Shift class. The same goes for testing the dbPersons.php module and the Person class, as well as the dbBookings.php module and the Booking class.

However, this testing still leaves several critical inter-module interactions untested. In *Homeroom*, the interactions among Persons, Bookings, and Rooms are particularly critical. That is, a Person can have several closed Bookings over time, one for each past stay at the House. Each such Booking has an associated Room. Correspondingly, the database tables dbPersons, dbBookings, and dbRooms need to be updated whenever a Booking is either activated or.closed.

Here are two common situations that can be explicitly tested:

1. Activating a booking means assigning it to a room.

2. Closing an active booking means un-assigning that room and updating that person's past bookings.

Let's design a unit test for these two situations: one where a new booking is activated and the other where an active booking is closed. To make these tests, we can add new assertions to the dbBookings.php unit test that create and then test the proper implementation of each of these two cases.

1. For the first situation, we add statements that activate a pending booking with a specific room, and then test that that room actually has been assigned the id of that booking.

```
class dbRoomsTest extends TestCase{
  function testdbRooms(){
  // Setup -- insert two rooms into the database
    $room1=new Room("998", "2T", "3", "y", "clean", "", "");
    $room2=new Room("999", "Q", "2", "n", "reserved", "", "");
    $this->assertTrue(insert_dbRooms($room1));
    $this->assertTrue(insert_dbRooms($room2));

  // Test the retrieve and update functions
    $this->assertEquals(retrieve_dbRooms($room1->get_room_no(),
        "",""), $room1);
    $this->assertEquals(retrieve_dbRooms($room2->get_room_no(),
        "","")->get_room_no(),"999");
    $this->assertEquals(retrieve_dbRooms($room2->get_room_no(),
        "","")->get_status(),"clean");
    $this->assertEquals($room2->reserve_me(
        "13-07-26Alison2076942604"),$room2);
    $this->assertEquals(($room2->unbook_me(
        "13-07-26Alison2076942604"))->get_status(),"dirty");

  // Teardown -- test the delete functions
    $this->assertTrue(delete_dbRooms($room1->get_room_no()));
    $this->assertTrue(delete_dbRooms($room2->get_room_no()));
  }
}
```

FIGURE 6.17 A unit test for the dbRooms.php module.

2. For the second situation, we add statements that close an active booking, and then test that the person for whom the booking was closed has that booking's id in their list of prior bookings.

Figure 6.18 illustrates the statements that can be added to the dbBookings.php unit test in Figure 6.15 to model these situations for this particular integration test.

These statements actually appear in the unit test in the *Homeroom* `code base` so that readers can replicate this integration test if needed. Testing these kinds of interactions provides important assurance about the integrity of interactions among the dbPersons, dbBookings, and dbRooms tables in the database.

Many other interdependencies in CO-FOSS database development can lend themselves to an integration test of the sort described above. To identify these interdependencies and effectively create such tests, the developers need to

```
class dbBookingsTest extends TestCase {
  function testdbBookings() {
  // Setup -- create a booking and a person
    $today = date('y-m-d');
    $b = new Booking($today,"","Meghan2075551234","pending",
        "",array("Tiny"),array("Jean:father", "Teeny:sibling"),
        "", "", "", "Millie2073631234","Maine Med", "", "",
        "$10 per night", "","","","","new");
    $p = new Person("Jones", "Meghan", "female", "",
        "123 Cod St","Brunswick","ME", "04011", "2075551234",
        "","", "guest", "","Tiny", "98-01-01","Female","","");
    $this->assertTrue(insert_dbBookings($b));
    $this->assertTrue(insert_dbPersons($p));
  // Integration Test -- activate booking, check that room
    $this->assertEquals(($b->reserve_room("126",$today))
        ->get_date_submitted(),$today);
    $this->assertEquals(($b->book_room("126",$today))
        ->get_date_in(),$today);
    $bretrieved = retrieve_dbBookings($b->get_id());
    $r = retrieve_dbRooms($bretrieved->get_room_no(),$today,
        $bretrieved->get_id());
    $this->assertEquals($r->get_booking_id(),
        $bretrieved->get_id());
  // Integration Test -- close booking, check that person
    $this->assertEquals($bretrieved->check_out($today,""),
        $bretrieved);
    $bretrieved2 = retrieve_dbBookings($b->get_id());
    $this->assertEquals($bretrieved2->get_status(), "closed");
    $this->assertEquals($bretrieved2->get_date_out(),$today);
    $pretrieved = retrieve_dbPersons($p->get_id());
    $this->assertContains($bretrieved->get_id(),
        $pretrieved->get_prior_bookings());
  // Teardown -- restore dbBookings and dbPersons tables
    $this->assertTrue(delete_dbBookings($b->get_id()));
    $this->assertTrue(delete_dbPersons($p->get_id()));
  }
}
```

FIGURE 6.18 An integration test for dbPersons.php, dbBookings.php, and dbRooms.php.

understand the interactions that are suggested by the use cases in the project's design document. Later in the process, when the use cases are implemented at the user interface level, new insights and opportunities for integration testing will also emerge.

6.4 CLIENT REVIEW AND ISSUE TRACKING

Once the database modules have been fully implemented and tested, it is important to conduct a client review.

What can the clients contribute in a review at this stage, when they may not have seen much more of the user interface beyond the login page? What issues can the client post for the developers when they find problems as they review the database elements? We address these questions below.

6.4.1 Client Review

To prepare for client review, the integration-tested domain and database code should be pushed to the sandbox server so that developers and clients can see the same partially-developed product. An initial database should also be uploaded onto the database server, including a few entries in each of the tables corresponding to the database modules. In *Homeroom*, for example, these tables include dbBookings, dbPersons, dbRoomLogs, and dbRooms.

So what can the client test with the software at this early stage? We recommend that the developers provide one very central example of an interaction among database modules that will provide the client with a concrete illustration of how the system will work for them. For example, in *Homeroom* the client can be provided with an example of the use case where they make a Booking for a guest (Person) in a specific Room.

Since there may be no actual user interface, the developers will need to demonstrate how that interface would work by sketching one or two screenshots of the user interface and showing the client what will happen in the database when that particular use case is completed. These sketches may be available among the use cases in the project's design document. For example, in *Homeroom*, developers would show how booking a room for a guest will work, using the "Book a Room" use case as a starting point. Developers would show clients how booking a room would look from the database standpoint by displaying the database before and after the booking is made.

This demonstration should elicit some valuable feedback from the clients about that use case since they can visualize completing it themselves. Moreover, this exercise also provides some perspective to the client on what to expect (and what not to expect) from the software before the user interface development begins.

This review step may not be very exciting for the client, who at this point in the project is ready to begin interacting with the software itself. Nevertheless, keeping the client review process both active and narrowly focused on a single

```
#1 reported by hargu...@whitman.edu on 2 Mar 2015 at 6:47
Calendar Week Management Disappears When Adding After
Removing A Week

#2 reported by al...@bowdoin.edu on 5 Mar 2015 at 6:25
When listing volunteers available, software doesn't filter out
house from family room.

#3 reported by al...@bowdoin.edu on 5 Mar 2015 at 6:29
long list from volunteer search runs off the screen.  Add a
scrollbar.

#4 reported by al...@bowdoin.edu on 17 Mar 2015 at 5:09
Add an "Are You Sure?" dialog box to each delete function

#5 reported by l...@whitman.edu on 22 Apr 2015 at 6:27
Adding date range option for volunteer history report

#6 reported by beeb...@whitman.edu on 22 Apr 2015 at 6:45
Default password doesn't get assigned for new volunteers
```

FIGURE 6.19 The first 6 issues posted for the 2015 *Homebase* project.

user activity may provide better quality feedback than presenting a broader summary of the details of all the database tables.

In a CO-FOSS project that is adding new features to an existing product, client review of the database development stage can be more interactive, since some/much of the original user interface may still be accessible. With that interface, the client will be able to confirm that original features have not been compromised by the addition of database modules and tables to support new features.

6.4.2 Issue Tracking

At this point in the project, issue posting should begin to be active, especially among the developers. Issue posting by clients, on the other hand, may not begin to take off until they have had a chance to begin working with the user interface. For example, Figure 6.19 shows a list of the first 6 issues posted for the 2015 *Homebase* project.[9]

[9] A complete list of the issues posted, owned, and closed for this project can be viewed at `the project's GitHub site`. Note that these issues were originally posted on Google Code, and then were migrated to GitHub later in 2015 when Google Code was discontinued. However, the content and full tracking of these issues was retained in that migration.

Recall that the 2015 *Homebase* project adapted and extended an earlier version for a different client with almost identical needs. Thus, these first six issues all related to an experience with some aspect of the user interface. Also, these six issues were posted by four different developers, three of whom were students with no prior issue-posting experience.

Notice also the dates when the issues were posted. Although the project began on January 15, 2015, the first issue wasn't posted until over a month later, on March 2. In fact, the clients didn't begin posting issues until April 23 (issue #14), after they had begun to review the new user interface in detail. This project was completed on May 7, 2015. This is a typical timeline for issue posting on a new CO-FOSS project. In our experience, clients don't really become active issue posters and commenters until late in the project.

The idea of *issue tracking* refers to the temporal series of comments that occur in response to the posting of an issue, from the time it is opened until the time it is closed. The last comment in the series should reflect resolution of the issue by changing its status from "open" to "closed."

All developers and clients should be authorized to open, comment on, and close an issue, although we recommend that some convention for client sign-off be adopted before closing an issue. That convention could simply be a requirement that the client review a proposed "fix" from the developer and do the closing, rather than allowing the developer who owns it to close it. That ensures client review and acceptance of the proposed fix prior to closing.

We have also found that opening some issues should be done using a simple framework, or template, for describing the issue clearly. While many issues don't require this level of structure, some are detailed enough to require a careful description so that the developer can understand it clearly. This framework we used in our later CO-FOSS projects is shown in Figure 6.20.

```
Brief description of the issue:
...
What steps will reproduce issue?
1.
2.
3.
...
What is the expected output? What do you see instead?
```

FIGURE 6.20 Simple framework for posting a new issue.

For clients and developers to open or comment on an issue, they must first be registered collaborators in the project repository where the code base is stored, as discussed in Chapter 4. In GitHub, for example, new collaborators can be registered by the project owner by using the **Settings** tab on the project page.

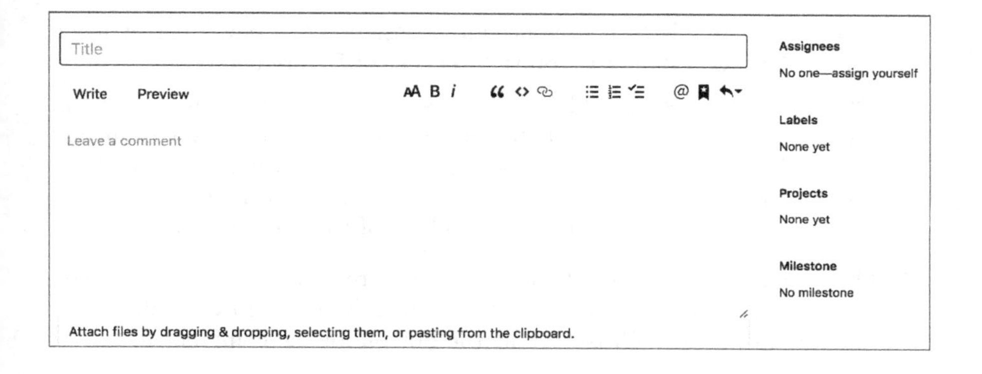

FIGURE 6.21 Form for posting a new issue on a GitHub project.

Recall the simple issue-tracking protocol for CO-FOSS projects that we introduced in Chapter 5. Using this protocol, let's look at the key steps in tracking a single issue, using as an example the tracking of issue #5 during the 2015 *Homebase* project.

1. An issue can be opened by a developer or a client during the Review step of any CO-FOSS cycle. In a GitHub project, the developer or client can open an issue by logging into the project and hitting the **New Issue** tab at the top of the page. A form like the one shown in Figure 6.21 should appear.

   When done filling out this form, hit **Submit New Issue** tab at the bottom of the form.

2. The issue is posted as "open" with a definitive title, a brief description, and an example. Once submitted, the issue is added to the top of the list of open issues for the project, with the current date and name of the team member who posted it. All team members are notified of this new posting. The person posting the issue may or may not become its owner.

3. Each open issue is reviewed by the lead developer, who determines whether it needs immediate correction, overlaps (and can be combined) with another open issue, can be deferred until later in the development process, or is not relevant to the project. For example, Issue #5 in the list in Figure 6.19 was later determined to be not relevant to this project, and thus was closed. The **Labels** tab at the right of Figure 6.21 can be used to help characterize the issue, such as **Priority-Medium** or **duplicate** or **Type-enhancement**.

4. The lead developer assigns issues that need immediate correction to the other developers in an equitable fashion. Each issue is thus "owned" by

a developer. In GitHub, an issue can be assigned to a developer by using the **Assignees** tab at the top right corner of Figure 6.21.

During the process of addressing an issue, the owner or another team member can report its status to the rest of the team by filling out and submitting a comment form that appears when viewing the issue. Submitting that comment appends it to the end of the issue's discussion thread and also notifies all team members of that submission.

5. Once an issue is corrected by its owner, it is reviewed by the person who posted it and, if approved, the issue is then "closed." In GitHub, this is done by hitting the **Close** tab at the bottom of the comment form.

Because all comments on an issue are retained in chronological order, developers and clients can return at a later date and review the status of any issue, especially the ones that are open. Moreover, issues that were closed earlier can be reopened if a team member determines that more work needs to be done to completely fix the issue. Below is a brief summary of the history of issue #5 in the 2015 *Homebase* project.

```
Issue reported by l...@whitman.edu on 22 Apr 2015 at 6:27
  Currently, everyone will show up in the volunteer report
  from the beginning of time up to now.

  Joanna and Sue would like to select the date range for the
  report, either annually (from January 1st) or  month (from
  the first day of a month).

Comment by beeb...@whitman.edu on 27 Apr 2015 at 5:29
  Issue 8 has been merged into this issue.

Comment by l...@whitman.edu on 29 Apr 2015 at 6:56
  fixed 4/29/15
```

This example shows the issue summary appearing on April 22 and two later comments being added on April 27 and April 29, at which time the issue was closed. While this is not a particularly interesting issue, it does reflect the kinds of tracking that are available to a project through its code repository.

Often, in fact, individual commits of new code whose comments mention a specific issue # will become automatically merged into the discussion thread for that issue. Thus, developers can see who actually made the fix and tested it. This kind of issue tracking information becomes invaluable as the project itself becomes more complex and client participation becomes more proactive, as we shall see in the next chapter.

6.5 SUMMARY

This chapter has covered a lot of ground. It began by introducing database principles along with their realization in the setting of SQL, the widely-used Structured Query Language. It covered the important topics of table creation, keys, normalization, and queries that accomplish elementary operations on database tables.

In a CO-FOSS project, the creation, integration, and testing of its database modules is made feasible by the layering principle. This chapter explains the correspondence between database tables, their actions, and the underlying domain classes that define the project's vocablulary. We introduce and illustrate each database module required by a project using examples from our own experience.

Unit testing and integration testing of the database modules are also introduced and illustrated using examples from our own projects. Finally, this chapter revisits the important step of client review and issue tracking in the setting of database development.

6.6 MILESTONE 6

1. Create a new database table corresponding to each database module developed for your CO-FOSS project. Each table should have a complete definition of its columns and a handful of initial entries that will facilitate testing.

2. Complete the initial development of all the database modules for your CO-FOSS project. Each module should have a Create, Replace, Update, and Delete (CRUD) function for manipulating individual rows of data. Use the protocol described in this chapter for designing each CRUD function.

3. For each database module whose code is being adapted from another open source project, perform all the refactoring needed to remove unnecessary or redundant functionality.

4. Complete the unit testing and integration testing of all database modules.

5. Conduct an appropriate client review, encouraging clients to identify and post any issues that arise out of this review.

6. Evaluate these new issues and assign ownership to individual developers for any issues that need to be resolved during the upcoming CO-FOSS cycle.

CHAPTER 7

User Interface Development

"What you see is what you get."
—*Flip Wilson*

The user interface of a CO-FOSS product is the main point of interaction between the client and the software. The interface provides the glue that holds all the other elements together—not only in the code base but also in the development process. That is, all developer-client discussions about a particular functionality, bug, or new feature begin and end at the user interface.

The user interface is also the place where system integrity and usability are established and maintained. Any Web-based software artifact that has a clumsy user interface, fails to ensure database security and integrity, or does not enable different users to exercise the system simultaneously, is relatively weak and useless.

This chapter introduces both the principles of user interface design and the practice of user interface development. Our main sources of examples are the completed CO-FOSS projects *Homebase*, *Homeroom*, and *Homeplate*. Because our projects are coded using PHP, JavaScript, MySQLi, and HTML/CSS, developers using other platforms should supplement this discussion with examples from other language-specific sources, such as:

Java/Spring `https://www.tutorialspoint.com/spring/spring_web_mvc_framework.htm`

Python/Django `https://www.tutorialspoint.com/django/django_overview.htm`

Ruby on Rails `https://www.tutorialspoint.com/ruby-on-rails/rails-framework.htm`

Nevertheless, the principles discussed and illustrated here apply to a wide variety of Web-based CO-FOSS projects.[1]

7.1 PRINCIPLES

The design and development of a good user interface is a special craft that requires both skill and experience. Being a good programmer is a necessary but not sufficient requirement for becoming a good user interface designer or developer. A good user interface follows several aesthetic and functional principles that guide its development process.

Many books and articles have been written about user interface design (e.g., see [41]). The Web contains many examples of excellent user interfaces, but unfortunately it also contains many more examples of bad user interfaces. For the remainder of this chapter, we shall use the following principles as our guide for what we mean by a *good user interface*.

1. **Task-Oriented** There is a clear mapping between the steps of each use case in the design document and a page or group of related pages in the user interface.

2. **Language** The language of the interface must be consistent with the language of the domain and the user. All labels and user options that appear on individual pages must be consistent with their counterparts in the design document and the domain classes.

3. **Simplicity** No page should contain too much information or too little. Each page should be pleasant to view, yet its functionality should not be buried by elaborate stylistics.

4. **Navigability** Navigation within a page and between related pages should be simple, explicit, and intuitive.

5. **Visual consistency** All pages in the user interface should have a similar look and navigation style.

6. **Discoverability** Each page should provide the user with clear feedback on what has just been done, what can be done next, how to do it, and how to undo it.

7. **Data integrity** The types and valid values for individual user data entries must be clearly indicated. The software should validity-check all data at the time of entry and the user should be required to correct errors before the entries are saved in the database.

[1]To facilitate hands-on engagement with these examples, readers are encouraged to download and install the `demo version` of Homebase (see `https://npfi.org/homebase-install/`) on their own servers.

8. **Client-server integrity** The activities of several different users who happen to be using the system at the same time must be kept separate and independent.

9. **Security** An individual user should have access to the system's functionality, but only that functionality for which he/she is authorized.

10. **Documentation** Every page or group of pages in the user interface should be linked to a step-by-step instruction that teaches a user how that page can be used to accomplish a task.

When designing an individual element of the user interface, for example a menu or a form that displays and receives information from the user, developers must ensure that all these principles are followed. If an element of the user interface cannot provide that assurance, the software that underlies the entire user interface is potentially compromised.

The following sections introduce design strategies that, when used together, can help developers embrace the first nine of these principles. The last principle, **documentation**, requires so much discussion that it is covered separately in Chapter 8.

7.1.1 Model-View-Controller Pattern

The code underlying a user interface is both complex and loaded with details. It may contain a mix of HTML and additional programming elements called *scripts* that define a user interface that is simultaneously pleasant to view and responsive to users' needs. Breaking the user interface design into separate conceptual components can help the developer manage this inherent complexity.

A widely-used strategy for managing the user interface design is known as the *Model-View-Controller (MVC)* pattern (see Figure 7.1). The MVC pattern separates the code underlying each component (use case) of a user interface into three distinct conceptual pieces:

The Model Code underlying the application's logic and underlying data,

The View Code presenting that component of the interface to the user, and

The Controller Code controlling the user's input, output, and navigational functionality for that component.

Together, these three pieces form a "MVC triple" for that component.

This separation promotes independent coding, testing, and maintenance of each component. It also provides documentary clarity for later developers who are modifying or reusing individual components of a user interface.

In a layered architecture, the code in the model is encapsulated within the database and domain layers, while the code in the view and controller are encapsulated within the user interface layer.

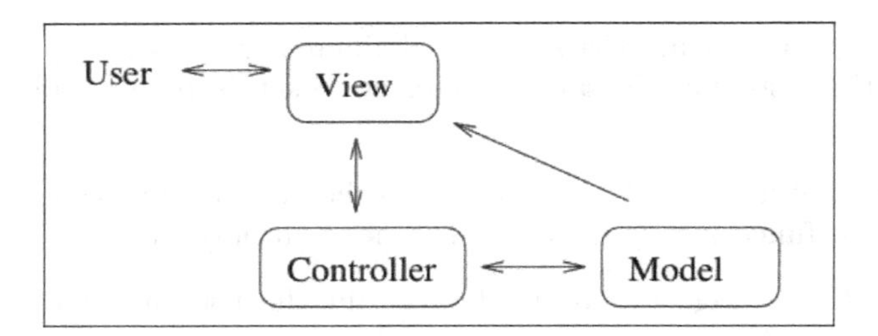

FIGURE 7.1 The Model-View-Controller pattern.

A complete user interface for a software product can be organized as a collection of MVC triples, each triple responsible for implementing a particular user activity or use case within the product.

In a single MVC triple, the *model* contains the system's session-specific data, or *state*, that exists for a single user activity, along with all the functionality in the database and domain layers that can manipulate that data for that activity. This includes the active variables in the current user's session, the relevant database tables, and their related database modules and domain classes. When information in the database is requested or changed, the model can thus fulfill that request or perform that change.

The *view* for a single MVC triple renders that user activity in a way that supports user interaction according to its underlying use case. This rendering includes one or more forms, including graphics, text, and other widgets that enable information to be easily transmitted between the user and the controller. In Web-based applications, the view can be created using HTML, CSS, and an embedded scripting language like PHP, JavaScript, Python, Ruby, or Java Servlets, that displays information and facilitates user interactions. The view plays a major role in enforcing design principles 2, 3, 4, 5, and 6 listed above.

The *controller* for a single MVC triple receives user input via the view and initiates a response by making transformations on the data in the underlying model. In this sense, the controller maintains and interprets the current *state* of an interaction between the user and the system. To accomplish this, the controller maintains so-called SESSION, GET, and POST variables; verifies user input; and updates information in the underlying model.[2]. In this sense, a controller implements the system's steps as it responds to the user's input during completion of a particular use case. The controller plays a major role in enforcing design principles 7, 8, and 9 listed above.

The following three MVC examples illustrate how every *view* in a user interface has an underlying module that creates it, and then interacts with

[2]Briefly, the SESSION variables distinguish this particular user's session from all others accessing the running software at some instant in time. The GET variables are set by the system to help call up a particular view, while the POST variables are set by the user when entering data into a particular view. More detailed information about SESSION, GET, and POST variables is provided in Section 7.2.1.

related modules that embody its associated *model* and *controller*. These three examples are taken from three different CO-FOSS projects, *Homebase*, *Homeroom*, and *Homeplate*.

Altogether, these examples illustrate how the MVC pattern can provide a powerful organizational tool for any client-server application. Using this paradigm also helps partition programming tasks among several team members who are simultaneously developing different views in a user interface.

MVC Example 1: Editing a Shift in *Homebase*

As a first example of an MVC triple, let's look at the *view* that appears when a user edits a particular shift in *Homebase*, as shown in Figure 7.2.

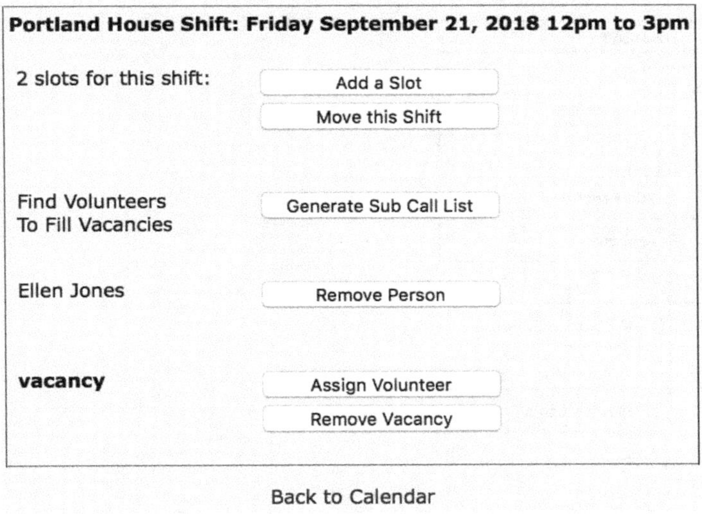

Portland House Shift: Friday September 21, 2018 12pm to 3pm

2 slots for this shift: Add a Slot

Move this Shift

Find Volunteers To Fill Vacancies: Generate Sub Call List

Ellen Jones: Remove Person

vacancy: Assign Volunteer

Remove Vacancy

Back to Calendar

FIGURE 7.2 The Edit Shift view in *Homebase*.

The *model* that underlies the view shown in Figure 7.2 includes all the active variables in the user's session, all the contents of the dbShifts and dbPersons database tables, and all the underlying functionality in the classes and modules that can manipulate that data.

The *view* shown in Figure 7.2 is rendered by code that is activated by the link `editShift.php?shift=18-09-21:12-3&venue=portland`. That is, the view is rendered by the editShift.php, editShift.inc, and styles.css modules using the GET variables `shift=18-09-21:12-3` and `venue=portland`.

The *controller* for this view is shared between the editShift.php and editShift.inc modules. They handle GET and POST information and use it to update the active variables and database tables inside the model for this SESSION. The controller also includes code that supports user navigation by transferring control between this view and other views in the *Homebase* user interface.

For all the details, the reader should visit the complete code for these modules, which can be seen at `https://npfi.org/homebase-code/`.

MVC Example 2: Editing a Person in Homeroom

As a second example of an MVC triple, let's look at the *view* that appears when a user edits a particular person in *Homeroom*, as shown in Figure 7.3.

Person Edit Form
Here you can edit (or delete) a person in the database.

(* denotes required fields)

* First Name:	Katelyn
* Last Name:	Jones
Gender:	
Employer:	
* Address:	208 Fern Rd.
* City:	Ashborn
State, Zip:	VA , 20147
* Primary Phone:	(703) 270 - 3896
Alternate Phone:	( ) -
Email:	jonesey@aol.com

* Role(s) for this person:
- [x] Primary Guest
- [ ] Volunteer
- [ ] Manager

Prior Bookings (date checked out):
Feb 11, 2016
Feb 24, 2016

Create new booking

Manager's Notes:

Hit Submit to submit these edits.

Check this box and then hit Delete to delete this entry.

FIGURE 7.3 The Person Edit view in *Homeroom*.

The *model* that underlies the view shown in Figure 7.3 includes all the active variables in the user's session, all the contents of the dbBookings and dbPersons database tables, and all the underlying functionality in the classes and modules that can manipulate that data.

The *view* shown in Figure 7.3 is rendered by code that is activated by the link `personEdit.php?id=Katelyn7032703896`. That is, the view is rendered by the personEdit.php, personForm.inc, and styles.css modules using the GET variable `id=Katelyn7032703896`.

The *controller* for this view is shared between the personEdit.php and personValidate.inc modules. They handle GET and POST information and use it to update the active variables and database tables inside the model for this SESSION. The controller also includes code that supports user navigation by transferring control between this view and other views in the *Homeroom* user interface.

For all the details in this example, readers should visit the complete code for these modules in the *Homeroom* code base.

MVC Example 3: Editing a Stop in *Homeplate*

As a third example of an MVC triple, let's look at the *view* that appears when a user edits a particular stop in *Homeplate*, as shown in Figure 7.4.

Food Lion Kitties Xing (1330)

Area: Bluffton
Date: Monday, September 10, 2018

Data Entry

Meat Weight: 15 lbs.
Deli Weight: 15 lbs.
Bakery Weight: 200 lbs.
Grocery Weight: 0 lbs.
Dairy Weight: 0 lbs.
Produce Weight: 500 lbs.

Total Weight: 730 lbs.

Additional notes:

Save *Hit Save to re-total and save these weights and notes.*

Return to Route

FIGURE 7.4 The Stop view in *Homeplate*.

The *model* that underlies the view shown in Figure 7.4 includes all the active variables in the user's session, all the contents of the dbStops database table, and all the underlying functionality in the classes and modules that can manipulate that data.

The *view* shown in Figure 7.4 is rendered by code that is activated by the link `viewStop2.php?client_type=donor&stop_id=18-09-10-SUN....` That is, the view is rendered by the viewStop2.php and styles.css modules using the GET variables `client_type=donor` and `stop_id=18-09-10-SUN....`

The *controller* for this view is also embedded in the viewStop2.php module. It handles GET and POST information and uses it to update the active variables and database tables inside the model for this SESSION. The controller also includes code that supports user navigation by transferring control between this view and other views in the *Homeplate* user interface.

Readers can review the complete code for these modules by visiting `the` *Homeplate* `code base`.

7.1.2 Linkages among MVC triples

To see the connections between individual views within a single use case, one can either read the use case in the design document or exercise the sandbox version to follow the links among the views themselves.

For example, the Shift view in Figure 7.2 shows the link **Back to Calendar**, which will take the user over to the calendar view for a different MVC triple in the *Homebase* user interface. For another example, the Stop view in Figure 7.4 shows the link **Return to Route**, which will take the user over to the route view for a different MVC triple in the *Homeplate* user interface.

There is one high-level MVC triple that controls user navigation among all the use cases. That is the triple underlying the user's main menu view shown in Figure 7.5. Each part of this figure shows one of the main menus in *Homebase*, *Homeroom*, and *Homeplate*, respectively. Users can navigate to begin any particular use case by clicking on that use case's name, such as **guests: add**, in the *Homeroom* main menu.

(a)
home | about | help | calendars: house, guest chef, activity | around the house
master schedule | volunteers: search, add, screenings | reports | logout

(b)
home | about | bookings: pending, search, new
guests: view, search, add | log | data
Room Logs | help | logout

(c)
home | routes | tablets | master schedules | volunteers |
affiliates | areas | donors and recipients | reports | logout

FIGURE 7.5 The main menu views in (a) *Homebase*, (b) *Homeroom*, and (c) *Homeplate*.

The underlying source code for this view in each of these CO-FOSS products is a variant of the same source code for the other two. For example, Figure 7.6 provides a skeleton of that code which appears in the **header.php** script in *Homebase*. That script contains both the view and the controller for the main menu shown in part **(a)** of Figure 7.5.

```
1  //they're logged in and session variables are set.
2  if ($_SESSION['access_level'] == 0) {
3    echo(' <a href="'.$path.'personEdit.php?id='.'new'.
4        '">apply</a>');
5    echo(' | <a href="' . $path . 'logout.php">logout</a><br>');
6  }
7  else {
8    ...
9    if ($_SESSION['access_level'] >= 1) {
10     echo('<a href="' . $path . 'index.php">home</a>');
11     echo(' | <a href="' . $path . 'about.php">about</a>');
12     echo(' | <a href="' . $path . 'help.php?helpPage=' .
13       $current_page . '" target="_BLANK">help</a>');
14     echo(' | calendars: <a href="' . $path .
15       calendar.php?venue='.$_SESSION['venue'].'">house, </a>');
16       ...
17   }
18   if ($_SESSION['access_level'] >= 2) {
19     echo('<br><a href="' . $path . 'viewSchedule.php?venue='.
20       $_SESSION['venue'].'">master schedule</a>');
21     echo(' | volunteers: <a href="' . $path .
22       'personSearch.php">search</a>,
23       <a href="personEdit.php?id=' . 'new' . '">add, </a>
24       <a href="viewScreenings.php?type=new">screenings</a>');
25     echo(' | <a href="' . $path . 'reports.php?venue='.
26       $_SESSION['venue'].'">reports</a>');
27   }
28   echo(' | <a href="' . $path . 'logout.php">logout</a><br>');
29 }
```

FIGURE 7.6 Part of the view and controller for the main menu MVC in *Homebase.*

The underlying logic in Figure 7.6 renders a series of HTML links in the form `<a href= ... >` that, when selected, will take the user to the opening view for one particular use case. For example, line 23 shows `<a href=personEdit.php ... >`, which will take the user to the opening view of the MVC triple related to the use case for adding a volunteer in the dbPersons database table. We will refer back to more of this code in the next section, where we discuss system security.

The MVC design for the main menu in these three CO-FOSS products provides a fine example of code reuse. That is, students who developed *Homeroom* in 2011 simply reused and adapted the main menu code from *Homebase* rather than reinventing it from scratch. Similarly, students developing *Homeplate* in 2012 again reused and adapted the same main menu code. The open source nature of these products facilitates such code reuse.

7.1.3 User-Level Security

As we read in Chapter 6, a software architecture can define the database tables so that they promote data security and integrity. Two user interface elements must also exist to ensure database security:

1. Require users to log in (their encrypted passwords are stored in the database), and
2. Define user access levels and use them to limit access to activities reserved for managers.

User Login and Password Encryption

It is conventional to store a user's password in the database in an encrypted form. The reason for this is that someone else gaining access to the database (independently from the application) can't see the password—that person only sees the encryption. In general, the *encryption principle* is simply that only the password owner knows the unencrypted version of his/her own password.

In *Homebase*, *Homeroom*, and *Homeplate*, each user's password is stored in the dbPerson's table in encrypted form, using the so-called `md5` encryption algorithm.[3] When the user enters his/her password, it is sent to the server, encrypted, and compared with the encrypted value stored in that person's database entry.[4] A match allows the user's login to succeed, while a mismatch prevents the user from seeing any of the other underlying views.

Password verification occurs in essentially the same way for each of *Homebase*, *Homeroom*, and *Homeplate*. The view for password verification for all three is the login form shown in Figure 2.10. The underlying code for the view and controller for password verification in *Homebase* appear in the `login_form.php` module, which is summarized in Figure 7.7. The

[3] In the 10 years since *Homebase* was developed, the `md5` encryption algorithm has been found to be vulnerable to malicious attacks. For a discussion of more modern approaches to password encryption in PHP, see `https://npfi.org/md5-vulnerability/`.

[4] Note the security vulnerability revealed by this approach, since the password is transmitted from the client to the server in plain text before it is encrypted on the server side. To mitigate against this vulnerability, the server can provide transport layer security via SSL or TLS encryption (see `https://en.wikipedia.org/wiki/Transport_Layer_Security`).

complete `login_form.php` module can be found at `https://npfi.org/homebase-code/`.

Figure 7.7 shows how the view and controller work. The view is rendered as a simple HTML table (lines 2-10) with three rows, the Username, the Password, and the Submit button. On lines 12-16, the controller retrieves the user's password and login name from the user's POST, encrypts the password, retrieves the user's record from the database, and compares the password stored for that person in the database with the user's encrypted password. If they are equal, the login succeeds (the controller sets the SESSION access level in lines 17-25)); otherwise it fails with an error message to the user (lines 28-29).

```
// the essential view for the login form
  echo('<p><table><form method="post"><input type="hidden"
        name="_submit_check" value="true">
    <tr><td>Username:</td>
        <td><input type="text" name="user"></td></tr>
    <tr><td>Password:</td><td><input type="password"
        name="pass"></td></tr>
    <tr><td colspan="2" align="center"><input type="submit"
        name="Login" value="Login"></td></tr>
    </table>');
// the essential controller for the login form
  $db_pass = md5($_POST['pass']);
  $db_id = $_POST['user'];
  $person = retrieve_person($db_id);
  if ($person) { //avoids null results
    if ($person->get_password() == $db_pass) {
        $_SESSION['logged_in'] = 1;
        if ($person->get_status() == "applicant")
            $_SESSION['access_level'] = 0;
        else if (in_array('manager', $person->get_type()))
            $_SESSION['access_level'] = 2;
        else
            $_SESSION['access_level'] = 1;
        echo "<script type='text/JavaScript'>
            window.location = 'index.php';</script>";
    }
    else {
        echo('<div align="left"><p class="error">
            Error: invalid username/password...
```

FIGURE 7.7 The View and Controller for the *Homebase* login form.

User Access Levels

Once logged in, a user's session is assigned an appropriate access level that restricts the user from seeing/modifying parts of the database to which they should not have access. In *Homebase*, for example, three different access levels are defined as follows.

Access Level 2 Only persons with type "manager" can access the dbPersons table. Also, only managers have write access to the dbWeeks, dbDates, dbShifts, and dbMasterSchedule tables, which underlie the MVC triples that generate a new calendar week and schedule volunteers for that week.

Access Levels 1 and 2 Only persons with type "volunteer" or "manager" can access the dbWeeks and dbShifts tables. This allows them to view the weekly calendar and update any shift for which they are scheduled. Individual volunteers may also have read/write access to their own entry in the dbPersons table, so that they can update their own profile data when it changes.

Access Level 0 Site visitors have read access only to the volunteer application form, which allows them restricted access to a single row in the dbPersons table (for entering their own profile data). Visitors must login as "guest," giving them access to no other main menu items.

Looking back at Figure 7.6, we can see that the controller for displaying elements of the main user menu for *Homebase* enforces these access levels by showing only those elements that should be visible to persons with a particular access level. For example, lines 18-26 enforce that only users logged in as a manager can see the **master schedule** and **reports** menu items.

Enforcement of Access Levels

Ensuring system security via the global `$_SESSION['access_level']` variable combines well the system-wide convention for requiring all users to log in. Security is further enhanced inside the view and controller of each MVC triple by checking that the access level of the current user is appropriate to the use case being exercised.

For example, a person logged into *Homebase* as a volunteer (access level 1) will never see the option to add a slot or move a shift when viewing the form shown in Figure 7.2, since these options are reserved for managers (access level 2). This is enforced by the code shown in Figure 7.8 line 2, which is part of the overall editShift.php module shown in `the` *Homebase* `code base`.

However, this code ensures that logged-in managers will have the option of adding a new slot or moving a shift to a different time slot. If a manager selects one of these options, either `_submit_add_slot` or `_submit_move_shift` will be added to the `$_POST` array accordingly. These variables are used later to control what action the model will perform in response to that selection.

```
//   ONLY A MANAGER CAN ADD SLOTS OR MOVE A SHIFT
if($_SESSION['access_level']>=2) {
  echo "<tr><td valign=\"top\"><br> "
    .do_slot_num($persons, $shift->num_vacancies()).
    "</td><td>";
  echo ("<form method=\"POST\" style='margin-bottom:0;'>
    <input type='hidden' name='_submit_add_slot' value='1'>
    <br><input type='submit' value='Add a Slot'
    style='width: 150px' name='submit' ></form>");
  echo ("<form method='POST' style='margin-bottom:0;'>
    <input type='hidden' name='_submit_move_shift' value='1'>
    <input type='submit' value='Move this Shift'
    style='width: 150px' name='submit' > </form>");
}
```

FIGURE 7.8 Ensuring security in *Homebase* using `$_POST` and `$_SESSION` variables.

7.1.4 Protection against Outside Attacks

Some users are malicious, in the sense that they intend to gain access to the software and do harm to the database. Here, we discuss two strategies at the user interface level that can defend the software from malicious attacks:

1. avoiding against SQL injection attacks, and
2. protecting against cross-site scripting attacks.

While the following discussion is by no means a complete or thorough treatment of software security, it should make readers aware of the importance of security in their own CO-FOSS projects.

Avoiding SQL Injection Attacks

SQL injection was identified in Chapter 6 as one of the common vulnerabilities of a database. SQL injection is a code injection technique that exploits a security vulnerability to corrupt or destroy information stored in the database. The vulnerability is present when user input is inadequately filtered for string literal escape characters embedded inside database queries.

For example, by reading the `connect` function in Figure 6.5 a malicious hacker immediately learns the name of the database, the root user, and the root password. Thus, if the following SQL query were "injected" into another query, it would remove the entire `homebasedb` database from the server:

```
DROP DATABASE homebasedb;
```

How can SQL code such as this be injected into another query? The main avenue for this activity is the user interface, where a user is either logging in (entering a username and password) or filling out an HTML form. If user input is not properly screened, it can contain an embedded snippet of malicious SQL code. Here's an example:

Consider the following PHP code, which is used in an authentication procedure to validate a user logging in as `userName`:

```
mysqli_query($con, "SELECT * FROM users
    WHERE name = '" . userName . "';")
```

Now suppose that the user logs in by entering `' or '1'='1` as the userName. Now the above `mysql_query` call will actually execute the following query:

```
SELECT * FROM users WHERE name = '' OR '1'='1';
```

This query will retrieve all entries from the users table because its WHERE clause is always true.

Fortunately, measures can be taken to help prevent these kinds of attacks. First, user input data can be filtered by the function `mysql_real_escape_string()`, so that single quotes will be escaped to prevent them from being "injected" into the text of an SQL query. In the above example, the following rewriting of the call will escape all instances of single quotes from the user's input before executing the query.

```
mysqli_query($con, "SELECT * FROM users
    where name='" . mysql_real_escape_string($userName) . "';")
```

Second, the PHP `mysqli_query` function does not allow more than one query to be executed in a single call. Thus, injecting the above DROP query into a `mysql_query` call would not drop the database.

Avoiding Cross-Site Scripting Attacks

Cross-site scripting is another common vulnerability of a software system. This is a technique that injects external information into Web pages viewed by other users. This vulnerability is present when input from an external source is inadequately filtered.

To help prevent such an attack, the code should filter all external data coming in from users. That is, it should assume that all user-entered data are invalid until they can be proven valid. If a user is supplying his/her last name, the code should check that it contains only alphabetic characters, hyphens, and spaces. Examples of this filtering are given in various of our user interface scripts, such as `personEdit.php` in *Homebase*.

In these and other examples, the PHP functions `preg_replace`, `str_replace()`, and `htmlentities()` can be used to help screen user input to prevent cross-site scripting attacks. Several instances of these functions, such as the following, occur in the personEdit.php script in *Homebase*, *Homeroom* and *Homeplate*:

```
$clean_phone1 = preg_replace("/[^0-9]/", "", $phone1);
```

This line removes all non-decimal digits from the user's input of a phone number, leaving only a string of decimal digits.

We conclude this section with an observation. That is, a fundamental distinction of FOSS is that the source code can be read by anyone, friend or foe. Does this make FOSS more likely than proprietary software to be attacked by cross-site scripting or SQL injection?

Definitely not. The entry point for these types of attacks is the user interface. Thus, an attack can be carried out with or without any specific knowledge of the source code, as long as the underlying code itself has security vulnerabilities. For example, `a recent study` of US voting systems' proprietary software found a very large number of security vulnerabilities.

Software customers seem to understand this, since (recall from Chapter 1) open source software is often selected over proprietary software because of its superior security. The argument that, by hiding its source code, proprietary software is inherently more secure is viewed as false by most security experts.

Thus, it is always prudent to understand and use best practices for ensuring security when developing software, whether it is open source or proprietary. Taking defensive measures such as the ones suggested above is not a waste of time for any software developer.

7.2 PRACTICE

In this section, we discuss the details of an MVC approach to user interface development, using examples from our experience with CO-FOSS development. We also discuss the utility of JavaScript and `the jQuery library`[5] as valuable tools for implementing key user interface elements, such as calendar datepickers and scroll bars.

7.2.1 Sessions, Query Strings, and Global Variables

As noted above, separating the activities of several different simultaneous users is aided by the notion of a *session*. Each individual user who logs in to the system initiates a unique *session*. Associated with that user's login is a set of *session variables* stored on the server that distinguishes that user from

[5]jQuery is a widely-used open source JavaScript library designed to simplify client-side scripting of HTML.

all other user sessions that are active at the same time. Here's how it works in PHP.[6]

A session is created by a call to the **session_start()** function. If a session is already active for that user (that is, the user hadn't logged out) when a new page is loaded, then **session_start()** simply resumes that session. When the **session_cache_expire(minutes)** function is called, it extends the time available to the user before the session expires. For example, when:

```
session_start();
session_cache_expire(30);
```

appears at the beginning of a PHP user interface script, it 1) ensures that the session is running and 2) extends the run time for the session by 30 minutes.

During the life of a session, its session variables are stored on the server as a global array named **$_SESSION**. A session is officially terminated when the session˙unset() function is called or the session times out. The **session_unset()** function is called when the user logs out of the system (e.g., see the `logout.php script` in *Homebase*).

Each separate user login is assigned a unique value stored in the session variable `$_SESSION['id']`. Other application-dependent variables can be added to that `$_SESSION` array at the time the user logs in. For example, in *Homebase*, execution of the login_form.php script causes the following session variables to be established in that array.

`$_SESSION['_id']` gives the user's id as stored in the dbPersons database table. A visitor to the site can login as "guest," which becomes his/her id for the duration of that session.

$_SESSION['access_level'] defines this user's level of access to various functions and data.

$_SESSION['f_name'] stores the user's first name.

$_SESSION['l_name'] stores the user's last name.

$_SESSION['venue'] stores the venue for this session (portland or bangor)

$_SESSION['type'] stores the user's type (manager, volunteer, or guest)

User navigation between views in a user interface relies on two primary vehicles: the links in the main menu and other links that are embedded within individual views. The controllers in their respective MVC triples generate these links from their respective scripts, as discussed in Section 7.1.2.

The HTML *form* displays a view that provides the framework for capturing user-entered data for the controller to process. Underlying the form are the so-called `$_GET` array and `$_POST` array. The `$_GET` array provides "input" to the

[6] Python, Ruby, and Java Servlets have similar conventions.

script via the URL that links to that view, and the `$_POST` array's individual variables define the current values entered by the user on that page.

To illustrate these ideas, look again at the shift view displayed in Figure 7.2. That view provides several options for editing a particular shift—adding or removing a volunteer, for example—on a particular day. This view is achieved when the user selects this shift on a calendar display, which activates the script `editShift.php` with $_GET variables `shift=18-09-21:12-3` and `venue=portland`.

The editShift.php script generates this view using these values `18-09-21:12-3` and `portland` along with the data for that shift from the dbShifts table. This allows editShift.php to show the user that the shift has two slots, one vacant and the other filled by the volunteer `Ellen Jones`.

This view also provides several user options: adding a third (vacant) slot, moving the shift to another time slot, generating the shift's sub call list, removing `Ellen Jones` from the shift, assigning another volunteer to the vacant slot, or removing the vacant slot from the shift. Selecting one of these options causes a new query string to be generated and control to pass to a corresponding PHP script to display a new view.

For example, look at the code in Figure 7.9. This code is executed only if the shift has one or more vacancies. In that case, it displays a row in the shift view that allows the user to find volunteers to fill those vacancies, which spans lines 3-23.

The phrase beginning `action="subCallList.php..."` on lines 5-7 in Figure 7.9 is the means for transferring control to the subCallList.php script to display a new view when the user selects the button "Find Volunteers to Fill Vacancies" (lines 3-4). The `$_GET` variables passed to that script are `shift` and `venue`.

Lines 10-22 in Figure 7.9 reveal how, depending on whether the shift already has a sub call list, the text in the button on the right will vary ("View Sub Call List" or "Generate Sub Call List," respectively), as will the value added to the `$_POST` array (`_submit_generate_scl` or `_submit_view_scl` respectively).

This information is all that is needed for the subCallList.php script to display the view that displays the current sub call list for this shift.

7.2.2 Working with Scripts and HTML

The user interface modules in a CO-FOSS product may be the easiest to read and understand, since each MVC triple corresponds to a particular view in the user interface. Moreover, navigation among these triples corresponds strongly with the steps in the use cases that underly the product's design. Reading the user interface code also helps clarify the relationships among the layers in the code – the domain classes, the database modules, and the user interface.

However, mastering the code in a MVC triple presents challenges to new developers who have not seen how three different languages – PHP, JavaScript,

```
//   LIST ALL VOLUNTEERS AVAILABLE FOR THIS SHIFT
if($shift->num_vacancies()>0) {
  echo("<tr><td valign='top'><br> Find Volunteers
      <br> To Fill Vacancies</td><td>
      <form method='POST' action='subCallList.php?
          shift=".$shiftid."&venue=".$venue."'
          style='margin-bottom:0;'>
      <input type='hidden' name='_shiftid'
          value='".$shiftid."'>");
  if(!$shift->has_sub_call_list() &&
      !(select_dbSCL($shift->get_id()) instanceof SCL)) {
      echo "<input type='hidden' name='_submit_generate_scl'
          value='1'><br><input type='submit'
          value='Generate Sub Call List' name='submit'
          style='width: 150px'>";
  }
  else {
      echo "<input type='hidden' name='_submit_view_scl'
         value='1'><br><input type='submit'
         value='View Sub Call List' name='submit'
         style='width: 150px'>";
  }
  echo "</form><br></td></tr>";
}
```

FIGURE 7.9 Controlling navigation using $_POST variables.

and HTML – can be coordinated to render an effective user interface. This discussion is intended to disentangle the details of this coordination beyond what we have shown in the foregoing sections.

Scripting Example 1: Editing a Shift

As an example, let's look at the code shown in Figure 7.10, which provides a summary overview of the entire editShift.php module.

When comparing this with its corresponding view in Figure 7.2, we notice first how the code lines up with that view. Embedding a PHP script within an HTML page requires enclosing it between the tags <?PHP and ?> shown on lines 1 and 45 of Figure 7.10. This PHP script implements the view and controller elements that support user-system interaction when a particular shift is edited.

The elements of an HTML form are rendered in the view in the order in which they appear in the code. The particular form in Figure 7.10 is rendered

```
1  <?PHP
2    $venue = $_GET['venue'];
3    $shiftid=$_GET['shift'] . ":" . $venue;
4    include_once('editShift.inc'); // additional functions
5    if($shiftid=="") { echo "<p>No Shift ID Supplied. </p>"; }
6    else {
7      // use model to retrieve shift from the database
8      $shift=select_dbShifts($shiftid); ...
9      // use controller functions to respond to user's POST
10     if (!process_fill_vacancy($_POST,$shift,$venue) &&
11       !process_add_volunteer($_POST,$shift,$venue) &&
12       !process_move_shift($_POST, $shift) ... {
13       ... // create the view in HTML
14       $persons=$shift->get_persons();
15       echo "<br><table align='center' border='1px'><tr><td><b>"
16         .get_shift_name_from_id($shiftid)."</b></td></tr>";
17       if($_SESSION['access_level']>=2) {
18         echo "<tr><td valign='top'><br> "
19           .do_slot_num($persons, $shift->num_vacancies()).
20           "</td><td>";
21         echo ("<form method='POST' style='margin-bottom:0;'>
22           <input type='hidden' name='_submit_add_slot'
23           value='1'><br><input type='submit' value='Add a Slot'
24           style='width: 150px' name='submit' >
25           </form>"); ...
26         echo "<br></td></tr>";
27       }
28       if($shift->num_vacancies()>0) {
29         echo("<tr><td><br> Find Volunteers</td>
30           <td><form method='POST' action='subCallList.php?
31             shift=".$shiftid."&venue=".$venue."'">
32           <input type='hidden' name='_shiftid'
33             value='".$shiftid."'>");   ...
34         echo "</form><br></td></tr>";
35       }
36       // invoke functions to show the rest of the view
37       echo display_filled_slots($persons);
38       echo display_vacant_slots($shift->num_vacancies());
39       echo "</td></tr></table>";
40       echo "<p align='center'><a href='calendar.php?id=".
41         substr($shiftid,0,8).":".$venue."&edit=true&venue=".
42         $venue."'>Back to Calendar</a>";
43     }
44   }
45 ?>
```

FIGURE 7.10 Excerpts from editShift.php view and controller module.

as an HTML table, which is denoted by the tag pair `<table>` and `</table>` in Figure 7.10 lines 15 and 39. The table has a series of rows and columns, which are denoted by the tag pairs `<tr>` and `</tr>` and `<td>` and `</td>` respectively.

As noted previously, the arrays `$_GET` and `$_POST` contain values used by the controller to share information with the view and the model. For example, the assignment `$shiftid=$_GET['shift']` on line 3 retrieves the id of the particular shift to be displayed in this view, which is the string `'18-09-21:12-3'` stored in the `$_GET` array.

A number of auxiliary functions, such as `process_add_volunteer` on line 11, help implement parts of this view, and thus receive information from the user via the `$_POST` array in order to function properly. These functions are grouped separately in the `editShift.inc` module, and are included in this script by the statement on line 4.

The PHP and HTML code in Figure 7.10 thus combines to render the user's view shown in Figure 7.2.

Scripting Example 2: Managing a Sub Call List

For another example, Figure 7.11 shows some of the subCallList.php module, which contains the view and controller for the MVC triple that manages user interactions with a Sub Call List (SCL). This module is activated whenever the user hits the Generate Sub Call List button in Figure 7.2.

A reading of this code reveals a simple structure, which generates/retrieves a SCL (line 8) and then calls one of two functions depending on what the user has just requested:

1. viewing the shift's SCL (line 19), or
2. saving changes and then viewing the current SCL (lines 16 and 19).

The functions `process_edit_scl` and `view_scl` are also part of the controller for subCallList.php module, but are not shown in Figure 7.11. Either call returns the `$id` of the resulting SCL or not, in which case the program will display either the newly-modified SCL (calling `view_scl` line 18) or a list of active SCLs (calling `do_scl_index` on line 25).

To gain more insight into this code, we should exercise that part of the GUI where the user is working with a Sub Call List. We can start by viewing the form shown in Figure 7.12. This form appears to support only the third user action from the shift view shown in Figure 7.2, where the user makes changes to the current shift's SCL. Notice also, however, that this form provides a link back to that shift's view, so maybe we can understand the other two actions by looking at the shift view itself.

Recalling that the shift view is displayed by the editShift.php module, we see that this Shift's SCL can be viewed by selecting "Generate Sub Call List" in Figure 7.2. Once a Shift's SCL has been generated, it can be later accessed from the same shift view, but will now carry the message "View Sub Call List" instead.

```
1  <?PHP
2  include_once('database/dbSCL.php');
3  include_once('database/dbShifts.php');
4  include_once('database/dbLog.php');
5  include_once('database/dbPersons.php');
6  $id=$_GET['shift'];
7  $venue=$_GET['venue'];
8  generate_scl($id);  //creates a sub call list for this shift id
9  if(array_key_exists('_submit_generate_scl',$_POST)) {
10     $id=$_POST['_shiftid'];
11 }
12 else if (array_key_exists('_submit_view_scl',$_POST)) {
13     $id=$_POST['_shiftid'];
14 }
15 else if(array_key_exists('_submit_save_scl_changes',$_POST)) {
16     $id=process_edit_scl($_POST);
17 }
18 if($id) {
19     $id=view_scl($id,$venue);
20 else {
21     // The first 8 characters show the date
22     $yy_mm_dd = substr($_GET['shift'], 0, 8);
23     // Displays the option of going back to the Calendar.
24     back_to_calendar($yy_mm_dd, 700, $venue);
25     do_scl_index($id,$venue);
26 }
27 ?>
```

FIGURE 7.11 Underlying view and controller for managing a SubCallList.

So this little investigation of the code and its run-time behavior provides both a better understanding of the subCallList.php code and a starting point for exploring other views and controllers in the shift-editing process.

7.2.3 Reading Deeply

The above exercise with the editShift.php and subCallList.php code reveals some information about how the user interface relates to specific MVC triples. However, it does not reveal how specific functions in the underlying domain classes and database modules support a user activity.

To begin understanding these deeper relationships that occur during a user session, we can always start with a specific user action and explore how the user interface, domain classes, and database modules work together to

Back to Shift

Sub Call List for Portland House Shift: Friday September 28, 2018 9am to 12pm

1 sub needed for this shift.

Name	Phone	Date Called	Notes	Accepted
Mary Jones	7036423647			-
Nancy Jones	7032210332			-
Sarah Jones	7035425380			-
Star Jones	7036536759			-
Suzanne Jones	7037018778			-

Assign Subs / Save Changes

FIGURE 7.12 Using the SubCallList form.

implement that action. To illustrate, let's take another look at the Shift view for September 21, 2018 12-3pm, shown in Figure 7.2.

Let's assume we want to remove "Ellen Jones" from that shift, creating a second vacancy. If we are running *Homebase*, the URL will show us that the `editShift.php` module is active at the time this view appears in the browser, along with the $`GET variables identifying the shift's id and venue. So if we select "Remove Person" beside the "Ellen Jones" entry on this form, the form reappears with two vacancies rather than one. What happened to cause this?

When we look at the entire editShift.php code that underlies Figure 7.10, we see a large module, though it appears to be well organized. It has the following major sections:

A main section, which is a block of HTML code that presents the view of the Shift form (see Figure 7.10).

A collection of controller functions (see the auxiliary module `editShift.inc` for details)—`do_slot_num`, `display_filled_slots`, and `display_vacant_slots`—that assist with the display of a shift.

A series of additional controller functions (again, see `editShift.inc` for details) that govern the management of vacancies in a shift—`process_fill_vacancy`, `process_unfill_shift`, `process_clear_shift`, `process_add_slot`, `process_move_shift`, and `process_ignore_slot`. Notice that each of these functions returns `true` or `false` depending on whether it succeeds or fails. Success requires the presence of a user request to perform that action and a successful completion of that action.

There are more controller functions (again, see editShift.inc for details) that assist with the identification and selection of a volunteer to fill a vacancy—get_available_volunteer_options, get_all_volunteer_options, and process_add_volunteer.

A helper function, fix_SCL, brings a shift's SCL into agreement with its number of vacancies after a vacancy has been filled or created.

Notice also that editShift.inc uses elements of the underlying model (especially dbShifts.php, dbPersons.php, dbSCL.php, and dbLog.php) as resources for carrying out its various view and controller functions.

So removing "Ellen Jones" from the September 21, 2018 12-3pm shift is triggered in the main section of editShift.php, where the following code determines whether or not the user has requested that action:

```
if (process_unfill_shift($_POST,$shift,$venue))
    $shift=select_dbShifts($shiftid);
else if (process_add_slot($_POST,$shift,$venue))
    $shift=select_dbShifts($shiftid);
else if (process_clear_shift($_POST,$shift,$venue))
    $shift = select_dbShifts($shiftid);
else if (process_ignore_slot($_POST,$shift,$venue))
    $shift=select_dbShifts($shiftid);
```

This code makes four tests. The first test calls process_unfill_shift to determine whether or not the user has requested removal of a person from a slot in this shift and, if so, then performs that removal. The arguments $_POST, $shift, and $venue are passed in this call.

The first argument is an array of current user actions, including the action '_submit_filled_slot_0', and the second is the Shift object itself, which includes the id = "18-09-21:12-3". The persons scheduled for a shift are stored as an array, so "Ellen Jones" occupies slot 0 in that array.

Figure 7.13 shows a code snippet from the function process_unfill_shift where the removal of "Ellen Jones" actually takes place. This code's outer loop is searching through the array of slots in the shift to find out if the user has selected any person for removal (lines 3-17). Since "Ellen Jones" is in slot 0, _submit_filled_slot_0 should appear in the $_POST array.

At that point, the inner loop (lines 6-11) builds the new array $p2 that contains everyone who had been in the shift except "Ellen Jones." Finally, this new array is reassigned to the $shift (line 12) and the shift's vacancy count is incremented by 1 (line 13). This change is permanently recorded in the dbShifts database table (line 14) before the function returns true.

To understand the underlying model for this MVC triple, we can dig further into the dbShifts.php module in the database layer, where the update_dbShifts function resides. In the spirit of our discussions in Chapter 6, Table 7.1 summarizes the major CRUD functions in dbShifts.php.

```
update_dbShifts($shift);
$persons=$shift->get_persons();$p2 = array();
for($i=0;$i<count($persons);++$i) {
    $p2=array();
    if(array_key_exists('_submit_filled_slot_'.$i, $post)) {
        for($j=0;$j<count($persons);++$j) {
            if($i!=$j)
                $p2[]=$persons[$j];
            else
                $name=$persons[$j];
        }
        $shift->assign_persons($p2);
        $shift->add_vacancy();
        update_dbShifts($shift);
        ...
        return true;
    }
}
return false;
```

FIGURE 7.13 Code snippet for removing a person from a Shift.

What happens when an individual row in the dbShifts database table is updated? The code for `update_dbShifts` is deceptively simple, having a pair of calls that removes the row whose key matches that of shift $s and then inserts a new row using $s again:

```
delete_dbShifts($s);
insert_dbShifts($s);
```

So to get to the bottom of an update, we need to examine the code for `delete_dbShifts` and `insert_dbShifts` in succession. Let's persist with this

TABLE 7.1 CRUD Functions in the dbShifts module

Function	Purpose
`insert_dbShifts($s)`	Inserts Shift $s into dbShifts as a new row
`select_dbShifts($id)`	Returns the Shift with id = $id from dbShifts
`update_dbShifts($s)`	Replaces the row corresponding to the id of Shift $s by $s itself
`delete_dbShifts($s)`	Removes the row corresponding to Shift $s from dbShifts

example because it reflects some principles that are useful when evaluating other MVC triples.

First, the code for **delete_dbShifts** has the essential steps shown in Figure 7.14. There, line 1 establishes a connection to the database, while lines 2 and 3 delete a row from the dbShifts table whose id matches the id in Shift **$s**. Lines 4-8 check to see if this deletion has been completed successfully. A **false** result indicates failure to delete the row.

```
1  $con=connect();
2  $query = "DELETE FROM dbShifts WHERE id=\"".$s->get_id()."\"";
3  $result = mysqli_query($con,$query);
4  if (!$result) {
5      echo "unable to delete from dbShifts " .
6           $s->get_id() . mysqli_error($con);
7      mysqli_close($con);
8      return false;
9  }
10 mysqli_close($con);
11 return true;
```

FIGURE 7.14 Essential steps for deleting a Shift from the dbShifts table.

Second, the code for **insert_dbShifts** has the essential steps shown in Figure 7.15. Insertion of a new row with `id == $s->get_id()` is done on lines 2-8, and then a check is made on lines 9-13 to be sure that the insertion was made. Lines 1 and 14 open and close the connection to the database in the same way that was done in lines 1 and 10 of the previous example.

So this exercise has examined in detail the code underlying the MVC triple where a user removes a person from a shift. The triple includes, along with the dbShifts database table, modules from all three layers: editShift.php and editShift.inc from the user interface layer, Shift.php from the domain layer, and dbShifts.php from the database layer.

Important to this exercise is the disciplined use of a consistent module-naming convention, which enables easy retrieval of all the modules in a particular MVC triple for a particular use case, such as editing a shift. This discipline becomes especially important when we need to dig through the code to fix errors, add new features, or make the code available for others to reuse.

7.2.4 Using JavaScript and jQuery UI to Improve the User Interface

In 2008, our first CO-FOSS project teams used PHP, MySQL, and HTML as the programming tools for developing user interfaces. Later teams discov-

```
1   $con=connect();
2   $query = "INSERT INTO dbShifts VALUES ('" . $s->get_id()
3       .  "','" .$s->get_start_time() ."','" .$s->get_end_time()
4       . "','" .$s->get_venue() ."'," .$s->num_vacancies() . ",'"
5       . implode("*", $s->get_persons()) . "','"
6       . implode("*", $s->get_removed_persons()) . "','"
7       . $s->get_sub_call_list() ."','" . $s->get_notes() ."')";
8   $result = mysqli_query($con,$query);
9   if (!$result) {
10      echo "unable to insert into dbShifts " .
11          $s->get_id() . mysqli_error($con);
12      mysqli_close($con);
13      return false;
14  }
15  mysqli_close($con);
16  return true;
```

FIGURE 7.15 Essential steps for inserting a Shift into the dbShifts table.

ered the beauty of embedding JavaScript and jQuery elements to simplify the coding and create cleaner and more functional user interfaces.[7]

`JavaScript` was invented in 1995 as a programming language for use in combination with HTML to create dynamic and interactive client-side web applications. JavaScript was soon extended to support server-side applications and has been used by Microsoft to support its ASP and .NET Internet applications. Since the mid-2000s, other server-side JavaScript implementations have been introduced, such as `Node.js` in 2009.

`jQuery` is a JavaScript library invented in 2006 to simplify the client-side scripting of HTML pages. It is free open source software, and is the most widely-deployed JavaScript library among all Web applications. jQuery is supported on all major browsers.

`jQuery UI` is a JavaScript library containing a set of user interface elements built on top of the jQuery library. For example, the `jQuery UI library` includes a calendar datepicker, a menu, a progress bar, and many other pre-packaged elements that commonly appear in Web pages but would be cumbersome to program using straight PHP, HTML, and CSS alone.

A good example of jQuery UI's utility is reflected in the user's experience when entering a calendar date into an HTML form within a particular MVC triple's view. In 2008 and 2011, before jQuery UI had become well-known,

[7]Today, if we were to begin a CO-FOSS project from scratch, we might make different choices altogether. Choosing a different language and platform for a new CO-FOSS project must always be balanced against the availability and value of reusing legacy code. This choice is not always simple.

our *Homebase* and *Homeroom* developers tediously coded the following user's form for entering a calendar date:

A sketch of the code for this form is shown in Figure 7.16. As shown, the entire date is rendered using two HTML selects and a text input box.

```
<?php
   echo '<select name = "patient_birth_month">';
   $months = array("January","February","March","April",
     "May","June","July","August",
     "September","October","November","December");
   echo("<option> </option>");
   for ($i = 1 ; $i <= 12 ; $i ++)
     echo ("<option value=".$i.">".$months[$i-1]."</option>");
   echo '</select>';
   echo '<select name = "patient_birth_day">';
   echo("<option>  </option>");
   for ($i = 1; $i <= 31 ; $i ++)
     echo ("<option>".$i."</option>");
   echo '</select>';
   echo("<input type='text' size ='6' maxLength = '4'
     name='patient_birth_year' value='".date('Y')."'/><br />");
?>
```

FIGURE 7.16 Coding calendar date using HTML selects.

Homebase and *Homeroom* were later updated by other student teams, who replaced the code in Figure 7.16 by a jQuery UI widget called a "datepicker," which renders a user form like the one shown below:

Not only does this form offer a better user experience, but it also greatly simplifies the underlying code, as shown in Figure 7.17. Note here that the date is received from the user as a single string, properly formatted for entry into the database without any need for further coding.

```
<script>
$(function() {
    $( "#patient_birthdate" ).datepicker();
});
</script>
...
$formattedDate = date("F j, Y",strtotime($patient_birthdate));
echo '<input type="text" id="patient_birthdate"
    name="patient_birthdate"
    value="'. $formattedDate.'" size="15" />';
```

FIGURE 7.17 Coding calendar date using a jQuery UI datepicker widget.

Since the jQuery UI tools are open source, they can be freely downloaded and used in any CO-FOSS application. When this is done, an enormous amount of programming and testing tedium can be saved, and the quality of the resulting user interface can be substantially improved. For detailed guidance on the jQuery UI, please visit `https://npfi.org/jquery-tutorial/`.

7.2.5 Responsive User Interfaces

Recent years have seen the rapid evolution of Web-based software that can be effectively accessed and used on either a desktop, a laptop, a mobile phone, or a tablet interchangeably. This genre of software applications is called **responsive** because its user interface adapts gracefully to the particular device from which the user is accessing it, as shown in Figure 7.18.

The desktop/laptop picture in Figure 7.18 shows a single page of a user interface divided into four parts; a gray main menu at the top, a black content area in the middle, and two other gray sidebars on the left and right of the content area. When this page is adapted for viewing on a tablet or phone, it must become scrollable, its left and right sidebars must be moved to the bottom of the scrolling area, and its menu must be compressed to maintain its functionality. A responsive application will perform these adaptations automatically, depending on the type of device on which the user is viewing it. By contrast, a non-responsive application will present the page the same way on all types of devices, leaving the tablet or phone user with a dysfunctional page most of the time.

Making applications responsive makes especially good sense when they are used “in the field” as well as in the office. Such applications need to support different types of use by different users in a variety of locations. Responsiveness is also consistent with the idea of client-server architecture, since it supports

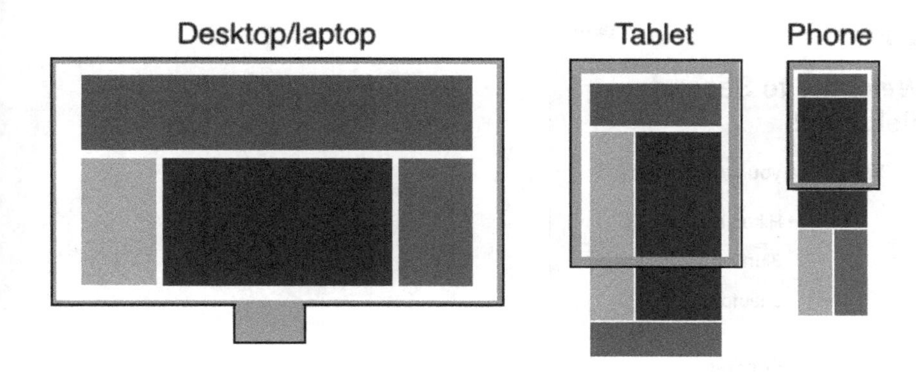

FIGURE 7.18 A responsive user interface.

several simultaneous users gracefully accessing a shared database over the Internet from several different locations.

Responsive applicatons have been available to a limited extent on mobile devices (tablets and phones) for several years. These applications are traditionally implemented for a particular mobile platform, such as IOS (for iPhones and iPads) or Android Java (for Android tablets and smartphones). They are downloaded from a mobile app library, such as Apple's App Store or Google's Play Store – some are free and others are paid.

However, these applications do not typically run on client-server platforms by way of a Web browser. Instead, they have stand-alone user interfaces and may or may not have direct database access via a Wi-fi connection to the Web. For example, in the *Homeplate* project, the application *Homeplate Mobile* runs on an Android tablet specially designed for use by volunteers working on a food rescue truck for Second Helpings. *Homeplate Mobile* has been running effectively ever since it was developed in 2012. The home screen for *Homeplate Mobile* is shown in Figure 7.19.

Second Helpings could replace *Homeplate Mobile* by a Web app with a responsive user interface that runs on the tablet using a Web browser. They are considering this change for three main reasons:

1. At least once a week, volunteer drivers need to find a wi-fi hotspot in order to upload their data to the *Homeplate* server and download new assigned routes for the upcoming week. This is often not convenient for the drivers.

2. While the *Homeplate Mobile* user interface is easy to use, it only runs on an Android tablet registered to Second Helpings. It would be much more convenient for drivers to access this application using their own smartphones or tablets, whether they be iPhones, iPads, or Android phones or tablets.

FIGURE 7.19 The *Homeplate Mobile* home screen.

3. *Homeplate Mobile*'s user interface is not responsive. It is designed for a 10" Android tablet, so it cannot even be fully viewed on a 7" tablet, let alone a smartphone.

For these reasons, a few drivers have begun to record their data on a clipboard, which must later be entered manually into the *Homeplate* user interface on the server. This is regressing back toward the days before *Homeplate*, where manual data entry was the only way to keep records on the truck.

But *Homeplate* is a client-server application whose database and reporting functions still work well. So developers are considering a redesign for *Homeplate Mobile* so that it will run in a Web browser on any mobile device. A responsive redesign would solve all three issues listed above.

Responsive user interface design

Designing a responsive user interface may be challenging for a layered architecture, since the responsive user interface layer must be developed on top of an existing domain and database layer (the *model*, in MVC terms). Moreover, if we think of the user interface as a collection of MVC triples, we need to focus on redesigning the respective views in these triples. But since the views are often intertwined into the same module as their controllers, the redesign is not by any means trivial.

Looking back at Figure 7.18, it should be apparent that the amount of information that can be presented in the view for a cell phone user is much smaller than what can be presented in the same view for a tablet user or a desktop user. So, although it is tempting to imagine creating a responsive user interface by just shrinking each view in the desktop version to fit onto the screen of a tablet or a cell phone, that would be a simplistic approach.

For example, Figure 7.19 shows us that a view on a mobile app provides less information, and uses a larger font, than the same view on a laptop or desktop. Moreover, the user navigation options are much simpler on a mobile device, and the navigation buttons themselves are large enough to be selected by a normal one-finger touch (not a mouse click). Pop-up screen-based keyboards should also be provided whenever the view calls for textual user input.

For more complete examples of responsive user interfaces, we know of two different CO-FOSS applications that were developed by computer science students and an instructor at Dickinson College. These are called *FarmData* and *AnimalData*, and they were developed to support mission-critical needs of a client that is the `Dickinson College Farm`. From its `Sourceforge site`, FarmData is described as follows:

> FarmData is a web-based database system for entering and reporting crop production records, including seeding, transplanting, harvest, cover crop, compost, fertilization, irrigation, pest scouting, spray activities, packing and distribution records and customer invoicing. The FARMDATA portal, accessible by both smartphone and desktop computer, uses "smart forms" that minimize incorrect records and save farmer time by making calculations and remembering important data such as field sizes, spray and seeding rates.

The current version of FarmData was deployed in 2016 and is now in productive use not only at the Dickinson College Farm but also at several other farms. To log in as a guest and experience the responsiveness of the FarmData user interface, readers can visit `https://farmdata.dickinson.edu/guest.php`. AnimalData is designed with the same philosophy, but it applies to the management of farm animals rather than crops. AnimalData is in an earlier stage of development, so at this time only an `experimental version` is in use at the Farm.

One way to design a responsive user interface is to cast each view as a nest of HTML $< div >$s and $< ul >$s that expands to fill the entire screen. In case the view is deeper than the screen size, the table's rows are vertically scrollable. Also, each row's column entries equally divide the screen width, so that each entry is as wide as the others in that row. An example of this layout is shown in Figure 7.20, using the experimental version of AnimalData.

The MVC triple underlying this view is a collection of HTML, JavaScript, and PHP, and the SQL database is managed using the PDO abstraction layer, as discussed in Section 6.2.6. The scrollbar on the right side of Figure 7.20 allows users with smartphones to see the entire page by simply scrolling. Each row is proportionately sized according to its number of entries. Labels on rows,

FIGURE 7.20 A responsive user interface view.

buttons, and other widgets carry large fonts for ease of reading on smaller mobile devices.

The HTML code underlying the first two lines below the FEED PURCHASE INPUT FORM title shown in Figure 7.20 is shown in Figure 7.21. The first half of the code in Figure 7.21 renders a date and the second half renders a select labeled **Feed Major Type**. The model and controller for this view are in separate modules.

Finally, we note that a number of tools that support responsive user interface design have recently emerged, and some of them are open source. We are not familiar enough with these tools to make an informed recommendation at this time. However, interested readers should search the Internet actively, since the development of responsive user interfaces is a rapidly-emerging field.

7.3 TESTING, DEBUGGING, AND REFACTORING

Testing the user interface in a CO-FOSS project is challenging. Each MVC triple, such as the ones shown in Section 7.1.1, has a small group of 2-3 modules that underlie a specific view and its related controller. Those modules combine to present that view, process user inputs for that view, and update the database from those inputs.

```
<div class="ui-field-contain">
<label for="feed_input_date">Date:</label>
<fieldset class="ui-grid-b" id="feed_input_date">
</fieldset>
</div>
<div class="ui-field-contain">
<label for="feed_input_type">Feed Major Type:</label>
<select name="feed_input_type" id="feed_input_type"
    onchange="update_feed_subtype('feed_input', true, false);">
</select>
</div>
```

FIGURE 7.21 HTML code underlying part of the view in Figure 7.20.

When we tested the underlying domain classes and database modules, we designed a unit test for each module separately, and then integration-tested those modules that interact with each other. We could continue that strategy by developing unit tests for each MVC triple that underlies each particular view in the user interface. For example, we could develop a unit test for the MVC triple that underlies the *Homebase* shift view shown in Figure 7.2.

Following that, we could fully test each use case by designing a unit test that covers the interactions among its three or four different views. The question that test would address is, "Do those interactions correctly and fully support a user executing all the steps in that use case?"

Taken together, our complete set of unit tests for all the use cases would provide a degree of "integration testing" or "acceptance testing" for the entire system in the eyes of the user. Implicit in this strategy is the assumption that unit testing the use cases takes place only after the unit and integration testing of all the domain classes and database modules has been completed.

However, trying to develop unit tests for a complete user interface is tedious business. As we have seen, the view and controller modules in each MVC triple contain a mix of PHP, HTML, CSS, and JavaScript code that creates the view and controls the system's responses to each user action. So each triple's accompanying unit test would need to contain assertions that validate the underlying effect of performing each user action allowed by its view.

Because of the inherent complexity of designing and maintaining effective user interface tests, the idea of "automated user interface testing" has received quite a bit of attention in recent years. Several `automated testing tools`, such as `Selenium`, have been developed as well. However, the whole idea of automated user interface testing is more suitable for more complex and mature software applications than those addressed by our CO-FOSS projects.

In our CO-FOSS projects, we use a simpler, yet effective, approach to testing the user interface. It is effective in the sense that it ensures that the user interface works well and is relatively bug-free by the end of the project. Our approach asks the client to exercise each view that appears in the new user interface. Guided by the requirements document, the client can corroborate that all the steps in each use case are covered and navigation among the steps is cleanly and effectively implemented.

For example, to test the view shown in Figure 7.2 the client would exercise each of the user actions and observe the result, asking the question, "Does the result reflect correct fulfillment of that action?" Since clients have a vested interest in the software working well for them, they are happy (if not excited!) to try out the new pages and critically share their experiences with the developers. The issue tracking activity introduced in Section 6.4 provides the primary medium for client-developer communication, not only for reporting bugs but also for making other suggestions about the details of the user interface.

The discussions below are based on our experiences testing various user interface elements in our own CO-FOSS projects with our clients. The goal for our tests is to ensure that each MVC triple and its related user interface form delivers the functionality required by the client and its use case. By combining these results with those from prior unit and integration testing of the domain classes and database modules, the entire development team, especially the client, can gain a high level of confidence in the quality and usability of the entire CO-FOSS product.

7.3.1 Testing a User Interface

Let's begin by looking at the *Homebase* `code base`. Its `tests` directory already has unit and integration tests for all the domain classes and database modules (as guided by Chapters 5 and 6). Now we wish to complete the testing for the entire CO-FOSS product by actively engaging the clients with all the views in the user interface and responding to their feedback by fixing bugs and cleaning up other usability issues.

This activity is organized around use cases. For an example, consider **Use Case 4: Editing the Calendar** in the *Homebase* `design document`. The key views underlying that use case are shown in Figures 17-20 of that document. The *Homebase* code has three underlying MVC triples that support these views. These are listed in Table 7.2, along with their respective view and controller modules:

TABLE 7.2 The three views in the Editing the Calendar use case

MVC triple	View and controller modules
Calendar (Figures 17 and 18)	calendar.css, calendar.inc, calendar.php
Edit Shift (Figure 19)	editShift.inc and editShift.php
Sub Call List (Figure 20)	subCallList.php

Key to effective user interface testing is the existence of database tables with sufficient data that will support each of the use cases being tested. For example, testing Use Case 4 in *Homebase* requires that the database contains enough scheduling information to populate an entire calendar week in the Portland venue.

To test this use case, the user should separately test each of these views, as well as the transitions among them. For example, to test the Edit Shift view, the user should exercise all the buttons and other options on the form shown in Figure 7.2: "Add a Slot," "Move this Shift," "Generate Sub Call List," "Remove Person," "Assign Volunteer," "Remove Vacancy," and "Back to Calendar."

Each of the options in the other two views should be similarly tested. In addition, all the interactions among the three views should be tested, such as the link **Back to Calendar** at the bottom of the Edit Shift view shown in Figure 7.2.

We also need to test the links between the views. For example, the Edit Shift view in *Homebase* Use Case 4 has the link "Back to Calendar" in Figure 7.2. When this link is selected, the user should be sent over to the Calendar view that appears in Figure 7.22.

(view this week) (manage weeks) Next Week >>

Portland House Calendar: September 17, 2018 to September 23, 2018 (week 37: odd)

	17 Monday	18 Tuesday	19 Wednesday	20 Thursday	21 Friday	22 Saturday	23 Sunday	
9am 10am 11am	Jane Jones Cathy Jones Cheryl Jones	Stacey Jones	Aynne Jones Charlie Jones	Cathy Jones Meg Jones	Sally Jones Becky Jones	**Vacancies (1)**	Gaye Jones	9am 10am 11am
12pm 1pm 2pm	Cheryl Jones **Vacancies (1)**	Cindy Jones **Vacancies (1)**	John Jones **Vacancies (1)**	Marjorie Jones **Vacancies (1)**	Ellen Jones **Vacancies (1)**		**Vacancies (1)**	12pm 1pm 2pm
3pm 4pm 5pm	Robin Jones Claire Jones	Betsy Jones	Amy Jones Ann Jones	Nancy Jones Suzanne Jones	Elaine Jones **Vacancies (1)**		**Vacancies (1)**	3pm 4pm 5pm
6pm 7pm 8pm - 9pm	Nonie Jones **Vacancies (1)**	Kara Jones Daniel Jones	Marilee Jones Claudia Jones	Jody Jones Allyson Jones				6pm 7pm 8pm - 9pm
night					**Vacancies (1)**	**Vacancies (1)**		night
manager notes								manager notes

Save changes to all notes

FIGURE 7.22 The Calendar view inside *Homebase* Use Case 4.

The number of user options here is large, but most of them are navigational in nature. Moreover, each shift box on the calendar is a logical copy of all the other shift boxes, so testing the links in one or two is a sufficient shortcut for testing the links in all of them. For example, we can check that navigating to the view in Figure 7.22 from the view in Figure 7.2 by hitting

Back to Calendar, shows that the Friday September 18 12-3 shift is populated identically in both views.

The Calendar view itself also allows the user to navigate among the following five activities:

1. Editing a shift (discussed above)
2. Adding or changing notes for a shift
3. Adding or changing manager notes for an entire day
4. Navigating to the next week's Calendar view
5. Adding new weeks to the calendar (this is a different use case)

To test these options, the user must select each link and confirm that its suggested navigation is actually accomplished. That is:

1. To edit a shift on the Calendar view, the user must select that shift, which should transfer over to the Edit Shift view for that shift, as shown in Figure 7.2.
2. To add or change notes for a shift, the user edits the shift's notes and then selects the "Save changes to all notes" button at the bottom of the Calendar view. The calendar should now show the changes.
3. To add or change manager notes for an entire day, the user types into the manager notes line and then selects the **Save changes to all notes** button at the bottom of the Calendar view. The calendar should now show the changes.
4. To navigate to the next week's Calendar view, the user should select **Next Week** at the top and observe that the next week's Calendar view replaces the current one.
5. To add a new week to the calendar, the user should select "manage weeks" at the top and confirm that a new view appears that initiates a different use case (not discussed here).

This is tedious and repetitive work, but it is the only practical way that users can validate the effectiveness of the new software and report issues. The added benefit here is that users are gaining practical experience working with the new software and a new level of understanding about both its strengths and its limitations.

Organizing the Testing Process

The above discussion reveals two key points to consider when organizing the testing process for the user interface.

1. Creating a sandbox database, and
2. Sequencing the testing of use cases.

First, preparation for testing the user interface requires creating a rather robust database before clients can begin to test the use cases. In our projects, we have relied on the instructor creating a sandbox database in advance, as described in Section 2.5. An example sandbox setup for *Homebase* can be found at `its GitHub site`. The sandbox database iitself can be prepared by extracting a small sample of "live" data from the client's current system; that data may be in spreadsheet form or in an existing database.

When doing this, developers must be careful to anonymize personal and sensitive organizational data in the interest of protecting privacy. In the *Homebase* sandbox database mentioned above, for example, we have anonymized the names, addresses, phones, and emails of all persons in the database so that real persons cannot be identified by this information. This is an easy step that can be done off-line using stylized SQL queries in a global replace fashion.

Second, the sequence in which clients test the different use cases should reflect certain global interdependencies among the use cases. For instance, in *Homebase* we should not begin testing **Use Case 4: Editing the Calendar** until we have a fully-populated calendar week for testing. But populating a calendar week is done by running **Use Case 3: Generate new calendar weeks** (see the *Homebase* `design document` again). So it makes practical sense for the client to test Use Case 3 before testing Use Case 4. Sequencing of other use case tests may also be governed by these types of interdependencies.

7.3.2 Refactoring: Removing a Layering Violation

In Chapter 5, we introduced the idea of refactoring the source code to improve its organization, clarity, and functionality. Now that we are testing and debugging the user interface for our CO-FOSS project, we should find new opportunities for refactoring. In this section, we discuss a refactoring opportunity that appeared in *Homebase*, one that could only have been discovered during the user interface testing stage of the project.

At this stage, the three layers of code in our project are fully populated, and the domain, database, and user interface modules are clearly delineated. The user interface layer should have only the modules that drive the views and controllers for individual MVC triples, such as editShift.php and editShift.inc discussed in Section 7.3.1.

Chapter 2 introduced the *layering principle* of software architecture, which suggests that no user interface code (view or controller module) should creep into the domain or database layer. It also suggests that no domain class definitions or direct database table querying code should appear in any view or controller module in the user interface layer. If any such code appears, we have a layering violation, which in turn provides an opportunity for refactoring.

```
1  function generate_scl($id) {
2    $shift=select_dbShifts($id);
3    ...
4    $con=connect();
5    $query="SELECT * FROM dbPersons WHERE
6      (status = 'active' AND type LIKE '%sub%'
7        AND availability LIKE
8          '%".$day.":".$time.":".$venue."%')
9        ORDER BY last_name,first_name";
10   $persons_result=mysqli_query($con,$query);
11   mysqli_close($con);
12   for($i=0;$i<mysqli_num_rows($persons_result);++$i) {
13     $row=mysqli_fetch_row($persons_result);
14     $id_and_name=$row[0]."+".$row[3]."+".$row[4];
15     if (!in_array($id_and_name, $shift_persons) &&
16       !in_array($id_and_name, $shift->get_removed_persons())) {
17         $persons[] = array($row[0], $row[3], $row[4],
18         $row[9]." ".$row[10],$row[11]." ".$row[12],"","","?");
19     }
20   }
21   $new_scl=new SCL($id, $persons, "open", $vacancies,
22   get_sub_call_list_timestamp($id));
23   ...
```

FIGURE 7.23 Layering Violation: a user interface module directly querying the database.

A review of the *Homebase* code base uncovers a couple of layering violations that were left undetected by its developers. One of these appears in the `generate_scl` function, which is embedded in the subCallList.php module (see Figure 7.11). The offending code in that function is shown in Figure 7.23. This code builds an array of all active persons who have "sub" in their type, and are available for a particular day, time, and venue. Out of this list, the code goes on to build a sub call list out of all these persons. All well and good. The code works, right?

Wrong. This is poor code in two respects. First, its layering violation appears as a direct query into the dbPersons database table (lines 4-11) – such queries should all be housed in the appropriate database module (dbPersons.php in this case). Second, it has some bad smells in the way it builds the sub call list itself (lines 12-20).

To remove the layering violation, we need to replace that code by a call to an appropriate function in the dbPersons.php module that accomplishes this same task. The task is to find all persons in the database with type "sub," active status, and availability for the same day, shift, and venue as the current shift (given by the $id parameter).

So we should first ask, "Is there a function already defined in the dbPersons.php module that can deliver this result?" Looking at that module, we see two candidate functions that do this sort of thing:

`getall_available($type,$day,$shift,$venue)` Find all persons with $type who are available on a given $day, $shift, and $venue.

`getonlythose_dbPersons($type,$status,$name,$day,$shift,$venue)`

Find all persons with a given $type and $status and have names like $name who are available on a given $day, $shift, and $venue.

The second seems better matched with our needs than the first, since it will filter for `$status = 'active'` and will not filter by name if we set `$name = ''` in the call. So, returning to the code in Figure 7.23, we can replace all of lines 4-11 by the single line:

```
$new_persons = getonlythose_dbPersons('sub', 'active', '',
     $day, $time, $venue);
```

This brings us to dealing with the bad smells in lines 12-20 in Figure 7.23. That code works with the array `$persons_result` that has one row for each person retrieved from the database with the required availability. In turn, each row is itself an integer-indexed array of features for a single person: $row[0] is the person's id, $row[3] is that person's first name, $row[4] is that person's last name, and so on. Are you picking up the smell yet?

The rest of lines 12-20 do some additional filtering and then construct a new array called $persons (line 17), which contains only those persons and features needed to pass to the constructor for a $new˙sub˙call˙list in line 21. Those features are basically the person's id, first name, last name, primary phone, and secondary phone.

But here's the good news! The new function call that we used to replace lines 4-11 returns an array of Person objects. So rather than using the smelly integer-indexed entries in the $row array, we can use a Person's getter functions to access these individual features. Cutting to the chase, we can replace lines 12-20 by the cleaner code shown in Figure 7.24, which produces a much-improved `generate_scl` function overall.

Notice on lines 6-15 of Figure 7.24 the use of getter functions in place of the integer-indexed array references to retrieve the person's needed features. We can breathe more easily now!

```
function generate_scl($id) {
  $shift=select_dbShifts($id);
  ...
  $new_persons = getonlythose_dbPersons('sub','active','',
    $day,$time,$venue);
  foreach($new_persons as $per) {
    $id_and_name=$per->get_id()."+".$per->get_first_name()."+".
      $per->get_last_name();
    if (!in_array($id_and_name, $shift_persons) &&
      !in_array($id_and_name, $shift->get_removed_persons())) {
      $persons[]=array($per->get_id(),
      $per->get_first_name(), $per->get_last_name(),
      $per->get_phone1()." ".$per->get_phone1type(),
      $per->get_phone2()." ".$per->get_phone2type(),"","","?");
    }
  }
  $new_scl=new SCL($id, $persons, "open", $vacancies,
    get_sub_call_list_timestamp($id));
  ...
```

FIGURE 7.24 Layering Violation fixed and bad smell removed.

7.4 ADDING A NEW FEATURE: ALL LAYERS IMPACTED

Adding a new feature to an existing CO-FOSS product impacts all layers of the code. But it is smart to begin at the user interface layer, focusing particularly on additions to the view and the underlying functionality that will be required by the new feature. Thus, one or more MVC triples can be modified or created to implement the new feature.

When deciding how to add the new feature, developers should raise the following questions about each of the MVC triples involved:

1. How should the *view* be modified so that the user can easily identify the new feature?

2. How should the *controller* be modified so that the user can easily use the new feature?

3. How should the underlying *model* (domain and database modules) be modified to fulfill these new needs of the modified view and controller?

To illustrate this process in detail, we discuss the addition of a new feature to *Homebase* made in 2015, focusing on how these three questions were addressed.

This new feature requires *Homebase* users to keep track of the status of each volunteer as “active,” “applicant,” “on leave,” or “former.” This addition allows *Homebase* to exclude volunteers who are not "active" from appearing on a sub call list or a list of volunteers available for scheduling on the master schedule or the calendar, thus ensuring that only active volunteers are scheduled into a shift. Prior to 2015, *Homebase* had been listing *all* volunteers on these lists, which introduced some errors into the scheduling process.

To begin adding this feature, developers first addressed question 3 above by adding a new `status` field to the Person class. They unit-tested the modified constructor and new getter function, following the guidance in Chapter 5. They then added this field to the dbPersons table and made modifications to the affected CRUD functions in the dbPersons.php database module. They unit-tested that database change, following guidance in Chapter 6. This work created confidence that the added `status` field in the model would provide basic support for modifying the necessary MVC triples in the user interface.

Developers answered questions 1 and 2 by identifying five different MVC triples to be changed, as summarized in Table 7.3.

TABLE 7.3 MVC steps for adding a new feature

MVC triple	Module(s)	Changes
Edit Person	personEdit.php personForm.inc	add a `status` field
Search for Persons	personSearch.php	add `status` search option
Schedule Person	editMasterSchedule.php	list only "active" volunteers
Edit Shift	editShift.php editShift.inc	list only "active" volunteers
Sub Call List	subCallList.php	list only "active" volunteers

The following sections discuss each of these changes individually. (We note, in passing, that if this new feature were implemented by a 5-student development team, each developer could independently implement one of the 5 MVC changes described below.)

Changing the Edit Person MVC Triple

The desired change to the Edit Person view is shown in Figure 7.25. There we see a new select box labeled **Status**, with options "active," "applicant," "on leave," or "former."

This view was implemented by adding a new HTML select to the personForm.inc module, as shown in Figure 7.26.

The `if` statements in Figure 7.26 preload the select box so that it reflects the person's current status prior to the change. If the user selects a different option, the controller records that information in the `$_POST['status']` variable. When the user submits this form, all user changes are recorded in that person's entry in the dbPersons table.

The personEdit.php module was changed to provide an appropriate controller for the new status field in the Persons.php class and the dbPersons

Status: active

Position type: (check one or more)

☑ House Volunteer ☐ Weekend Manager ☐ Flex shift ☐ Manager

Availability:

Monday	Tuesday	Wednesday	Thursday	Friday	Saturday	Sunday
☐ 9-12	☐ 9-12	☐ 9-12	☐ 9-12	☐ 9-12	☑ 10-1	☑ 9-12
☐ 12-3	☐ 12-3	☐ 12-3	☐ 12-3	☐ 12-3	☑ 1-4	☑ 2-5

FIGURE 7.25 Showing a person's status in the Edit Person view.

```
1   echo ('<p>Status:');
2   echo('<select name="status">');
3   echo ('<option value="applicant"');
4   if ($person->get_status() == 'applicant')
5       echo (' SELECTED');
6   echo('>applicant</option>'); echo ('<option value="active"');
7   if ($person->get_status() == 'active')
8       echo (' SELECTED'); echo('>active</option>');
9   echo ('<option value="LOA"');
10  if ($person->get_status() == 'LOA')
11      echo (' SELECTED'); echo('>on leave</option>');
12  echo ('<option value="former"');
13  if ($person->get_status() == 'former')
14      echo (' SELECTED'); echo('>former</option>');
15  echo('</select>');
```

FIGURE 7.26 Coding to show a person's status in the Edit Person view.

database table. For example, Figure 7.27 shows how the construction of a new person was modified (line 4) to reflect this addition. To properly update the person's changed status in the dbPersons database table, line 8 calls the `update_person` function in the dbPersons.php module.

Changing the Search for Persons MVC Triple

Searching for volunteers who have a particular status was implemented by modifying the search for people view with an additional select titled **Status:**, as shown in Figure 7.28. Figure 7.29 shows a result that the user sees when selecting all persons with status= "applicant."

```
$newperson = new Person($first_name, $last_name, $location,
    $address, $city, $state, $zip, $clean_phone1, $phone1type,
    $clean_phone2,$phone2type, $email, $type, $screening_type,
    $screening_status,  $_POST['status'], $employer, $position,
    $credithours, $commitment, $motivation, $specialties,
    $convictions, $availability, $schedule, $hours,
    $birthday, $start_date, $howdidyouhear, $notes, $pass);
$result = update_person($newperson);
```

FIGURE 7.27 Updating a database entry with the new status field.

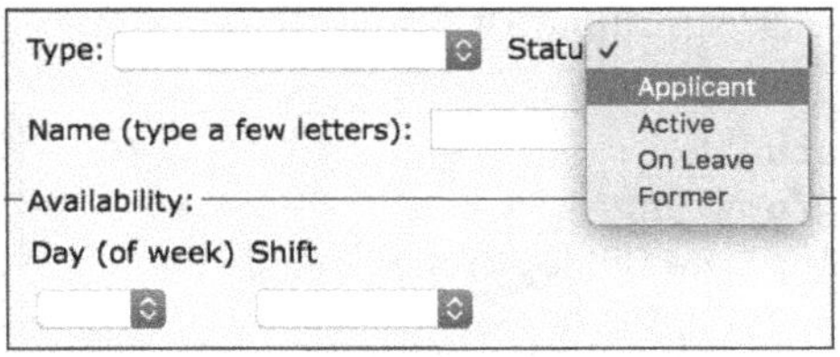

FIGURE 7.28 Searching for "applicant" status.

The code that renders this view is shown in lines 1-7 of Figure 7.30, while the code that retrieves all persons with a given status from the database is shown in lines 9-11. This latter code assumes that the definition of the function `getonlythose_dbPersons` (inside the dbPersons.php module in the database layer) has already been appropriately modified. To view that definition, please see the `dbPersons.php module`.

Changing the Schedule Person MVC Triple

Changing the editMasterSchedule.php module so that only "active" volunteers are listed as candidates for filling a vacancy required modifying the query that retrieves all volunteers who are available on that particular day and time. This view is shown in Figure 7.31.

The code in editMasterSchedule.php that underlies the first select in this view is summarized in Figure 7.32, where we see a call to the function `get_available_volunteer_options`. This function performs the required filtering and returns a series of HTML select options in the form of a character string listing the names and ids of all available volunteers for that particular shift. To view the entire editMasterSchedule.php module, readers should check the `editMasterSchedule.php module` in the *Homebase* code base.

Search Results:

Found 5 applicant persons (select one for more info).

Name	Phone	E-mail	Availability
Andrew Jones	703-289-9841	jonesey@yahoo.com	
Arne Jones	703-772-3073	jonesey@yahoo.com	Mon:9-12:po
Elizabeth Jones	703-756-5794	jonesey@yahoo.com	Wed:3-6:, Fr
Ellen Jones	703-797-9667	jonesey@yahoo.com	Tue:9-12:por
Eva Jones	703-749-9251	jonesey@yahoo.com	Mon:9-12:po

FIGURE 7.29 Search results for status = "applicant".

```
echo('Status:<select name="s_status">' .
    '<option value="" SELECTED></option>' .
    '<option value="applicant">Applicant</option>' .
    '<option value="active">Active</option>' .
    '<option value="LOA">On Leave</option>' .
    '<option value="former">Former</option>' .
    '</select>');
...
$result = getonlythose_dbPersons($_POST['s_type'],
    $_POST['s_status'], $_POST['s_name']), $_POST['s_day'],
    $_POST['s_shift'], $_SESSION['venue']);
```

FIGURE 7.30 searchPeople.php code for selecting a person's type.

As an aside, readers might have noticed a layering violation in the editMasterSchedule.php module, since its function `get_available_volunteer_options` contains a database query being executed from the user interface layer. This violation presents another opportunity for refactoring.

Changing the Edit Shift MVC Triple

Changes to editing a shift on the calendar were implemented so that only active volunteers were listed for selection, as shown in the modified view in Figure 7.33.

The code in the editShift.inc module that underlies this view is summarized in Figure 7.34. It calls the function `get_available_volunteer_options`, which is part of the editShift.inc module.

Filling a vacancy for Portland House Master Schedule, odd Thursdays, 12pm to 3pm

Select an active volunteer with this availability

Select from all active volunteers

Add Volunteer

Back to Shift

FIGURE 7.31 Listing only "active" volunteers when filling a vacancy.

```
...
echo "<select name='scheduled_vol'>
    <option value='0'>Select an active volunteer
        with this availability</option>"
    . get_available_volunteer_options($msentry,
        get_persons($msentry->get_id())) .
    "</select><br><br>" ;
...
```

FIGURE 7.32 Changing editMasterSchedule.php to list "active" volunteers.

This function performs the required filtering and returns a series of HTML select options in the form of a character string listing the names and ids of all available volunteers for that particular shift. To view the entire `get_available_volunteer_options` function, readers should check the `editShift.inc module` in the *Homebase* code base.

Changing the Sub Call List MVC Triple

The final change needed to implement the addition of a status field for *Homebase* volunteers is to modify the code that generates a sub call list so that it shows only active subs for a given shift, rather than all subs. The operative view that triggers this action is shown in Figure 7.2.

Whenever the user hits **Generate Sub Call List** in that view, control passes from the EditShift.php module to the `generate_scl` function inside the subCallList.php module, which displays the newly-generated sub call list, as shown in Figure 7.11. Figure 7.24 (line 4) shows that the `generate_scl` function generates the new sub call list for only the active volunteers.

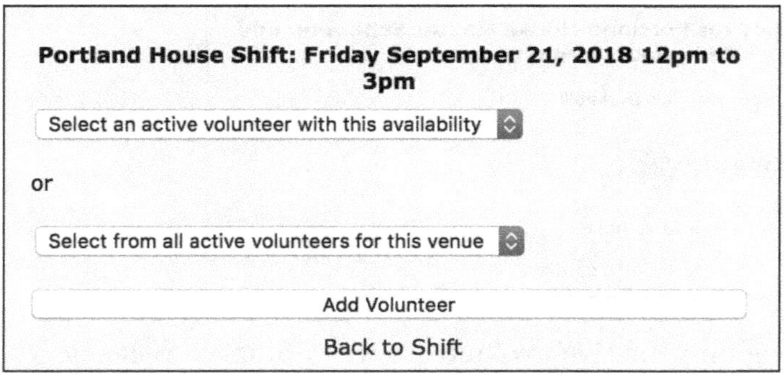

Portland House Shift: Friday September 21, 2018 12pm to 3pm

Select an active volunteer with this availability

or

Select from all active volunteers for this venue

Add Volunteer

Back to Shift

FIGURE 7.33 Selecting only active volunteers for filling a calendar vacancy.

```
...
echo "<select name='scheduled_vol'>
    <option value='0'>Select an active volunteer
        with this availability</option>"
    . get_available_volunteer_options($shifttime,
        $shift->get_day(), $shift->get_persons(),$venue)
    . "</select><br>"
...
```

FIGURE 7.34 Code for selecting only active volunteers.

7.5 CLIENT REVIEW AND ISSUE TRACKING

At this stage in the CO-FOSS project, the client should begin feeling some ownership of the software by exercising the new user interface and becoming active with the issue tracker. This was the case, for example, with the 2015 *Homebase* project. Figure 7.35 summarizes the issues posted during the user interface development and testing phase of that project. As we see there, the clients became active in commenting on issues (issue 10) and reporting new issues themselves (issues 14, 15, and 16). Student developers resolved each issue either by working alone or by working together (issue 14).

Overall, the issue tracking shown in Figure 7.35 reflects more critical feedback from clients than was apparent earlier in the project, as shown in Figure 6.19 of Chapter 6. This happened because clients were now engaged with the user interface for *Homebase*, and they had also become comfortable posting on the issue tracker itself.

For a complete list of all the issues reported for the 2015 *Homebase* project and how they were resolved, readers should review the complete issue track at its `GitHub repository`. When reviewing this list, recall that the 1-semester development period for this project ended on June 6, 2015 with a final presentation by student developers. That is why student participation on the issue tracker ended abruptly on May 25. However, the client continued to report issues after the semester ended, and those later issues were resolved by a professional developer. We return to this topic in Chapters 8 and 9.

In the remainder of this section, we dive more deeply into the work conducted by developers in responding to issues posted by clients, including their interactions during the process of issue resolution. Hopefully, this presentation will provide some insight into what types of bugs and issues can occur in a CO-FOSS project and how the developers and clients worked together to clarify and address the issues.

7.5.1 A User Interface Bug

The most straightforward kind of bug to detect and correct is one that can be isolated within a single module within a MVC triple in the user interface. The following bug report was posted by Gina (the client) for the developer (Alex) during development of the 2010 version of *Homebase*:[8]

> Hi Alex,
>
> On the calendar, beginning in January the button to click to save notes is missing. The last week it appears on is December 21–27.
>
> Thank you!
>
> Gina

To recreate this error, Alex tried to edit the calendar page for the same week reported by the user. He used a sandbox version of *Homebase*, so as not to interfere with the "live" version.

At the bottom of each current and future calendar week (but no previous weeks), the button "Save changes to all notes" should appear, as it does in Figure 7.22. For some reason, this button was not appearing on any calendar week after the week of December 21–27 of the current year.

The first step in fixing this issue is to locate the MVC triple and the module(s) that display and control the calendar view, which are `calendar.php` and `calendar.inc`. We can find this module easily by reading the URL at the top of the page when the calendar view appears. This is what the URL looks like, for instance, for the calendar view in Figure 7.22.

```
calendar.php?id=18-09-17:portland&venue=portland
```

[8]In 2010, when this bug was reported, client-developer communication took place via ordinary email exchanges, rather than issue trackers. Today, such communication would be done using a project-based issue tracker.

```
#7 reported by beeb...@whitman.edu 22 Apr 2015 at 6:47
Night shifts needed for Volunteer Log

#8 reported by beeb...@whitman.edu 22 Apr 2015 at 6:48
Date constraint for Volunteer History Report

#9 reported by hargu...@whitman.edu 22 Apr 2015 at 6:51
-1 Vacancies for Family Room on Sunday April 5, database issue?

#10 reported by beeb...@whitman.edu 22 Apr 2015 at 8:16
Improvements on Birthdays report
  comment by sczekal...@rmhprovidence.org 23 Apr 2015 at 5:23
  Joanne Tainsh birthday entered as 1929, came up as 2029.

#11 reported by beeb...@whitman.edu 22 Apr 2015 at 8:16
Exporting reports

#12 reported by beeb...@whitman.edu 22 Apr 2015 at 8:17
Optional person field in reports

#13 reported by beeb...@whitman.edu 22 Apr 2015 at 8:22
Inactive volunteers require a schedule

#14 reported by jpow...@rmhprovidence.org 23 Apr 2015 at 2:31
Volunteer Availability
  comment by beeb...@whitman.edu 11 May 2015 at 5:31
  In the master schedule, people appear who should not.  Molly
  Jones' schedule is for odd Fridays 9-1 in the House, but she
  is available for odd Wednesdays 9-1.  maybe a problem in
  editMasterSchedule.php or dbMasterSchedule function?

  comment by wa...@whitman.edu 11 May 2015 at 9:12
  I modified editMasterSchedule.php, and fixed this bug. This
  should now work. Thank you beeb... for your explanation!

#15 reported by jpow...@rmhprovidence.org 23 Apr 2015 at 2:34
Volunteer cannot be entered without assigning a shift.

#16 reported by jpow...@rmhprovidence.org 23 Apr 2015 at 3:06
Some volunteers are not able to be scheduled for vacant shifts
```

FIGURE 7.35 Issues 7-16 posted for the 2015 *Homebase* project.

Next, we look for that part of the calendar.php code that displays the button "Save changes to all notes." The particular date on which the calendar is being edited plays a role in this determination, since users are not permitted to edit calendar shifts for weeks that have already gone by.

So the display of this button is controlled by determining if the current date precedes the calendar week being edited. If the calendar week is fully in the past, this button should not appear.

We can begin examining the calendar.php module for a code snippet that determines whether or not to display this button. Since the button is at the bottom of the view, we expect that snippet to appear near the bottom of the calendar.php module. It does appear there, as shown in Figure 7.36.

```
1  if ($edit==true &&
2        ! ($days[6]->get_year()<$year ||
3            ($days[6]->get_year()>=$year &&
4                 $days[6]->get_day_of_year()<$doy)
5            ) &&
6        $_SESSION['access_level']>=2)
```

FIGURE 7.36 Locating a bug in the calendar.php module.

At first glance, this is pretty ugly code, so let's parse it a bit. The condition `$edit==true` tests whether the calendar is in "edit" mode or "view mode," respectively; users cannot edit a calendar week in view mode. The condition `$_SESSION['access_level']>=2` tests to be sure that the logged-in user has edit access to the calendar notes, as discussed in Section 7.1.3.

Now for the gritty middle part of this code in lines 2-4. A calendar week is an array of 7 days, indexed from 0 to 6, and so the reference `$days[6]` is talking about the last day of the week. The references `$year` and `$doy` are talking about the current year and day (0 to 365) of the current date.

So literally this code is displaying the "Save changes to all notes" button whenever all three of the following conditions are met:

1. The user is editing, not viewing, a calendar week,
2. The calendar week being edited does not precede the current date, and
3. The user is authorized to change the notes.

Clearly, the fault must lie with condition 2, since otherwise the error would not appear only in certain future weeks. Looking at the logic of condition 2, we can break it down a little more finely to read, display "Save changes to all notes" whenever:

```
the calendar week being edited is:
  not (a previous year or
      (the current year or later and
        its day of the year precedes today)
  )
```

Now the error becomes more apparent, since the clause "or later" allows the earlier part of a future year to "precede" today, which is incorrect! This would explain why the "Save changes to all notes" button disappears at the beginning of next year but not for the the last week of the current year. To correct that error, we need to eliminate that clause so that the text reads:

```
the calendar week being edited is:
  not (a previous year or
      (the current year and
        its day of the year precedes today)
  )
```

So to fix the code, we need only to change the `>=` operator in line 3 of Figure 7.36 to `==`, then view the calendar editing form to be sure we have corrected the error properly. Viewing the form again confirms that the bug is fixed.

7.5.2 A Multi-Layer Bug

Many issues in complex software systems require the developer to navigate among layers in the code to find the bug and fix it. Moreover, the developer can utilize the organization of the user interface into MVC triples to further assist with this navigation.

To illustrate this idea, here is another actual bug reported by the client Gina during the development of an earlier version of *Homebase*. Its resolution required developers to dive into the code at the database, domain, and user interface levels of the architecture. They also exploited the conceptual cohesion among the MVC triples that underlie the Editing a Calendar use case, as discussed in Section 7.3.1.

> Hello!
>
> I think we may have discovered another bug: a volunteer created a sub call list and began making calls, she saved the information in the sub call list. I went in and added notes under the shifts which then deleted her sub call lists and the information. If it would be helpful, I can try to recreate this with you over the phone.
>
> Thanks and no hurry
>
> Gina

The first response from the developer was to confirm and understand the error more precisely by trying to recreate it on the sandbox version of *Homebase*. This exercise invoked activated three views specified by the following URLs (the date reflects the week for which the exercise was being conducted, and is arbitrary):

```
calendar.php?id=18-10-08:portland&venue=portland&edit=true
editShift.php?shift=18-10-12:12-3&venue=portland
subCallList.php?shift=18-10-12:12-3:portland&venue=portland
```

Now when we return to the calendar.php module and edit and save the notes for some other shift, the sub call list associated with the shift id `18-10-12:12-3:portland` suddenly vanishes.

Three MVC triples underly these three views. Let's look at the first triple, whose view and controller lie in the calendar.php and calendar.inc modules, which are responsible for saving the shift, day, and guest chef notes. The function `process_edit_notes` insidecalendar.inc is shown in Figure 7.37.

```
1  function process_edit_notes($week, $venue, $post) {
2    $days = $week->get_dates();
3    for ($i = 0; $i < 7; ++$i) {
4      $shifts = $days[$i]->get_shifts();
5      foreach ($shifts as $key => $shift) {
6        $note = trim(str_replace('\"', '\\\"',
7          str_replace('\'', '\\\'', htmlentities(
8                $post['shift_notes_' . $shift->get_id()]))));
9          $shift->set_notes($note);
10         update_dbShifts($shift);
11     }
12     $mgr_note = trim(str_replace('\"', '\\\"', str_replace(
13       '\'', '\\\'', htmlentities($post['mgr_notes' . $i]))));
14     $days[$i] = select_dbDates($days[$i]->get_id());
15     $days[$i]->set_mgr_notes($mgr_note);
16     update_dbDates($days[$i]);
17   }
18 }
```

FIGURE 7.37 The `process_edit_notes` function inside calendar.inc.

This shows that there are only two ways that the shift in question, and hence its sub call list, can be permanently changed. One is to execute the `update_dbShifts($shift)` call on line 10, and the other is to execute the `update_dbDates($days[$i])` call on line 16.

The update_dbShifts function is defined inside the dbShifts.php module, and it works by executing a delete˙dbShifts call followed by an insert˙dbShifts call on the same shift. No other side-effects seem to be taking place here.

The update_dbDates function inside the dbDates.php module exhibits a similar pattern—a delete followed by an insert of the same date. However, there is something else going on here, as shown in Figure 7.38.

```
1  function delete_dbDates($d) {
2  ...
3      $shifts=$d->get_shifts();
4      foreach ($shifts as $key => $value) {
5        $s = $d->get_shift($key);
6        delete_dbShifts($s);
7        delete_dbSCL(new SCL($s->get_id(),null,null,null,null));
8      }
9  }
10 function update_dbDates($d) {
11     if (! $d instanceof RMHdate)
12       die ("Invalid argument for dbDates->update_dbDates call");
13     delete_dbDates($d);
14     insert_dbDates($d);
15 }
```

FIGURE 7.38 Locating a bug in the dbDates module.

That is, while an update is accomplished by deleting and adding the same shift to dbShifts, each shift's sub call list is also deleted at the same time that shift is deleted (lines 6 and 7 in Figure 7.38). Once gone from the database, that sub call list cannot be re-added.

So the proper correction for this bug is to remove the delete˙dbSCL call (line 7 in Figure 7.38) from the delete_dbDates function, leaving each sub call list orphaned in the dbSCL table even though its corresponding shift has been deleted. Once the shift is re-inserted into the database to complete the update, it is reunited with its old sub call list (if it had one) that had been temporarily orphaned.

7.6 SUMMARY

This chapter has introduced the principles and practices of user interface development in the context of an active open source software project. It exposes the value of design principles like model-view-controller, session management, and software security. It also describes the impact of adding new features on the user interface modules and the underlying layers in the code base.

Because all the levels of a software architecture are interrelated, we can see that adding new features to a user interface often has a ripple effect on the underlying domain classes and database modules. Moreover, refactoring and debugging activities that begin with a user interface module also migrate to modules and classes at all other levels.

Finally, this discussion leads naturally to the development and proper maintenance of good developer and user documentation, which provides a focal point for the next chapter.

7.7 MILESTONE 7

1. Create one or more MVC triples for each of the use cases for your CO-FOSS project. Each triple should relate to a single view identified in the design document and be clearly linked to the other triples for its use case.

2. For the view and controller in each MVC triple, be sure that its layout and functionality are consistent with the first eight Principles of User Interface Design in Section 7.1.

3. For each MVC triple, add functionality to the model (domain and database modules) as needed to support its view and controller.

4. Enlist your clients to test each of the use cases by exercising each of its views and reporting any bugs or other concerns on the project's issue tracker.

5. Evaluate these new issues and assign ownership to individual developers for any issues that need to be resolved in order to complete the user interface for your CO-FOSS project.

CHAPTER 8

Preparing to Deploy

"The skill of writing is to create a context
in which other people can think."
—*Edwin Schlossberg*

This chapter is about preparing the software so that it will be well-used long after the development period has ended. This preparation includes creating user documentation, user forums, and other user support. Such preparation helps to ensure a successful handoff and integration of the software into the client's ongoing mission-critical activities.

The chapter begins by reviewing principles of good writing, especially as they apply to writing user documentation. We also discuss ways of integrating documentation within the code base, so that users have access to immediate support while they are working with the software.

User support is not only about documentation; it is also about feedback and responsiveness. We cover these other kinds of user support, which include user training, forums, feedback surveys, and ongoing communication between developers and users. The effectiveness with which these tools are used relies directly on how well they are set up and supported by the developers.

8.1 TECHNICAL WRITING

Throughout the life of a CO-FOSS project, developers should maintain good *developer documentation* in the code base, ensuring that future developers will understand every module and function by reading the code alongside its accompanying commentary. Section 3.1.6 identified techniques for writing and embedding good developer documentation during the coding process itself.

However, software development also requires writing good *user documentation.* Such documentation reinforces connections between the software and its users. User documentation can be viewed as a user manual of "how to's" for using the software to perform each of the tasks that the system supports. User documentation can also act as a tutorial for beginners when they encounter the software for the first time.

Writing good user documentation is a special kind of "technical writing." It follows stylistic and other standards that apply to good writing in general. When we think about writing good user documentation, we can begin by thinking about its overall structural characteristics:

Organization is the way the document is arranged. Organization also includes navigability and access. How easily can the user navigate among related sections of the document? For an on-line document, how easily can a relevant section of the document be accessed from a related view within the software's user interface itself?

Illustration is the interplay among text, images, and other media. Appropriate selection and placement of images can do a lot to clarify a written description for the reader. Some readers respond better to images, while others prefer text. If images and text are carefully interwoven, both types of readers will be well-served.

Style is the way words and media are combined to describe a software feature or user activity. It includes word order, sentence length, and usage of grammar, punctuation, spelling, examples, and various detailed presentation options such as font size and colors.

Tone shows the attitude of the writer toward the reader. For example, a *peer-to-peer* tone suggests using a level of language and subject matter understanding that the writer and reader share in common. A *mentor-student* tone, on the other hand, requires that the writer adapt to a level of language that is different from one's own.

Thus, good software documentation takes into account the audience of users for whom it is written. In addition, good software documentation should reach a high standard of writing quality. These two subjects are discussed more carefully below.

8.1.1 Writing for an Audience

The audience of users who read and utilize a software document typically varies from one software product to another. Even within the same product, some users will be more familiar with the application domain, better educated, more fluent in the native language of the application (typically English), more adept at using computers, and/or more familiar with the software itself than others. A well-written software document tries to address the diverse perspectives of all users — this is not a simple task.

To help address different users' levels of experience, two separate kinds of user documents can be developed: one that initiates *new users* to the software and another that provides *experienced users* with the finer details of using the software. The former document is sometimes called a *tutorial* while the latter is sometimes called a *user manual.* A third type of document, an *open discussion forum*, tends to address the needs of both novice and advanced users (for more details, see Section 9.4). All three types of documents are associated with a specific software product and are typically accessed on-line.

When the software is Web-based, it often comes with a "Help" tab on the main menu that facilitates convenient keyword-based searching for reference material. There are many examples of good documentation on the Web. Here are their two main characteristics:

1. Good documentation *directly answers any question* that a user may ask, no more and no less. For example, there is no point in showing an experienced user all the different views that may appear when logging in to use the software. On the other hand, a novice user will need precisely this detailed level of information.

2. Good documentation *covers all the situations* that users may encounter when using the software. For this reason, it is desirable for the document to be organized along the same lines that the software's use cases are organized. That is, we may start organizing a user manual by identifying one chapter with each use case, beginning with the login and logout procedures.

Thus, it is usually desirable to present on-line help in a way that parallels the way the user interacts with the software. Since each main menu item initiates a series of steps that the user must perform to accomplish a certain task, a corresponding help page for that menu item can describe how to accomplish those steps. That help page is enhanced by screen shots that the user can expect to encounter as those steps are completed. We illustrate this in more detail in Section 8.2.3.

A printed copy of a user manual may also be desirable. Some users prefer reading a printed copy while using the software, so that they can refer to it when they have a question about what to do next. Other users may prefer to "fiddle" with the software in order to become familiar with the nuances of each activity as they learn. To ensure consistency, the printed copy should always be generated as a byproduct of its online counterpart rather than written as a separate document.

For some applications, not all users may be native speakers of English. Thus, keeping documentation as simple as possible is very important. To this end, some technical fields have adopted a `Simplified English` that contains about 1,000 words, each with just one meaning. Being aware of any relevant Simplified English for your software's application domain will help you write text that users are likely to understand.[1]

[1]All our CO-FOSS projects are targeted strictly to an English-speaking audience. The development of multilingual software, called *internationalization and localization*, is an

When significant populations of users speak a different language from English, it may be necessary to produce the documentation in two or more different languages. This is called "localization," and it is a particularly common feature of more mature HFOSS products like `OpenMRS` or `WordPress` that have large international user communities.

On a different note, some users may have reading disorders or cognitive disabilities. Others may just read poorly. Text content will always pose problems for these users. Some of the ways in which documentation can be made more helpful to this wider audience are:

1. Supplement the text with illustrations. That is, do everything possible to clarify and simplify the text, and then go one step further by supplementing the text with a redundant (perhaps optional) graphic.

2. Reduce the text to a bare minimum on each page. Pages with a large amount of text can intimidate users with reading difficulties.

3. Be as literal as possible—avoid using metaphor or humor to illustrate a point. Some people with cognitive disabilities cannot distinguish between the literal meaning of an idea and its implied meaning.

Overall, good documentation assumes that readers are intelligent, but it does not assume that readers know the subject matter as well as the author. Therefore, explaining and illustrating basic concepts of the software are more helpful than insulting, as long as the explanations show respect for the reader.[2]

Surely no collection of user documentation for a software product will please all types of users. However, knowing and being sensitive to users' backgrounds and preferences will take the writer a long way toward producing documentation that will be helpful for most users.

8.1.2 Standards for Writing Quality

A high standard for writing quality begins with a clear and simple document. While developing a clear and simple document can be one of the most difficult of all writing tasks, it is a goal to which all software developers should aspire. In the words of Ernest Hemingway:

> "My aim is to put down on paper what I see and what I feel in the best and simplest way."

---

extensive subject. For more information on internationalization and localization, readers may want to visit `https://npfi.org/internationalization/` as a starting point.

[2]Some experts say writers should aim for "eighth grade level" when writing user documentation. For instance, popular magazines and news sources, such as *Time Magazine*, are written at about an eighth grade level so as to reach a general audience.

Conversely, unclear or complex software documentation can deflate the user's overall enthusiasm for the software. It can be especially off-putting for new users and people with reading disorders or cognitive disabilities.

The following guidelines enumerate what is needed to create clear and simple software documentation. A few examples are given here, and more extended examples appear in Section 8.2.3.

Organize your topic into a logical outline before you begin writing. This may be the most important guideline of all. You must think clearly about a topic in order to communicate it clearly. The organizing process is continuous, starting before any words are written and extending throughout the entire development process. In this sense, documentation writing is like code refactoring.

Put summary information first. Describe the purpose of each section before laying out any of its details.

Use the language of the application domain. Users relate more quickly to documentation if it uses words and phrases that are familiar and (in some cases) unique to the work environment where they are using the software. The key nouns and verbs in the documentation should match those that appear in the user interface, which in turn reflect the language of the design document and the application domain itself.

Be precise. Use precise words as opposed to more general variants. Provide enough detail so as not to keep the reader in the dark. Avoid using ambiguous words.

Use the present tense. Software use has no past or future. Everything happens in the moment as a direct result of some event, usually initiated by the user. As soon as an event occurs, the software has a specific reaction. So good technical writing uses the present tense to capture each step in an interaction.

Be direct. Use the second person and the imperative mood. That is, use "you" and phrases like "do this and then do that." The result is usually more precise and clear. Direct instructions can increase comprehension and place more of a sense of responsibility on the user.

Be concise. Avoid stating the obvious or using unnecessary padding, repetition, verbosity, or pomposity. Software users are not looking at the style of your prose; they are only looking for guidance that helps them perform a task. Wordiness is not a virtue in technical writing.

For example, instead of "The unwise leaving their login or password on their workspace may result in a dangerous breach of software security and therefore it is recommended that one retain their password in a safe and secure location in order to ensure data integrity," just say "Keep your password secure."

Use simple and short sentence structure. Avoid compound sentences and long phrases. Some suggest that a good maximum sentence length for technical documentation is somewhere between 15 and 20 words.

Use parallel sentence construction. Ensure that sentence structure is internally consistent.

For example, the inconsistent sentence "This algorithm is correct, improves memory usage and efficiency." can be better written as 'This algorithm is correct and improves both memory usage and efficiency."

Use the active voice. The active voice works better than the passive in technical writing because the focus of the sentence is the user's action rather than the object being acted upon.

Use positive terms. Use phrases that don't contain a negative element like "no" or "not."

For example, "impossible" is a positive construction as opposed to "not possible." Emphasize the way things are, rather than what they are not.

Punctuate properly. Use complete sentences terminated by a period. Use other punctuation appropriately and sparingly.

Check your spelling. Use an automated spell checker, but don't rely on it completely. Proofread the document to find correctly spelled words that are used incorrectly.

For example, the spelling error in "For score and seven years ago..." will not typically be picked up by a spell checker.

Use computer terms as needed, but only ones that are commonly known. For example, click, double-click, select, type, mouse, touchpad, browser, button, window, and file are all okay.

Avoid specifying gender. For example, using "he/she" or "one" or "they," rather than "he" or "she" alone, creates reading comfort among diverse groups of users that include both genders.

Clarify acronyms and abbreviations. Unfamiliar acronyms and abbreviations mean nothing to users. Expanding acronyms when they first appear allows users to learn their meaning. When the documentation has many acronyms, adding a hyperlinked glossary can be useful.

Avoid slang and jargon. Slang and jargon can be useful to people who understand it, but confusing to people who don't.

Avoid weak verbs. Don't use a form of the verb "to be" (is, are, was, were) when a more active verb would be more appropriate. Over-use of "to be" often forces writers to use the passive voice more than necessary.

Write cohesive paragraphs constructed around a single major idea. All of the ideas in a paragraph should relate back to the main point, which typically belongs in the first sentence.

Use two columns to describe the steps of a user-computer interaction. Label one column "User" and the other "System." Number the steps serially, which clearly indicates their order. For example, see Section 2.2.3.

Use short numbered lists (3–9 items) to list a sequence of steps. Lists are useful because they break up long sentences and they identify easy-to-digest information chunks. Again, see Section 2.2.3 for an example.

When making a list of steps, use the same form in each step. Start with a verb to establish the imperative mood. For example:

1. Open the file you want to rename.
2. Select Rename from the Edit menu.
3. Type the file's new name.
4. Click Save.

Add a graphic whenever it will help visualize a point. A picture is worth a thousand words. Many users will prefer a visual clue over a textual one. If practical, write instructions with both visual and textual content.

When followed carefully, these guidelines can help ensure that written user documentation is of high quality. Examples in later sections will illustrate these guidelines in greater detail.

8.2 USER DOCUMENTATION

This section provides an overview of the organization of user documentation that can accompany a software product. We pay particular attention to the organization of on-line documentation that is integrated with the software itself, presenting examples from our own CO-FOSS projects.

8.2.1 User Manuals, FAQs, and Demo Versions

Historically, user manuals and other documents for a software product were prepared and distributed to users in print form. More recently, these documents are available as downloadable PDF files, where users can either print them or use them on-line alongside the working version of the software itself. These documents can also be presented as a related set of Web pages that facilitates easy reading and navigation.

Whatever their medium, user manuals should be organized so that discussions about individual topics can be accessed randomly. This allows experienced users to search for any topic whenever the need arises and read about it directly. To facilitate searching, a user manual must be well-indexed. On-line user manuals typically include a search button that leads users directly to the topic of interest, as well as internal links that facilitate navigation among major sections.

A user manual should describe every feature of a software artifact. It should also provide relevant examples that assist users who want to exercise those features to accomplish useful tasks. A good user manual can also provide troubleshooting assistance. It is very important for a user manual to be kept up to date with the software itself. In this sense, the manual can be viewed as a contract describing what the software will do.

A user manual should always have a thorough and searchable index. It is also useful to provide a "Frequently Asked Questions" (FAQ) section at or near the front. So an overall organizational scheme for a user manual may have the following main elements:

Table of Contents This is like a table of contents in a textbook. However, it should also provide direct links to each of the sections so that users don't need to navigate manually. For an example, see the eBook version of this text.

Frequently Asked Questions (FAQ) This provides answers to questions about the software that occur so often that an average user is likely to find help here rather than diving more deeply into the manual. Typically, the length of a FAQ list is rather short. The list shows each question in **boldface** so as to make it easy to locate with a quick visual scan. A longer FAQ list is usually accompanied by a "search box" which can assist users to locate relevant questions more quickly.

Chapter Layout Each chapter in a user manual should be short and address a particular function of the software. If describing that function requires more text, the chapter should be divided into shorter sections focused on individual sub-topics. Hyperlinks among chapters and sections should be abundant, so that users can easily navigate among related topics as they read. Again, see the eBook version of this book for examples.

Examples and Illustrations Each topic discussed should be accompanied by a brief, illustrative example. Good examples always help to clarify abstract ideas.

Indexing and Searching A thorough index is also a necessity for a user manual. Its search box allows users to find information about topics that do not appear explicitly in the index.

Live Demo Some software documentation is accompanied by an on-line live demo, which gives the reader a chance to exercise the software itself without needing to obtain login credentials.

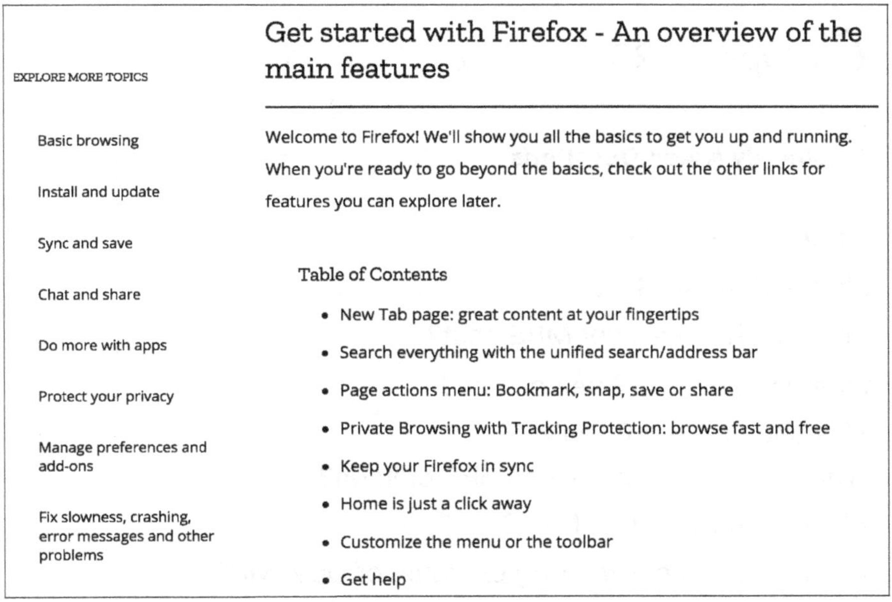

EXPLORE MORE TOPICS

Basic browsing

Install and update

Sync and save

Chat and share

Do more with apps

Protect your privacy

Manage preferences and add-ons

Fix slowness, crashing, error messages and other problems

Get started with Firefox - An overview of the main features

Welcome to Firefox! We'll show you all the basics to get you up and running. When you're ready to go beyond the basics, check out the other links for features you can explore later.

Table of Contents

- New Tab page: great content at your fingertips
- Search everything with the unified search/address bar
- Page actions menu: Bookmark, snap, save or share
- Private Browsing with Tracking Protection: browse fast and free
- Keep your Firefox in sync
- Home is just a click away
- Customize the menu or the toolbar
- Get help

FIGURE 8.1 First page of the *Firefox user manual*, including Help link.

Below are examples of particularly effective and useful on-line user manuals, FAQs, and live demos. These examples illustrate the writing and communication quality that developers of emergent CO-FOSS projects should aim to achieve.

Example: Firefox User Manual

The first page of the Firefox browser's user manual for beginners is shown in Figure 8.1 (see `https://support.mozilla.org/en-US/kb/get-started-firefox-overview-main-features` for more details). Its prominent features are a **Get help** tab at the bottom and a short list of links on the left to topics that are most useful for beginning users. The manual thus tries to lead users in the right direction without any fanfare.

Manually searching the expanded table of contents for this manual would be a daunting task, since it references hundreds of articles arranged in topical (non-alphabetical) order. To enable easier searching, the **Get help** tab at the bottom of the page is often the most practical avenue for finding a specific item in the manual.

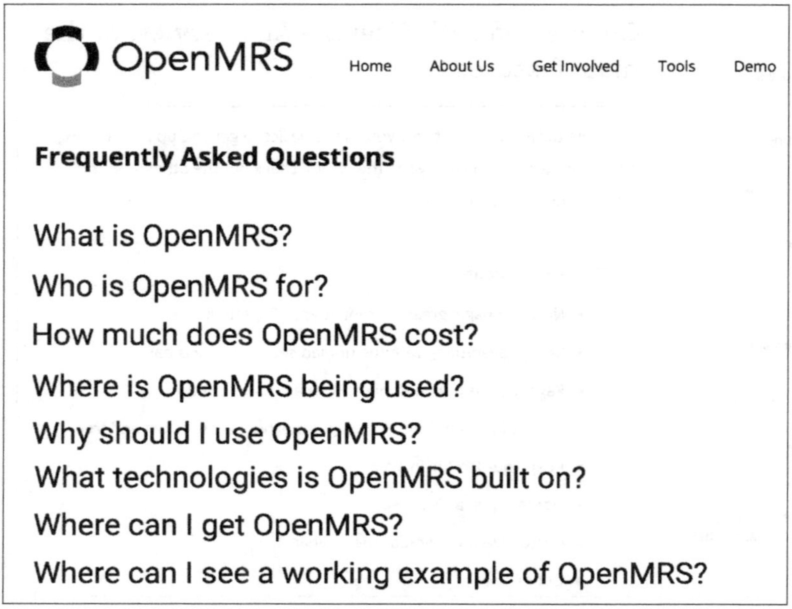

OpenMRS Home About Us Get Involved Tools Demo

Frequently Asked Questions

What is OpenMRS?
Who is OpenMRS for?
How much does OpenMRS cost?
Where is OpenMRS being used?
Why should I use OpenMRS?
What technologies is OpenMRS built on?
Where can I get OpenMRS?
Where can I see a working example of OpenMRS?

FIGURE 8.2 The Introductory OpenMRS FAQ List.

Example: OpenMRS FAQ and Demo

`OpenMRS` is an HFOSS medical records system that can be freely downloaded and customized for use by any organization providing medical services to a community of clients. Figure 8.2 shows a 2018 snapshot of the intrdoductory *OpenMRS* FAQ. Each entry in the this FAQ list is a question that has been raised by so many site visitors that it is answered right there.

Notice also in the top right corner of Figure 8.2 that an on-line demo for OpenMRS is available for uninitiated developers and users who want to gain some familiarity with its user interface and basic functionality. This demo's login page is shown in Figure 8.3. To exercise this demo live, readers can browse to `https://openmrs.org/demo/` and log in.

Example: Homebase Demo

A similar demo is available for our CO-FOSS product *Homebase* so that uninitiated developers and users can gain some familiarity with its user interface and basic functionality. This demo's login page is shown in Figure 8.4. To exercise this demo live, readers can browse to `https://npfi.org/homebase-demo/` and log in as a volunteer as shown in Figure 8.4.

The process of embedding a live demo of a CO-FOSS product within a Web site is straightforward, though it requires coordination with the Web site

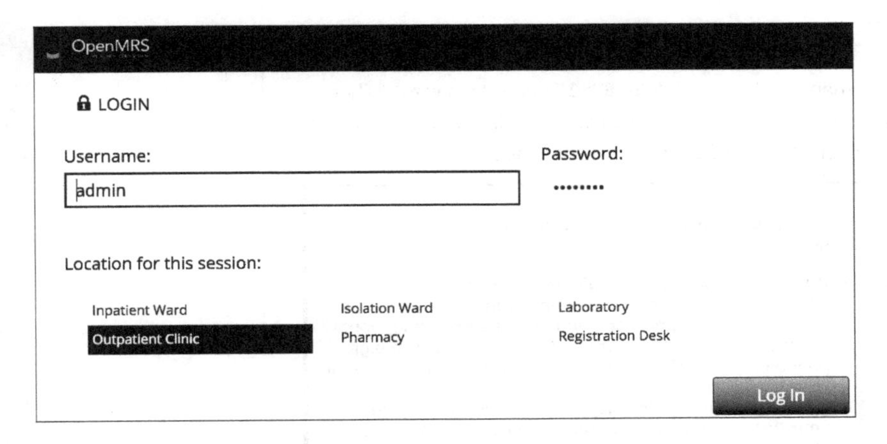

FIGURE 8.3 The OpenMRS on-line demo.

developer who will transfer the product's code base and demo database to the Web site's live server. Care must be taken in this process to anonymize the database to ensure that it does not reveal the identities of real persons.[3]

8.2.2 On-Line Help

On-line help is documentation integrated with the software itself, so that users can hit a "help" button whenever they need help with a particular feature. An on-line help page is usually organized so that it can be followed in a step-by-step manner, starting with the assumption that the user has never seen that particular feature before. It is organized as a collection of pages, with a table of contents that provides links to the individual pages.

The quality of an on-line help session is enhanced when the help pages themselves follow these organizational guidelines:

User access and navigation Most software systems provide users with a main menu, which identifies each of the major functions that they can perform. The organization of this menu provides guidance for organizing the on-line help pages themselves.

That is, identifying (at least) one help page for each entry in the main menu also identifies the major sections of the help document. On-line help should have a table of contents page that is organized in this way and contains links to each of the help pages themselves.

[3]Anonymizing a database so that it contains only fictitious persons and contact information is straightforward. The *Homebase* database does this, for example, by assigning all persons the surname "Jones" and bogus street addresses, phone numbers, and emails. Free tools are also available, such as `ARX`, for anonymizing a large database.

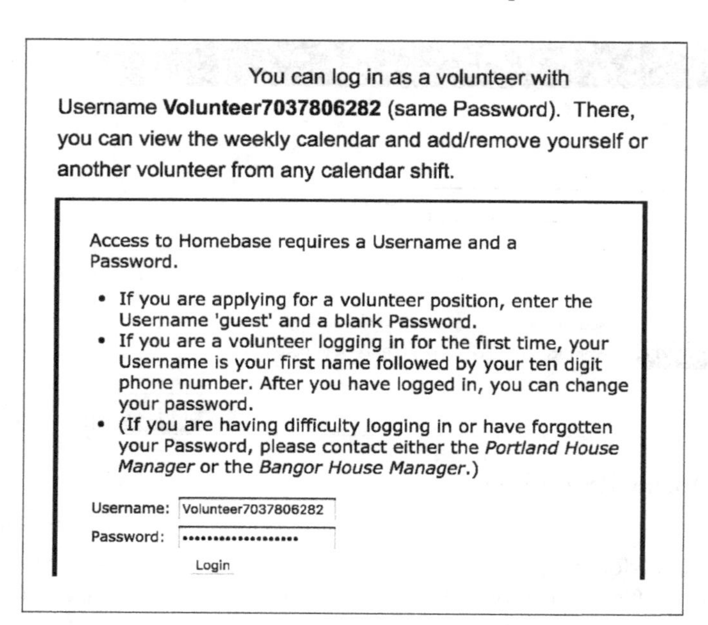

FIGURE 8.4 The Homebase on-line demo.

The table of contents and every help page should open in a separate window from the user's session, so as not to interrupt the user's interaction with the software itself while he/she is reading a help page. The correspondence between the table of contents and the main menu entries should be obvious to the user.

Step-by-step guidance Each on-line help page should address a single user task from beginning to end. Typically, a user task is accomplished as a series of steps (as in its original use case). Thus, it makes sense to present a task's help page as a series of steps to be carried out in order.

Each step should be described as a simple English-language sentence, using the principles of good writing introduced in Section 8.1. A step may often be accompanied by a visual picture (screen shot) that describes how that step is carried out.

Screen shots and thumbnails Each screen shot that accompanies a step in an on-line help page should be captured from a "live" software session using the anonymized database used for the demo version discussed in Section 8.2.1.

Because most users will not need to look at the full screen shot for each step on the help page, a "thumbnail" image (rendered at about 10% of the screen shot's full size) should appear by default. Thereby, only those users (maybe novice users) who need to view the full screen shot can click on the thumbnail to take a look at the full image, while others will be satisfied by reading the text alone.

Various free tools are available to facilitate the process of writing and generating on-line help pages using the guidelines suggested above. For a useful overview, visit `this Quora Digest article`. The on-line help pages that accompany our CO-FOSS projects were generated and integrated within the code base manually, as described in the next section.

8.2.3 Example: Homebase On-Line Help

The following discussion illustrates the above guidelines by illustrating the on-line help documentation that is incorporated into *Homebase*. Readers may consult the `code base` itself and the `on-line demo` for more details.

As a layered architecture, *Homebase* has domain, database, and GUI layers. Alongside these three functional layers, there are two other directories — a "tests" directory containing all the unit tests (discussed in Chapter 5) and a "tutorial" directory containing the on-line help pages.

Context-Sensitive Help

Each page in *Homebase* where the user is working has an associated on-line help page, which appears directly in a separate window when the user hits the **help** tab on the main menu shown in Figure 7.5(a).

For example, suppose the user wants to search for volunteers and lands on the search page shown in Figure 8.5.

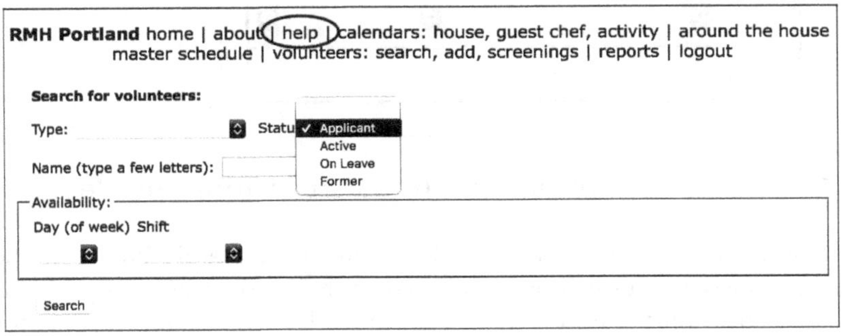

FIGURE 8.5 Context-sensitive help for the search page.

Hitting the **help** tab (circled) at the top of that page takes the user directly to the associated help page, whose first two steps are shown in Figure 8.6.[4]. So in this case, and many other similar cases, the user has no need to search the entire help table of contents to obtain direct help for the task at hand.

[4]To see the entire help page, log into the `Homebase demo` and hit "help.")

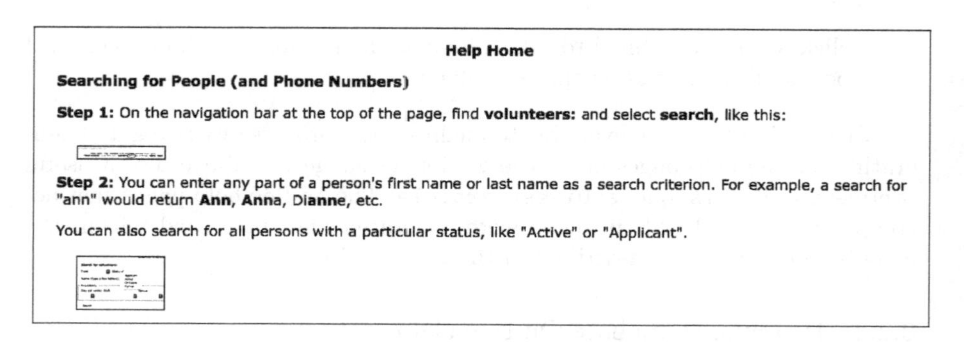

Help Home
Searching for People (and Phone Numbers)
Step 1: On the navigation bar at the top of the page, find volunteers: and select search, like this:
Step 2: You can enter any part of a person's first name or last name as a search criterion. For example, a search for "ann" would return Ann, Anna, Dianne, etc.
You can also search for all persons with a particular status, like "Active" or "Applicant".

FIGURE 8.6 The first two steps in the **Searching for People** help page.

Each step on a help page has an accompanying thumbnail image for a screenshot that illustrates that step. Hovering over a thumbnail enlarges that image to full size, as illustrated in Figure 8.7 for the thumbnail below step 2 in Figure 8.6.

Step 2: You ca
"ann" would re
You can also se
Search for volunteers:
Type:
Statu: ✓
Applicant
Active
On Leave
Former
Name (type a few letters):
Availability:
Day (of week) Shift
Venue
Search

FIGURE 8.7 Enlarged thumbnail in **Step 2** of **Searching for People**.

Searching for people is a pretty straightforward task overall, but it can be daunting for someone who has never worked with the software before. This help page thus gracefully initiates a new user into the process of searching for people using different search criteria. Performing this task two or three times should enable the user to search independently, without the need to refer to its associated help page in the future.

Help Table of Contents and Navigation

For broader access to the *Homebase* on-line help table of contents, users can hit the **Help Home** link at the top of any help page, such as the one shown in Figure 8.6. The help table of contents itself is shown in Figure 8.8.

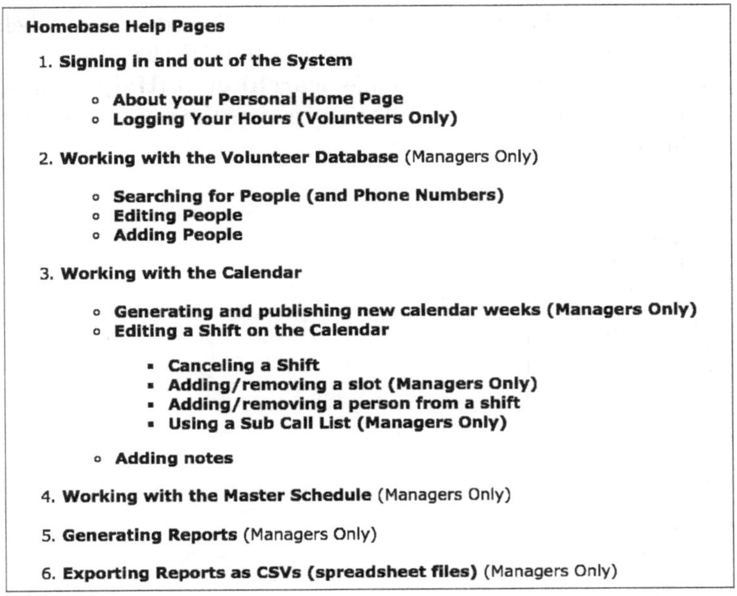

Homebase Help Pages

1. **Signing in and out of the System**
   - **About your Personal Home Page**
   - **Logging Your Hours (Volunteers Only)**
2. **Working with the Volunteer Database** (Managers Only)
   - **Searching for People (and Phone Numbers)**
   - **Editing People**
   - **Adding People**
3. **Working with the Calendar**
   - **Generating and publishing new calendar weeks (Managers Only)**
   - **Editing a Shift on the Calendar**
     - **Canceling a Shift**
     - **Adding/removing a slot (Managers Only)**
     - **Adding/removing a person from a shift**
     - **Using a Sub Call List (Managers Only)**
   - **Adding notes**
4. **Working with the Master Schedule** (Managers Only)
5. **Generating Reports** (Managers Only)
6. **Exporting Reports as CSVs (spreadsheet files)** (Managers Only)

FIGURE 8.8 The on-line help table of contents in *Homebase*.

The table of contents is organized into six major sections, with some sections linking to two or more help pages for a specific user function. For example, section 2 in Figure 8.8 is titled **Working with the Volunteer Database** and has links to three related help pages:

Searching for People (and Phone Numbers)

Editing People

Adding People

Selecting one of these links takes the user to its underlying help page in a separate window. For example, selecting the first link leads to the help page searching for people, whose first two steps are shown in Figure 8.6.

Help System Architecture

Each individual help page in the "tutorial" directory of the code base is associated with specific views in the user interface. The *Homebase* help pages are all coded in HTML with assistance from the jQuery UI library, and were written only after the user interface was fully developed.

To provide context-sensitive help such as this, the on-line help pages are integrated within the code base both physically and functionally. For example, in *Homebase*, the help files in the "tutorial" directory are listed in Figure 8.9.

Each file in the "tutorial" directory contains the HTML code for one help page, and its name matches that of the module in the GUI that renders its corresponding view. For example, the module **searchPersonHelp.inc.php** corresponds to the searchPerson.php module that renders the **person: search** page in the user interface.

FIGURE 8.9 Integrating help pages within the code base.

Figure 8.10 shows some of the HTML code in the help file named `searchPersonHelp.inc.php` that underlies Step 2 on the help page shown in Figure 8.6. The code on lines 12-19 that embeds a graphic image is particularly interesting, since it gives a clue about how to use HTML and jQuery tools to.display a thumbnail and enlarge it, as shown in Figure 8.7.

For organization, all the images for the tutorial pages are gathered in the subfolder "screenshots" within the "tutorial" folder. So line 16 reveals that the image itself is in the graphic file `searchpersonstep2.png`.

```
<p>
  <strong>Searching for People (and Phone Numbers)</strong>
  ...
<p>
  <B>Step 2:</B> You can enter any part of a person's first
  name or last name as a search criterion. For example, a
  search for "ann" would return <B>Ann</B>,
  <B>Ann</B>a, Di<B>ann</B>e, etc.
<p>
  You can also search for all persons with a particular status,
  like "Active" or "Applicant".<BR> <BR>
  <a href="tutorial/screenshots/searchpersonstep2.png"
    class="image" title="searchpersonstep2.png"
    target="tutorial/screenshots/searchpersonstep2.png">
        <img
    src="tutorial/screenshots/searchpersonstep2.png"
    width="10%" rel="popover" data-img=
    "tutorial/screenshots/searchpersonstep2.png"
    border="1px" align="middle"> </a>
  ...
```

FIGURE 8.10 HTML code for Step 2 in the help file searchPerson-Help.inc.php.

On line 17, the phrase `width="10%"` specifies the size of the thumbnail image to be displayed, while the phrase `rel="popover"` specifies that the image pop to full size when the mouse hovers over the thumbnail. The effect of this is shown in Figure 8.7.

The final point to make about help system architecture is to illustrate how the help pages are individually linked to the different user views that they are explaining. For example, how is the `searchPersonHelp.inc.php` page linked to the `personSearch.php` script that renders the view shown in Figure 8.5?

This linkage begins with the underlying code in the `header.php` file, where the `help` link in the main menu is displayed by the following line:

```
echo(' | <a href="' . $path . 'help.php?helpPage=' .
        $current_page . '" target="_BLANK">help</a>');
```

Now suppose the user is currently browsing the `personSearch.php` page and then clicks **help** in the main menu, as shown in Figure 8.5. This action opens a new page with the URL `help.php?helpPage=personSearch.php` in a separate window.

The code underlying this action appears in the script **help.php**, which uses an associative array to link each view with its associated help page. In this example, the view personSearch.php is linked with the help page searchPersonHelp.inc.php, by the array $assocHelp in the following steps:

```
$loc = 'personSearch.php';
$assocHelp[$loc]='searchPersonHelp.inc.php';
include('tutorial/'.$assocHelp[$loc]);
```

For more details about how this mechanism works, readers should consult the help.php script at **https://npfi.org/homebase-code/**.

8.3 OTHER USER SUPPORT

Other forms of user support include user training, feedback surveys, final presentations, and user forums. We briefly discuss each of these here, understanding full well that they may not be immediately or fully implementable at the end of a CO-FOSS project.

8.3.1 User Training

At the end of a new CO-FOSS project, live user training sessions can be organized for developers to work directly with new users to teach them about each function in the system. The goal of such a session is not only to train new users but also to further troubleshoot the new system for errors and other usability issues that had not occurred before. A user training session is also an excellent way to get feedback on the user documentation itself.

When *Homebase* was first completed in 2008, only the House Manager was familiar with the software. At that time, the need to establish staff buy-in had become an important goal. To that end, two training sessions were by the student developers. They used the on-line help pages to introduce staff members and volunteers to the software and its functionality.

The amount of user training conducted for our other CO-FOSS projects varied with the nature of the project and the user-friendliness of the software itself. In the case of *Homeplate*, the office administrator was familiar enough with its main user interface to not need further training. Moreover, the accompanying *Homeplate Mobile* Android app's user interface was so intuitive for the truck volunteers to use that they needed little or no training to record and upload data from their daily routes.

The same can be said for other *Homeplate*, *Homeroom*, and *BMAC-Warehouse* projects. All of these user interfaces had become very intuitive to assimilate and use, and any volunteer training needed by these systems was done informally by the managers who had mastered the software by the end of the project.

8.3.2 Feedback Surveys

At the end of the training session, it is also helpful to ask newly-trained users to provide feedback on their experience with the new system, and also to hear their thoughts about the general value of the new software in their working environment. Users usually provide frank and insightful advice, not only for debugging purposes but also in regard to the overall utility of the system and the possibilities for adding new functionality in the future.

Eleven persons attended the two workshops conducted by the developers for the original *Homebase* system in 2008. These included four RMH staff members and seven RMH volunteers. Workshop attendees completed a questionnaire where they rated the quality of the software and the chances that they would use it and be able to teach it to other volunteers in the future. The responses from all 11 participants are summarized in Table 8.1.

TABLE 8.1 *Homebase* User Questionnaire and Results

Question Evaluate *Homebase* with regard to:	**Average Response** **(1 = poor, ...** **5 = excellent)**
1. Logging in	4.7
2. Headings and logo	4.9
3. Searching for volunteers	5.0
4. Adding, deleting, and changing volunteers	4.7
5. Viewing the calendar	5.0
6. Changing the calendar (sub call lists)	5.0
7. Managing the calendar (generating weeks)	5.0
8. Editing and managing the master schedule	5.0
9. Understanding the on-line help	5.0
10. Logging out	5.0
11. Workshop effectiveness	5.0
12. Quality of client/developer collaboration	5.0

All 11 participants uniformly agreed that the software was "extremely user friendly" and that they could see themselves "easily teaching another volunteer" how to use the system. Several commented that the software should be useful at many other Ronald McDonald Houses and similar organizations in the future.

Some of our later CO-FOSS projects also gained user feedback more informally. The responses from users have always been positive and constructive. For example, several new features added to later versions of *Homebase*, *Homeroom*, and *Homeplate* were the results of suggestions made by users over the years since the software was first developed.

An important source of feedback for adding new features to an existing CO-FOSS project is its issue tracker. We have found that users post suggestions for new features freely and thoughtfully alongside any functional issues that they encounter with the software itself. We will return to this discussion in Chapter 9, which takes a longer view of the life of a CO-FOSS product.

8.3.3 Final Presentations

At the end of any 1-semester CO-FOSS projects course, it is natural to schedule a class meeting where student developers present the results of their project to the rest of the class and other visitors.

This final presentation provides an opportunity for the students to meet with their client, either face-to-face or via videoconference, to present their experience developing the software and discuss the entire project retrospectively. In short, the final presentation creates a sense of closure for the project and the course.

In most of our projects, the final presentation was attended by several visitors: other interested faculty, the client organization's director and staff, and a local software developer who would inherit the software following the presentation.

The presentation should have an agenda, so that two or more project presentations can have "equal time" and enough time will be left for questions and informal discussion. Table 8.2 shows the agenda for our 2015 course, which had 2 projects and 9 student developers.

TABLE 8.2 Agenda for a Final Presentation

11:00	Introductions	
11:05	RMHP Homebase Project	
	Student developers:	Adrienne, Connor, Xun, Phuong
	Clients:	Joanna and Sue (via videoconference)
11:30	BMAC Warehouse Project	
	Student developers:	Noah, Dylan, Moustafa, David, Luis
	Client:	Jeff
11:55	Informal discussions	

It is also useful to provide a framework to the students for organizing their presentation, one that would allow them to present their project to uninitiated observers and evaluate their outcomes. The framework we used for the 2015 course required that each presentation address the following questions.

1. What problem does your project solve?
2. How does your software work overall? (this is the live demo part)
3. What were the highlights of your project/experience?
4. What were the lowlights of your project/experience?

Both teams gave their presentations using PowerPoint and the overhead projector at the front of the room. The presentations were also screen-shared with the clients who were participating via videoconference, which brought them as close to being "in the room" as possible.

For the live demo, students displayed their CO-FOSS product using the sandbox server. This too was screen-shared with the clients who were connected via videoconference.

8.4 CLOSURE FOR STUDENTS

Once the student developers complete the project, their work will be done and they will either move on to other courses or else graduate. In either case, the students will need to achieve closure from this experience. This section briefly considers two activities that will contribute to this closure: self-assessment and leveraging the experience.

8.4.1 Self-Assessment

At the end of the project, it is important for student team members to reflect on their own experiences and accomplishments, as well as on the contributions of other teammates. This can be done via a self-assessment, which is best conducted confidentially so that students can make more candid assessments. For example, here are the four self-assessment questions addressed by each student in our 2015 course:

1. What was the quality of your team's overall effort this semester? Include such factors as communication, workload distribution, code sharing, etc.
2. What was the quality of your personal contributions to this effort? Identify 3 specific examples of your most important contributions.
3. How did the overall quality of your contributions compare with that of each of your teammates?
4. Were there any assignments during the semester in which you personally could have done better? Which one(s)?

The 9 students completed these self-assessments thoughtfully. Overall, each student was proud of his/her team's efforts and outcomes. Each one was also quite frank and introspective in evaluating their own work and that of their teammates.

Self assessments can also be helpful to the instructor, not only for helping to determine final course grades but also to gain a better understanding of how future CO-FOSS project courses can be improved.

8.4.2 Leveraging the CO-FOSS Experience

Student experience with a CO-FOSS project course is different from that provided by most other academic courses in the curriculum. That difference comes from both the community service aspect and the professional experience aspect of the project. Students can leverage this experience beyond the course in different ways.

For example, a student's code commits in a CO-FOSS project that sits on a public hosting service (such as GitHub) are part of their professional portfolio that they can present when applying for an internship, a job, or another FOSS project. Moreover, moving on from this CO-FOSS project to becoming a developer on a larger, more mature FOSS project can introduce a student to more automated processes, more complex software packages, and a larger developer and user community.

Whether or not the student moves from this experience to enter the software profession, this CO-FOSS experience provides unique opportunities for engaging first-hand in teamwork, community service, professional writing, and other important activities that are not normally included in academic courses. Acknowledging this experience in one's professional resume and job interviews is thus another important by-product of this course.

8.5 SUMMARY

This chapter has discussed the principles and practices of developing good user support for an open source software project.

Since it is a published work, user documentation must be well written. Thus, it must use principles of good (technical) writing. It must be clear, concise, and sensitive to a variety of user reading levels and skills.

Four prominent types of user documentation are particularly valuable—on-line help, user manuals, FAQs, and live demos. This chapter discusses and demonstrates the architecture of embedded on-line help pages within the software using *Homebase* as an example.

Finally, this chapter discusses other forms of support, such as user training, feedback surveys, developer self-assessments, and final presentations of the software itself. It concludes with some brief thoughts about project closure for the student team members.

8.6 MILESTONE 8

1. Consider the user documentation for your CO-FOSS project. If that documentation already exists in some form, briefly critique that documentation and rewrite it. If that documentation does not exist, write it. In either case, your written documentation should take the form of a user manual and should follow the principles of good writing introduced in this chapter.

2. Briefly characterize the audience of users who will use your software. Include in your characterization factors like reading level, cognitive skills, native language, and education level. Evaluate the quality of your

project's written documentation with respect to how well it matches the characteristics of its wider audience of users.

3. Following the architecture of on-line documentation for *Homebase* discussed in this chapter, develop on-line documentation for your CO-FOSS software. Ask your client to review it and suggest changes that will improve its usability.

4. Sketch the design of a training session that will introduce users to your software. Include a survey form that asks users to evaluate each of its features, along the lines suggested in Figure 8.1.

III

Deployment Stage

CHAPTER 9

Continuing the Journey

"It takes a whole village to raise a child."
—*African proverb*

Now that the CO-FOSS development project is complete, steps should be taken to install the software on the client's server and ensure its long-term viability, both for the client and for the larger open source community.

These steps include finding a developer, installing the software on the client's server, and releasing the code to a public repository for others to download, adapt, and reuse. These steps may also include organizing the product so that it can evolve into a more mature FOSS artifact.

This chapter discusses the key steps that the instructor can take to accomplish these goals. It also speculates on how a new CO-FOSS project can evolve into a more mature FOSS product, taking into account the growth of larger developer and user communities, the creation of public forums, and various other organizational considerations that surround such growth.

9.1 TRANSITIONING TO PROFESSIONAL SUPPORT

As the chief architect, lead programmer, and team leader for a newly-completed CO-FOSS product, the instructor must find a hosting service and a professional developer that will provide permanent support for the product.

A smooth transition of the product from its development stage into permanent use will be facilitated by the following preconditions:

1. The client and the hosting service both support this transition.
2. The hosting service already supports the client's Web site.
3. The new software is technically compatible with the client's server (Web site).
4. The professional developer can provide ongoing bug fixes at a reasonable cost to the client.

Item 4 above may pose the main challenge for a successful transition. That is, finding a professional software developer to install the new software on the client's server and provide regular bug-fixing and issue resolution can be challenging. We have found that many/most professional software developers are too occupied with their primary work commitments to provide such support.

However, a CO-FOSS project does provide opportunities for finding ongoing professional developer support. For example, a professional who has been working with the student team during the development phase may want to continue with the project through its transition into productive use. Moreover, the professional developer may also see the project as a recruiting opportunity, which is an additional incentive for providing ongoing support.

Another opportunity for finding ongoing professional support may come from the student development team itself. That is, a graduating senior may be able and willing to support the product after it has been deployed. While not having the experience of a more seasoned professional, that student does understand the code well and also knows the instructor who designed the project in the first place.

So these are two possible avenues through which ongoing support for a CO-FOSS product can be obtained.

9.1.1 The Hand-Off

The above four preconditions for a smooth transition reflect our own experiences. On the one hand, we have found that the client's own Web hosting company will easily embed the new CO-FOSS product into the client's Web site running on their server.

But on the other hand, we have also found that the Web hosting company does not normally provide programming resources that support cost-effective bug fixing and feature enhancement after the product has been deployed. That company's resources tend to be more focused on designing Web sites, which involves a different skill set from the one needed for fixing bugs in the source code. In addition, non-profit clients usually have limited technology budgets, which makes it unrealistic for them to pay a professional developer a market-competitive fee for making bug fxes and feature enhancements.

So most of our CO-FOSS projects' bug fixes and feature enhancements have been performed by `the Non-Profit FOSS Institute (NPFI)` on a *pro bono* basis as part of its mission. In one exceptional case, that programming support was provided for several years by the client's Web developer on a billable fee-for-service basis. But that support was finally dropped when the programmer relocated out of the area, after which NPFI resumed providing it.

Whenever one of our CO-FOSS projects was completed, its code base and initial database were turned over to the client's hosting company for installation on their Web server and integration with their Web site. All these Web sites were PHP/JavaScript/MySQL friendly, so the hand-off amounted to passing two files to the Web host — a MySQL database dump and a copy of the code base downloaded from GitHub — along with the security credentials that link the two together.[1]

Altogether, this hand-off step required a couple of email exchanges between and instructor and the Web hosting company, and then a few hours' work by the Web hosting company to create and upload the database and install the code base on the client's server. After installing the software, the company folded the cost of hosting the CO-FOSS product into the client's monthly hosting fee for the Web site itself, thus keeping the marginal cost of CO-FOSS product hosting support at a minimum.

9.1.2 Case Studies

The following paragraphs provide more details on the hand-off and ongoing support arrangements for each of our currently-installed and operational CO-FOSS products.

Homebase Hand-Off and Support

First developed in 2008 for the Ronald McDonald House in Portland, ME, the *Homebase* code base and database were installed on the client's server by `Artopa, Inc.`, a local Web developer that manages the House's Web site and server. All these services, including *Homebase* hosting support, are provided through a contractual arrangement between the RMH and Artopa.

Three different *Homebase* upgrades were made in 2011, 2013, and 2015 by different student teams as separate CO-FOSS projects. The 2015 upgrade extended *Homebase* for use at a second Ronald McDonald House in Bangor, ME in addition to the original RMH in Portland. Artopa managed the installation of these upgrades as part of their contractual arrangement with RMH.

All *Homebase* bug fixing and issue resolution are provided by NPFI, which also manages its source code repository and issue tracker on GitHub. Bugs are reported to NPFI by RMH staff, NPFI fixes the bugs, and NPFI passes the bug fixes to Artopa who installs them on the live server.

RMHP-Homebase Hand-Off and Support

Developed in 2015 for the Ronald McDonald House in Providence, RI, the code base and database were installed on the client's server by `Coursevector`, a Web developer located in Pennsylvania. Coursevector also manages the House's Web site. These services, including *Homebase* hosting support, are

[1] A similar level of simplicity should be achievable for installing a CO-FOSS product developed in Java, Python/Django, or Ruby on Rails. In those cases, the client's hosting service needs to support the Java, Python, or Ruby language and its associated SQL database implementation (MySQL, PostgreSQL, or SQLite). Beware, however, that the client's own Web hosting service may or may not support all of these languages. For example, the Godaddy hosting service does not currently support Ruby, while HostGator and DreamHost both do.

provided through a contractual arrangement between RMH Providence and Coursevector.

All the *RMHP-Homebase* bug fixing and issue resolution are provided by NPFI, who also manages the source code repository and issue tracker on GitHub. Bugs are reported by RMH Providence staff by posting them on the issue tracker. NPFI communicates bug fixes with Coursevector through the House's on-site technical support staff.

Homeroom Hand-Off and Support

Developed in 2011 by a student team for the Ronald McDonald House in Portland, ME, the *Homeroom* code base and database were installed by Artopa on the client's server. Later versions of *Homeroom* were developed by different student teams in 2013 and 2015 as separate CO-FOSS projects. A variant of Homeroom was developed in 2017 by NPFI for use at the Ronald McDonald House in Bangor, ME.

All these services are provided through a contractual arrangement between the RMH and Artopa. All *Homeroom* bug fixing and issue resolution has been provided by NPFI, which also manages the *Homeroom* code repository and issue tracker on GitHub. Bugs are reported to NPFI on the issue tracker by RMH staff, NPFI fixes the bugs, and NPFI passes the bug fixes to Artopa who installs them on the live server.

Homeplate Hand-Off and Support

Developed in 2012 for `Second Helpings, Inc.` in Beaufort County, SC, the *Homeplate* code base and database were installed on the client's server by `Progressive Technology`, a Web developer located in Bluffton, SC.

Progressive also developed the Second Helpings Web site, which is hosted by Godaddy and managed by a Second Helpings volunteer. So *Homeplate* hosting is also provided by Godaddy. Godaddy hosting support is provided for a nominal annual fee, and that includes excellent 24/7 technical support as well.

From 2013 to 2017, Progressive also provided bug fixes and feature enhancement for *Homeplate* on a fee-for-service basis. They discontinued this support in 2017, at which time NPFI began providing bug fixes and feature enhancements. NPFI also manages the *Homeplate* code repository and issue tracker on GitHub. Issues are posted by the Second Helpings office manager. NPFI fixes the bugs and then installs them on the live server.

BMAC-Warehouse Hand-Off and Support

Developed in 2015 for the `Blue Moutain Action Council` in Walla Walla, WA, the *BMAC-Warehouse* code base and database were installed on the client's server by `Vivio Technologies`, a local technology services company.

The BMAC Web site is hosted by Vivio Technologies, and the *BMAC-Warehouse* software is embedded within the Web site.

All *BMAC-Warehouse* bug fixing and issue resolution has been provided by NPFI, which also manages the source code repository and issue tracker on GitHub. Bugs are reported to NPFI by BMAC staff, NPFI fixes the bugs, and NPFI passes the bug fixes to Vivio who installs them on the live server.

9.2 PROJECT EVALUATION AND CODE RELEASE

The process of assessing, packaging, releasing, and distributing code to the larger open source community can be simple or complex, depending on the nature of the project. On the one hand, for a CO-FOSS project that fills a unique non-replicable software need for its client, the final version of its public source code repository provides a minimal release that may or may not be downloaded for reuse by others.

On the other hand, our experiences suggest that a CO-FOSS product can have a much wider potential market than the developers originally imagined, and so releasing code for this market may involve several additional steps that would help disseminate it for others to reuse.

Since the code base is in its infancy and relatively small, and since the client is in close communication with the developer, this is the moment for them to ask questions about the long-term reusability of the new CO-FOSS product and plan its release accordingly. Several key questions come to mind:

> "Who are the potential new clients for this product?"
>
> "Which open source license should govern its distribution for reuse?"
>
> "Where and how should the first release be published?"
>
> "How can we assess the maturity of this product as it evolves?"

The following sections address these questions in more detail.

9.2.1 Potential New Clients

Many CO-FOSS projects have been developed for specific clients that are part of a larger network of similar organizations with similar software needs. The following paragraphs discuss three larger networks that illustrate the potential for CO-FOSS project redeployment and dissemination.

Volunteer and Resource Scheduling

For example, *Homebase* is currently in use at the Ronald McDonald Houses in Portland, ME, Bangor, ME, and Providence, RI, while *Homeroom* is actively in use by two of those Houses. However, these are just three of more than 300 Ronald McDonald Houses throughout the nation, each one having a similar mission – to provide free overnight housing for families with children in nearby hospitals and to rely on volunteers to cover the day-to-day duties of running the House.

That said, every such Ronald McDonald House has a need for technology support that facilitates the tasks of volunteer scheduling and room scheduling, and each one fulfills that need differently. Some still do this scheduling manually. Others choose to pay for proprietary software that replicates the capabilities of *Homebase* and/or *Homeroom*, which carry little or no ongoing support cost.

But the potential for adapting *Homebase* and *Homeroom* for reuse in many more of these Houses is significant, since they all operate under the same licensing guidelines from the global Ronald McDonald House Charities. We are aware of several such Houses that would benefit from a new CO-FOSS project that would customize either *Homebase* or *Homeroom* to fit their specific scheduling needs.

Beyond the Ronald McDonald House network, many other local non-profits that rely heavily on the scheduling of volunteers are potential new clients for *Homebase*-like software as well. Many of these may be candidates for new CO-FOSS projects that use *Homebase* as a starting point.

Food Rescue and Redistribution

For another example, *Homeplate* is currently in use at Second Helpings in Hilton Head, SC to provide technology support for keeping track of food rescued and redistributed to needy clients throughout three counties in rural South Carolina. On the West coast, *BMAC-Warehouse* is currently in use by the Blue Mountain Action Council in Walla Walla, WA to keep track of food donated to their warehouse and redistributed to needy clients in that region. Both organizations are independently-managed non-profits who provide vital services to feed the hungry in their own local areas.

But these are just two of many such organizations throughout the nation that share the same mission: to rescue food and redistribute it to feed the hungry at low cost in their local community. `Feeding America` is a national network of over 200 food banks, each of which oversees a network of partner organizations that distribute food at no cost to fulfillment agencies in their community. For example, Second Helpings is a partner organization for the Low Country Food Bank in Charleston, SC, which is the Feeding America food bank serving coastal South Carolina. Similarly, the BMAC Warehouse is a partner with Second Harvest in Spokane, WA, which is the Feeding America food bank serving Eastern Washington.

So in all, there are hundreds of local food distribution organizations throughout the nation, like Second Helpings and the BMAC Warehouse, that partner with Feeding America food banks to deliver food to organizations that feed the hungry in their local communities. Our view is that most of these organizations do not have sufficient software support that would facilitate their day-to-day record keeping that is required by their own benefactors. Most, like Second Helpings and the Blue Mountain Action Council, are non-profit organizations that rely on volunteers and donations to keep their operations running. Many of these may be candidates for new CO-FOSS projects that use *Homepate* or *BMAC-Warehouse* as a starting point.

Agricultural Operations

As a third example, the CO-FOSS project FarmData, introduced in Chapters 1 and 7, was originally developed in 2015 by Dickinson College computing students for the Dickinson College Farm. FarmData has since been downloaded and installed at several additional farms. It also has an active Facebook group with over 150 members.

The potential for widespread dissemination of FarmData and its companion CO-FOSS project AnimalData (again see Chapters 1 and 7) to many more clients throughout the nation is enormous. Small and medium-slzed farms, of which there are about 2 million throughout the US, have increasing needs for cost-effective technology support to help manage their crop and animal operations. So there are a lot of opportunities for an undergraduate computing program to focus a student project on redeploying FarmData or AnimalData to fit the needs of a local small or medium-sized farm in their area. It is part of NPFI's mission to help facilitate the successful creation, completion, and deployment of such projects.

9.2.2 Licensing Choices

New CO-FOSS projects should always be licensed at their outset, before development begins. Many such projects, including our own, are initially licensed under the GNU General Public License (GPL). As we learned in Chapter 1, this license is restrictive, since it requires that derivatives of the software also be licensed under the GPL and cannot be relicensed under a proprietary license.

But once the project is completed and it begins evolving into a broader venture, the question of licensing might be reconsidered. For example, if a new CO-FOSS product could potentially be incorporated into an established proprietary product, it might be relicensed under a less restrictive license, such as the LGPL (Lesser GPL) or the MPL (Mozilla Public License). While they are less restrictive, the LGPL and MPL still protect the FOSS nature of the original software while permitting its future use in proprietary products.

Other non-GPL licenses, such as BSD and Apache, also allow developers to relicense derivatives of the software under less restrictive terms, even including a proprietary license. These are, however, still considered FOSS because they also protect users' freedom to use, study, share, and modify the original software.

The issue of FOSS licensing is a broad and complex one. Every individual situation differs. For example, the Open Energy Dashboard discussed in Chapter 2 is a CO-FOSS project licensed under a Mozilla Public License because it has components that are MPL-licensed. For more details and discussions of the various licenses, readers should review Chapter 1, consult the appropriate pages of the `FSF` and `OSI` Web sites, and even seek (free) legal advice.

Not surprisingly, some of the philosophical and political differences between the free software and open source software camps are played out in this particular decision process. As further guidance, readers may want to review `Richard Stallman's discussion` of when it is appropriate or not to use the LGPL, according to the free software philosophy.

9.2.3 Project Hosting Alternatives

Early in the life of an emergent CO-FOSS product, a decision should be made to use a Web-based hosting service for managing releases and encouraging communication among all of its developers and users.

GitHub

GitHub is fully appropriate for managing a new CO-FOSS project, since it supports version control and issue tracking. A more mature project requires additional tools to support the needs of its growing community of developers and users. To accommodate those additional needs, GitHub also supports `project milestones` and `project boards`. We have not used those features in our CO-FOSS projects.

However, several alternatives to GitHub are also available for hosting emerging and mature open source projects. These include GitLab, Bitbucket, and SourceForge. They all add support for larger developer and user communities in the form of group milestones, configurable issue boards, and many other functions. The following paragraphs summarize the major features of three such alternatives.

GitLab

`GitLab` is a powerful, secure, and feature-rich platform for managing software development and operations (DevOps). It supports group milestones, configurable issue boards and group issues, powerful branching tools, file locking, merge requests, custom notifications, and project roadmaps.

You can self-host GitLab on your own server or use hosted GitLab services for a price. You can import your GitHub repositories to your GitLab account.

Bitbucket

`Bitbucket` is a powerful, fully scalable and high-performance development platform designed for professional software development. Bitbucket is free for educational uses and open source projects with a small number of users. GitHub repositories can be easily imported to Bitbucket, which supports both Git and Subversion version control systems.

Bitbucket's features include pipelines, code search, diff views, pull requests, small project tracking, smart mirroring, issue tracking, and branch permissions for ensuring controlled access among developers. Bitbucket also supports Git Large File Storage (LFS). an unlimited number of private repositories, and seamless workflow integration that ensures continuous delivery.

SourceForge

`SourceForge` is a free open source software development and distribution platform built especially to support the maturation of open source projects. SourceForge began in 2000, when it was known as the Open Source Development Network (OSDN), a division of VA Linux Systems. It was set up to serve as a gateway for collaborative software development.

SourceForge offers code repositories, an open source directory, and tools for integrated issue tracking and project documentation. It also supports forums, blogs, and mailing lists. SourceForge is currently used to host over 500,000 projects and serves over 33 million monthly users and 4 million downloads per day. However, only a fraction of these projects are active.

SourceForge provides the following broad services:

Search: SourceForge contains thousands of open source software products in a wide variety of application categories. Powerful search features allow users to find software on SourceForge that fits their area of interest.

Community Building: Projects hosted on SourceForge are always looking for developers to join in their effort. Users can create a developer account and participate in fixing bugs by providing patches with one of the existing projects. They can also join in a project's forum, blog, or mailing list.

Project Hosting: Registered users can create their own development project. The registration process is open and accessible and provides a wide range of tools and services to support project development.

Among the services provided to the projects hosted by SourceForge are the following:

Code Hosting: The project's source code can be hosted on one or more of SourceForge's versioning servers: Git, Subversion, and Mercurial.

Web Hosting: SourceForge will host the project's Web site as a subdomain—i.e., with the URL myproject.SourceForge.net. It provides shell access for Web sites as well as traffic analytics.

Software Distribution: SourceForge provides easy-to-use tools to package your software for download and distribution. Its servers can detect the user's platform (Mac, Windows, Linux) and direct users to the appropriate version. Downloads are supported by a mirror network that spans five continents.

Issue Tracking: SourceForge provides extensive issue tracking and reporting services for each project. Tickets support Markdown for formatting and attaching files. Tickets can be organized with milestones, custom fields, and labels, can be efficiently searched, and can be organized by threaded discussion.

9.2.4 Maturity Assessment

After a new CO-FOSS project evolves into a full HFOSS product by gaining new developers and users beyond the original ones, it is important to be able to assess the maturity of the software itself. Several different models for measuring the maturity of an emerging software product have been proposed. Maturity models provide clear quantifiable measures for evaluating the maturity of a software project, open source or otherwise.[2]

Several factors are considered when evaluating the maturity of a young open source project and its community. Most of these are no different from those used for measuring the maturity of a proprietary project. These factors include quality assurance, scalability, security, performance, adoption, and support. However, some measures are specific to open source projects, such as community strength, community governance, support, and IT management.

The first measure of an open source project's quality is the strength of the community that surrounds it. A strong community can provide a wealth of diverse input. In comparison, a proprietary product can only benefit from the input of its owner's employees.

A second measure of quality includes the licensing terms and intellectual property management policies and controls in the project. As we saw in Chapter 1, several popular open source licenses have proven to be effective.

Finer-grained methodologies used for assessing an open source software product are defined in the `OpenSource Maturity Model (OMM)`. The OMM defines three basic maturity levels for open source software products:

1. Basic level, which is reached by adopting a few basic practices in the FOSS development process,

[2]Maturity models are not new in the software engineering industry. Capability Maturity Model Integration (CMMI) is one such example [1]. Originating in 1991, CMMI covers best-practices for planning, engineering, and managing product development and maintenance.

2. Intermediate level, which adds a product roadmap, relationships among developer and user communities, project monitoring, and quality assurance, and
3. Advanced level, which adds production integration, risk management, third party assessment, reputation, and contributions from software companies.

The following list identifies the key characteristics of the basic level of maturity under the OMM:

Documentation
Use of established widespread standards
Quality of test plan
Licenses
Technical platform
Number of commits and bug reports
Maintainability and stability
Configuration management
Project planning
Requirements management
Product roadmap

So the OMM provides a framework for assessing the quality of an open source project using a number of detailed criteria and related quantitative measures. However, a simpler way to make a quick judgment about project quality is to ask whether each core developer has at least 2–3 months experience on the project, the project has many more users than developers, and the following actions have been taken:

the source code has been placed in a repository for public download,

the project has established a public forum where users can post bug reports and queries about the system's adaptability to new users, and

the development team is open to taking on new members who may volunteer.

For a new project, it may take a few months to measure how many downloads have occurred, how many users have participated in the public forum, how many new developers have joined the team, and how many user posts have been made in bug and feature discussions. The results of these measures can provide good indicators about the future viability of the project.

Mitigating against these quality measures are measures of the risks involved in continuing a new open source project, as opposed to the alternative of simply letting the project go dormant. Questions that help assess the risk of continuing a project include:

Does the project have the (human and material) capital to go on?
Does it have the need/demand for new features or applications?
Are its developer and user communities sustainable for the projected life of the project?
Does the project have a stable and adaptable code base?
Is the code base running on a platform that is likely to remain stable for the foreseeable future?

A software developer can also use these sorts of assessment criteria to help decide whether or not to join an existing open source project. If the project is well rated in these areas, it may be more attractive for the developer to join and make useful contributions.

9.3 SOFTWARE MAINTENANCE AS A COMMUNITY ACTIVITY

A newly-established CO-FOSS project should already have an issue-tracking activity where the client has posted several new issues and the developers have addressed them. This activity should continue and grow alongside the software itself as the project matures and adds new users and developers. At that point, the project's issue tracker itself may evolve into one or more user-developer discussion forums, as suggested above and presented more carefully in Section 9.4.

In this section, we drill down on the kinds of interactions that can occur between the user and a professional developer after the CO-FOSS product has been installed on the client's server. First, we present a case study of an actual user-developer interaction that took place while one particular bug in *Homebase* was fixed in 2010, two years after it was first installed. Second, we summarize the overall experience of NPFI in fixing bugs and adding new features to support *Homebase*, *Homeroom*, *BMAC-Warehouse*, and *Homeplate* throughout the period 2010-2018.

9.3.1 Fixing Bugs: A Case Study

Most bugs appear at the time a user is interacting with the system through the user interface. Thus, the starting point for correcting a bug would naturally begin with a discussion between the user and the developer. To illustrate this communication in detail, we trace the activities of an actual debugging episode that occurred with the *Homebase* software.

When this episode took place in 2010, almost two years had passed since *Homebase* was initially installed. At that time, *Homebase* issue tracking was done by email exchanges – later that activity was transferred over to an issue board that accompanied its `GitHub source repository`.

In this episode, one developer, Alex, was maintaining the code base and another, Ellis, was managing the server where *Homebase* was running. Gina, one of the original clients, was the primary user of the software. The entire episode has a distinctive workflow in which all three parties played essential roles. This workflow is described in the following paragraphs.

User-Developer Discussion

The episode begins when the user discovers a problem when trying to search and edit an entry in the personnel database.

> **From: Gina**
> **Date: Friday, March 5, 2010 12:08 PM**
> **To: Alex**
>
> Hi Alex-
>
> Hope you are doing well.
>
> Homebase question: I now have three separate applications that have been submitted and were entered with two names separated by &. I am unable to open them and therefore cannot contact the applicants to let them know they need to resubmit without the "&" symbol. Is there any way to permanently remove these? They sit under the Open Applications.
>
> Heading off on vacation for a week so no hurry on responding.
>
> Best,
> Gina

Debugging Activities

The next step in this episode involves a series of developer activities to identify the bug, correct it, and test the correction.

To identify and fix this bug, the developer must first reproduce it using the "sandbox" version of the code base. This involves navigating to the **volunteers : add** menu in the user interface and adding a new person to the "sandbox" database whose first name contains an ampersand, such as "a & b applicant" or "John & Mary Jones." The result of this step is shown in Figure 9.1, using the **people: view** menu to list all persons in the database.

The next step in fixing the bug is to locate where the software fails when working with this particular data. If the developer selects the **edit** button beside the entry for "a & b applicant," the error message shown in Figure 9.2 is raised.

This message and its accompanying URL provide some important information:

> The module for the view that appears when this error occurs is `personEdit.php`, as shown in the URL.

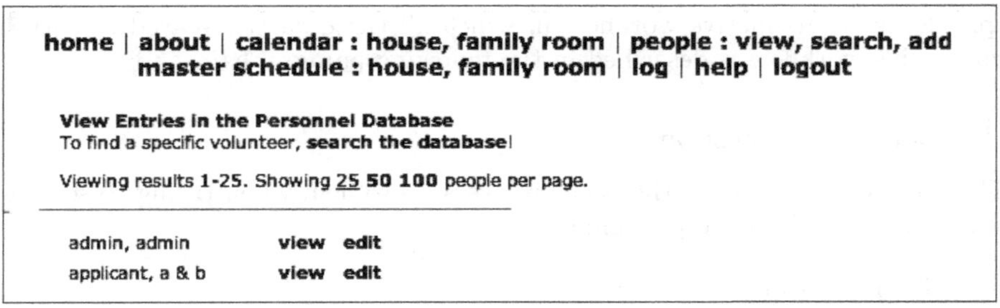

home | about | calendar : house, family room | people : view, search, add master schedule : house, family room | log | help | logout

View Entries in the Personnel Database
To find a specific volunteer, **search the database!**

Viewing results **1-25**. Showing 25 **50 100** people per page.

admin, admin **view edit**
applicant, a & b **view edit**

FIGURE 9.1 Reproducing the bug.

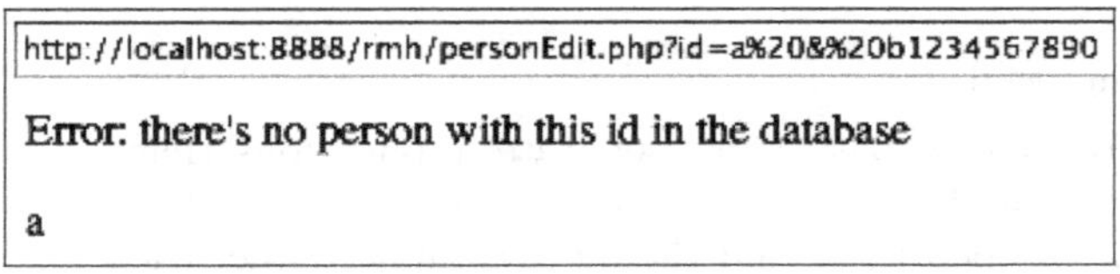

http://localhost:8888/rmh/personEdit.php?id=a%20&%20b1234567890

Error: there's no person with this id in the database

a

FIGURE 9.2 Locating the defect.

The URL shows that the argument passed to this module is `id=a%20&%20b1234567890`.[3]

The user interface module initiating the call is viewPerson.php.

The bug is exposed by the developer's knowledge about how HTML query strings are formed and punctuated. Specifically, the "&" character is used to separate individual arguments in the query, and so cannot be embedded as part of a single argument. Thus, this query string is parsed as two arguments, "a%20" and "%20b1234567890" rather than the single argument "a%20&%20b1234567890" that was intended.

There are a couple of options for fixing this bug, each of which involves tightening the requirements for storing the first name of a new Person in the database. Since each row in the dbPersons table has a primary key formed by concatenating a person's `first_name` with the primary phone number, the `first_name` should not contain the character "&."

Thus, when a user enters a first name for a new volunteer, like "John & Mary" or "a & b", all instances of "&" in that entry should be replaced by a reasonable substitute, like "and." The user interface module where this correction should be made is the point where the "&" is first entered, which is the personEdit.php module.

That module contains the function `process_form`, which processes new Person entries. The first line of that function reads:

[3]Recall that a user id is the concatenation of the user's first name and phone number, so in this case the user id being passed is a%20&%20b1234567890.

```
$first_name = trim(str_replace('\\\'','',
    htmlentities($_POST['first_name'])));
```

Literally, this line retrieves the value of `first_name` that was typed by the user and stored in the `$_POST` array, and then replaces special characters by their equivalent HTML representations (for example, blank characters are replaced by %20), removes all single quotes, and trims all leading and trailing blanks. The result is assigned to the `$first_name` variable, which becomes part of a new row added to the dbPersons table.

To correct the bug, the developer can modify this line so that it additionally replaces all instances of "&" by "and" before assigning the result to `$first_name`. The following replacement line of code is appropriate:

```
$first_name = trim(str_replace('\\\'','',
    htmlentities(str_replace('&','and',$_POST['first_name']))));
```

To test this correction, the developer should run editPerson.php and add a person to the database that has "&" in their first name, like "a & d." If all goes well, a new view of the persons list should show that new person's first name with "&" replaced by "and," as shown in Figure 9.3.

To complete testing the fix, the developer must be sure that selecting "view" or "edit" for a person with "and" in their first name (e.g., "a and d" in Figure 9.3) now navigates successfully. This is confirmed in the display shown in Figure 9.4.

Developer-Developer Discussion

Once the fix has been made and tested, the developer must pass this change over to the developer who is maintaining the live version of *Homebase* on the client's server.

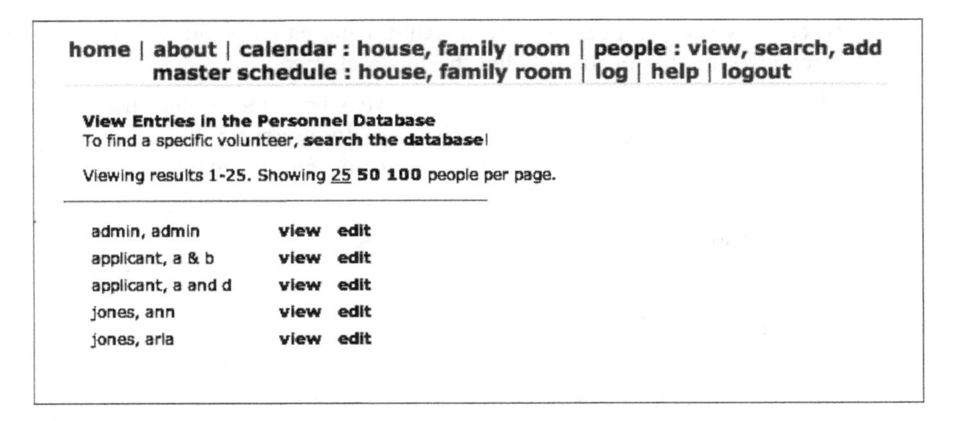

FIGURE 9.3 Designing the fix.

home | about | calendar : house, family room | people : view, search, add
master schedule : house, family room | log | help | logout

Personnel Edit Form
Here you can edit, delete, or change the password for a volunteer in the database.

* denotes required fields

First Name*:	a and d
Last Name*:	applicant
Address*:	a
City*:	a
State, Zip*:	ME, 11111

FIGURE 9.4 Testing the fix: editing a person.

In the e-mail below, Alex is passing the corrected module (editPerson.php) to the the developer, Ellis, who is managing that live version. He also advises the client that all current volunteers in the database who have "&" in their first names must be corrected manually.

From: Alex
Date: Wednesday, March 17, 2010 4:16 PM
To: Gina, Ellis

Hi Gina,

I hope you had a good vacation.

I have finally figured out your problem with the "&." I have changed the software so that if anyone enters "&" as part of their first name, like "Jack & Jill," it will store the name as "Jack and Jill" instead.

Attached is the file personEdit.php that Ellis can upload to the software to fix this problem when he has a chance.

Unfortunately, this doesn't fix any database entry that already has "&" in its first name. But I did use the **people: search** function in *Homebase* looking for "&" in the first name and found the following four entries:

Amos & Andy
Romeo & Juliet
Mutt & Jeff
Ding & Dong

Hopefully, Ellis can print out the database entry for each of these four persons and send it to you for re-entry.

Ellis, look in the table called dbPersons in the database, find these four rows, and just print them out and send them to Gina. Once she has re-entered these four persons' data using the corrected software, you should delete these four rows permanently from the dbPersons table.

Give me a call if you have any questions. In any case, keep me posted on how this works out.

Best,
Alex

In reply to this e-mail, Ellis offers to correct these entries directly in the database, a less time-consuming step than what Alex had suggested.

From: Ellis
Date: Wednesday, March 17, 2010 4:39 PM
To: Alex

Can I just edit these four rows in the table itself?

From: Alex
Date: Wednesday, March 17, 2010 7:00 PM
To: Ellis

Good idea, Ellis. If you want try that, it can save Gina from reentering these four persons' data.

There are two columns in the table that need the "&" replaced by "and," one named `id'` and the other named `first_name`. Let me know when that's done and I'll fix the rest. (That is, the passwords will need to be reset for these four persons.)

Thanks, Alex

From: Ellis
Date: Thursday, March 18, 2010 6:54 AM
To: Alex

I've uploaded the new file and have changed the four rows in the database.

Ellis

Closure

Now that the fix has been completed and installed, the developers wait for a reply from the user, who will exercise the new code. The episode ends with a final exchange between the developers and the user.

From: Alex
Date: Thursday, March 18, 2010 7:49 AM
To: Ellis, Gina

Thanks, Ellis. Everything looks okay.

Gina, let me know if you have problems accessing these four persons' information. In the future, there will be nobody in the database with "&" in their first name. It will show as "and" instead.

Alex

From: Gina
Date: Thursday, March 18, 2010 8:56 AM
To: Alex, Ellis

Thank you both so very much! Everything is working perfectly — you're good!

Gina

All's well that ends well.

9.3.2 Software Maintenance: A Multi-Year Developer Perspective

The above episode in the early life of *Homebase* is typical of many bug-fixing activities in the life of a new open source product. It is typical in the sense that the bug is first discovered by the user, reported to the developer, and addressed by the developer, and the fix is installed in the live version of the software.

However, many other types of issues can arise that require a broader perspective in order to address them properly. For example, the programming platform on which the software is built can experience an upgrade that depreciates one or more language features that were used in the software. For another example, the user can suggest adding a new feature that would create new functionality for the software.

Since it began supporting the live versions of *Homebase* in 2010, *Homeplate* in 2012, *Homeroom* in 2013, *BMAC-Warehouse* in 2015, and *RMHP-Homebase* in 2015, NPFI has had many interactions with the respective clients to fix bugs and add new features. The following discussion provides examples for each of these products, with the idea of illustrating the variety of situations that can arise soon after the software is first installed.

Homebase Maintenance: 2010-2018

For the most part, *Homebase* maintenance over these years has been done in two ways. Individual bugs have been fixed by NPFI in the style reflected in

Section 9.3.1, while new features have been added in groups by individual student teams as their semester project.

One major exception to this pattern occurred in the fall of 2015, right after the creation of *RMHP-Homebase* by a different student team earlier that year. At that time, the Ronald McDonald House in Portland, ME merged with its counterpart in Bangor, ME and wanted to have *Homebase* available for use in the Bangor house as well, but with a different database of volunteers.

Serendipitously, the architecture of *RMHP-Homebase* was perfectly suitable for use as a 2-House solution because it was already designed to serve two different venues. So NPFI took on this upgrade and installed a single version of *Homebase* to support both the Portland and Bangor houses, replacing the earlier 2013 version that supported only the Portland house. This upgrade would not have been so straightforward had it not been for the creation of *RMHP-Homebase* earlier in 2015.

A nice byproduct of this upgrade is that the newer version of *Homebase* installed at the Portland and Bangor houses also inherited many new features that had been added to create *RMHP-Homebase* by a different group of students for a different client.

Homeplate Maintenance: 2012-2018

Soon after it was completed and installed on the Second Helpings server in 2012, *Homeplate* maintenance – bug fixing and new feature addition – was transferred to Progressive Technology, a software developer and Web hosting company in Bluffton, SC. During the period 2012-2017, Progressive fixed several bugs and added some new features to both the *Homeplate* Web application and its companion *Homeplate Mobile* app that runs on Android tablets.

In 2017, Progressive was no longer able to support *Homeplate*, though Second Helpings still needed several new features added and bugs fixed. So in the fall of 2017, NPFI assumed responsibility for *Homeplate* maintenance and addressed these issues.

A list of *Homeplate*'s issues and their current status can be found at the project's `GitHub issue tracker`. Most of the more recent issues were resolved with a brief dialog between the client and the developer. For example, here is that dialog for `Issue #47`, where the developer added a new field to the client information form:

> **client commented on Dec 17, 2017:** Add new field 'Pest Ctrl Date' after Food Safe Date; make entry by dropdown calendar in original calendar format. Can move whole line to left. PDF sketch of layout is attached.
>
> **developer pushed a commit that referenced this issue on Dec 23, 2017:** fixed issue #47 - added pest ctrl date to client form and report

client commented on Dec 23, 2017: Looks good. Closed 12/23/2017.

The arrangement between this particular client and developer is for the client to post the issue; the developer to respond with either a fix, a question, or a determination not to fix; and the client to close the issue when it has been satisfactorily fixed.

Homeroom Maintenance: 2013-2018

While *Homeroom* was first implemented by a student team in 2011, it was not put into productive use until after it had been upgraded by another student team in 2013. In 2017, a second version of *Homeroom* was developed by NPFI as a variant of the original version for a different client. Both versions of *Homeroom* have been in productive use since then.

From August, 2016 to the present, *Homeroom* users posted 76 issues, all of which were resolved by NPFI soon after they were posted. For details, readers should review the *Homeroom* `issue tracker`. The two clients have also communicated additional issues with *Homeroom* to NPFI over the last two years. These issues were resolved by NPFI in September and December, 2016 and installed by Artopa, who manages the client's Web server. Here is a summary of the issues and upgrades that NPFI installed on *Homeroom* at that time.

1. New bookings now properly handle the primary guest as the first occupant and don't drop other occupants.

2. The Room Log shows all guests approved for each room, while clicking on an individual room shows only those guests who are marked "present." When you edit a booking, you can mark a guest "absent" or "present" and that will show up on the individual room view (but not the whole room log).

3. A family can have a second room assigned to them. Go to the Room Log and edit an active booking there. See the phrase "Add a second room for this family?" and choose a room to get that done. Later you can remove the second room using the same page, if needed.

4. Reports can now be generated for specific rooms – see the "data" tab at the top.

5. You can now search bookings for the Primary Guest's Last Name as well as First Name.

As readers can see, some of these items are bug fixes and others are new features added to improve room occupancy information and data reporting.

Both clients seem to be happy with *Homeroom* since its latest refinements were completed, as witnessed by the following note from the RMHC-Maine Development Director in May, 2018:

> Hi Alex,
>
> Thank you so much! I just tested out running a report in Data for 1/1/2017-1/1/2018 for room 126 and the page layout and content look great. I really appreciate you updating this in both Home-Room databases. It's going to save us SO much time in the future when pulling stats on specific rooms.
>
> Happy Friday and have a great weekend!
>
> Best, Alicia

This level of software customization is ideal – the software can be tuned to fit the exact needs of the client, no more and no less. The ability to achieve such customization is unique to the CO-FOSS development model.

BMAC-Warehouse Maintenance: 2015-2018

After *BMAC-Warehouse* was activated in May 2015, NPFI assumed responsibility for fixing bugs and adding features. NPFI also hosted this application on its own server until April 2016, when it was moved to BMAC's own server.

That move took place smoothly. NPFI transferred the code base and current database over to Viviotech, which was already hosting `BMAC's Web site`. At that time, *BMAC-Warehouse* was integrated into that Web site, where it has been in use ever since.

The *BMAC-Warehouse* `issue tracker` shows 19 issues raised during the beginning of its productive use in 2015. In addition, a list of issues was transmitted by the client to NPFI in June 2015 by email, and the issues were fixed during that summer. Then all was quiet until September 2018, when *BMAC-Warehouse* suddenly stopped working. A White Screen of Death (WSOD) appeared instead of the usual login page.

A quick review by NPFI identified the problem after learning that the BMAC Web server had just been upgraded to a new version of PHP (7.2) from version 5.3. Significantly, this new version had phased out support for PHP's MySQL database interface, and replaced it by MySQLi. At that time, the entire database layer of *BMAC-Warehouse* code was littered with MySQL queries, which would all have to be converted to MySQLi equivalents for this issue to be resolved.

Fortunately, NPFI had just converted the current version of BMAC-Warehouse (running on its sandbox server) to the PDO database abstraction layer, which was introduced in Chapter 6. Recall that using the PDO abstraction layer allows the PHP application to seamlessly switch between MySQLi, PostgreSQL, and SQLite without any recoding. This turned out to be exactly what was needed to get *BMAC-Warehouse* out of its current jam.

So NPFI retested the converted version using a copy of the current database, and then turned over the entire code base to Viviotech on September 7, 2018 for installing on the live BMAC server. The new version was installed on September 11, after which *BMAC-Warehouse* has been running smoothly ever since. Sometimes it's better to be lucky than smart!

RMHP-Homebase Maintenance: 2015-2018

RMHP-Homebase was put into productive use in May 2015. At that time, NPFI assumed responsibility for hosting, fixing bugs, and adding new features. By the end of that month, the first 18 issues had been posted by the client and resolved by the student developers, as discussed in Chapter 7.

Since May 2015, 31 additional issues were posted by the client and resolved by NPFI. Some were bugs and others were minor new features. These are summarized on the project's `issue board`. Overall, *RMHP-Homebase* has been running smoothly for the client.

RMHP-Homebase experienced three major maintenance events between May 2015 and December 2018. The first event involved transferring the software's hosting from NPFI's server over to the client's server in September, 2015. This was facilitated by the presence of a technology coordinator on the client's staff and a very responsive Web developer who managed the client's hosting configuration. In a period of a few days, NPFI posted the *RMHP-Homebase* source code and a dump of its active database on a shared drive. The developer downloaded those two files and installed the code and database on the client's server. Testing was done to be sure everything was clean and up to date, and the old version on the NPFI server was discontinued immediately.

The second event involved a rollout and staff training session for the new software, which was organized by the client in November 2015. The idea was to train the client's 200 or more volunteers to log in and self-schedule themselves into and out of a shift on the on-line calendar. This would distribute the tedium of volunteer calendar maintenance more widely, so as to relieve the house manager from doing it all.

The third major event took place in April 2018 and, like the one reported in the previous section, involved the need to respond to an update of the client's server from PHP version 5.3 to version 7.2. This response required recoding the PHP database modules to migrate from MySQL to MySQLi. This work was somewhat simplified by *RMHP-Homebase*'s layered architecture, which confined all the MySQL references to a handful of database modules, rather than spreading them throughout the code base.

9.4 CREATING A FORUM

At the conclusion of a CO-FOSS project, consideration should be given to establishing a user forum, especially if the project has more than one client and/or may soon attract new clients with similar software needs. We know of two such projects, `FarmData` and `The Open Energy Dashboard`. FarmData began as a single-client CO-FOSS project, while OED began with several clients. Now they are more mature and have several installations and users.

Upon its initial launch, a new CO-FOSS project's user forum can be viewed as an extension of its issue tracker, which was established earlier in the project and should now have several issues posted and resolved. Also, the client should now be comfortable posting issues and suggestions for new features. Thus, the tracker can begin to evolve into a multi-user forum.

A most critical element of community building for a newly-launched CO-FOSS project is the establishment of effective on-line forums where new developers and users can join in and discuss issues related to the development and effective use of the software on a broader scale.

Forums provide an immediate avenue through which a developer or user can report a bug or suggest a new feature for the software. They also provide timely information to developers about the status of all active bugs and other new features that are being considered for the next release. Thus, the developer and user forums play a vital role in ensuring that a new FOSS project becomes healthy and the young community of contributors and users remains up to date with the latest news about the project's progress.

In a more mature HFOSS project with many developers and users, an open discussion forum is designed to capture and publish questions "as is," providing an open channel through which experienced users and developers can address these questions directly. In general, forums tend to be more up-to-date on specific questions and issues than tutorials or user manuals.

On the other hand, the quality of information in forums can be more spotty compared with that found in on-line documentation and tutorials; a forum is only as good as the quality of questions and answers provided by its volunteer participants. Forum moderation can help retain good quality, and a certain level of "filtering skill" on behalf of the forum user can also help establish usefulness.

Forums are widely used by mature HFOSS projects. They are organized by topic, and individual topics are initiated by developers and users themselves. Anyone can access an open forum, and anyone can post a question or a response as well. Most forums are *moderated*, so that useless questions and answers can be filtered out before they are made available for general public access.

Examples of active HFOSS user forums for more mature CO-FOSS projects are given in the following sections.

9.4.1 Example: Wordpress Support Forums

`Wordpress` is an open source software platform that supports the development of content-management Web sites. Started in 2003, Wordpress has grown to become a leading Web site platform, and along with that it has a very large developer and user community.

To keep developers and users up to date with its rapid growth, Wordpress maintains several `support forums` for its developers (people who develop the Wordpress platform, its themes, and its plugins) and its users (people who create Wordpress Web sites). Each forum addresses a particular aspect of Wordpress development or use, as shown in Figure 9.5.

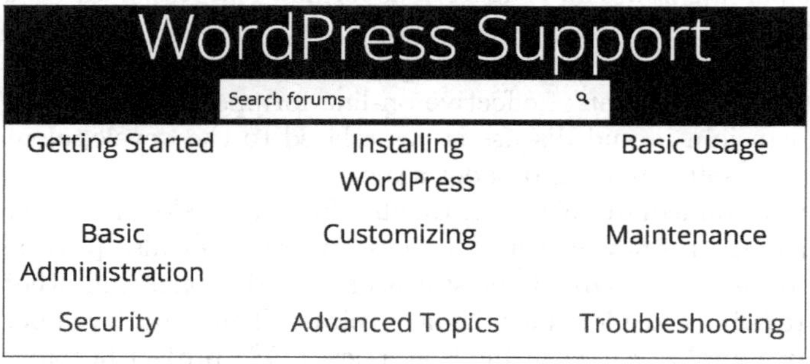

FIGURE 9.5 Points of access to the Wordpress forums.

Each Wordpress forum can be filtered in various ways by topic (most popular, no replies, non-support, resolved, and unresolved), so that developers and users can home in more precisely on the issues that concern them. For example, the user forum for installing Wordpress is shown in Figure 9.6.

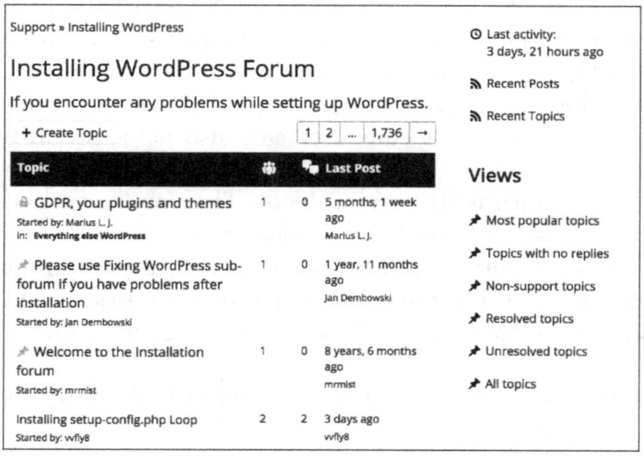

FIGURE 9.6 Snapshot of the Installing Wordpress Forum.

In addition, each individual Wordpress theme or plugin may have its own forum for its own developers and users. Thus, the granularity of direct support for a Wordpress issue resolution is quite fine overall.

9.4.2 Example: Firefox Forums

Like Wordpress, the `Firefox Support Forum` provides guidance for new and experienced users of Firefox. This forum's main page is shown in Figure 9.7.

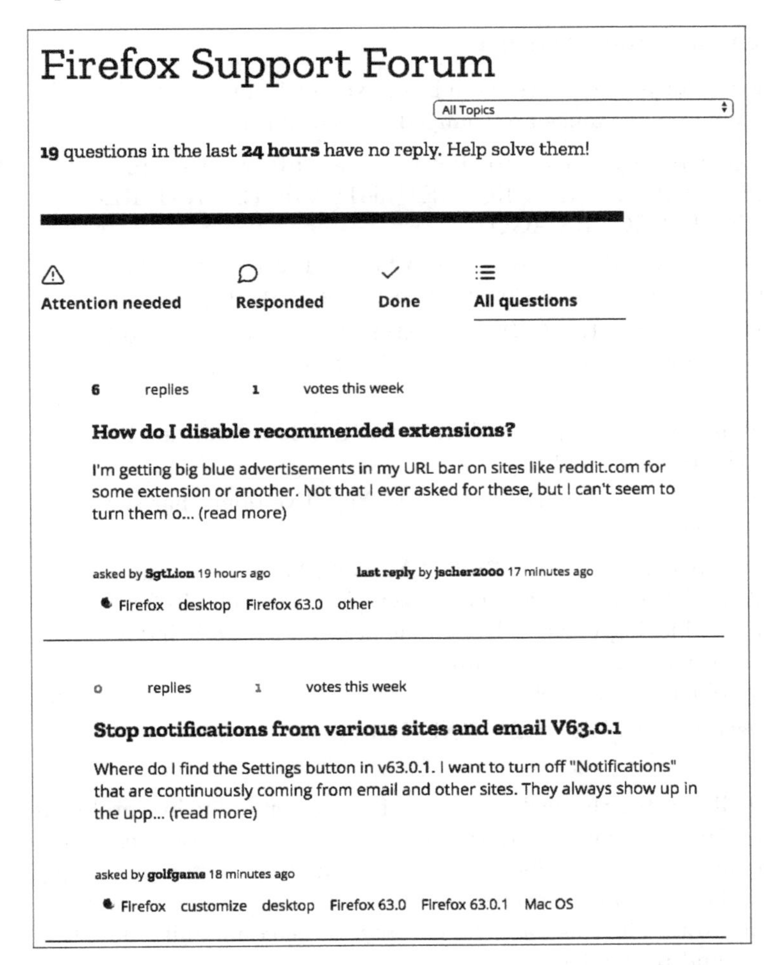

FIGURE 9.7 Accessing the Firefox user forum.

The threads in this forum appear chronologically, beginning with the most recent. Also shown is a snippet of each thread's discussion topic, its author, and the number of replies received to date. To narrow the search, users are encouraged to log in and "Ask a New Question" right at the top of the forum. This provides a Google-like list of discussion threads that are most relevant to the user's topic of interest.

For example, if a user wants to learn about disabling extensions, they would type, "How do I disable extensions?" In reply, the Support forum would likely retrieve the first item in Figure 9.7 but not any others.

9.4.3 An Example Forum Exchange

To illustrate how a forum can work, here is an excerpt from an actual Firefox user forum exchange (the names are changed to ensure privacy):

> **Firefox extremely slow to load pages after startup** 8 REPLIES 11 HAVE THIS PROBLEM 5411 VIEWS LAST REPLY BY BILL 3 HOURS AGO
>
> **Matthew Posted 8/14/18, 6:05 AM** For the past few days, Firefox has been extremely slow to load pages at startup. ...
>
> **Fred Posted 8/14/18, 9:48 AM** Start Firefox in Safe Mode web link by holding down the ¡Shift¿ (Mac=Options) key, and then starting Firefox.
>
> A small dialog should appear. Click Start In Safe Mode (not Refresh). Is the problem still there?
>
> **Matthew Posted 8/14/18, 9:49 AM** Yep, I've done that several times (including with a fresh profile).
>
> **Matthew Posted 8/16/18, 7:20 AM** Well, Windows updated last night, and this morning this problem is gone and Firefox is back to being blazingly fast. I have no idea if it was the Windows update, or perhaps an update to my security apps, but it evidently wasn't a problem with Firefox.
>
> **Fred Posted 8/16/18, 5:22 PM** That was very good work. Well done.

This thread illustrates the utility of user forums for reporting problems where developers and other users can respond and help fix the problem. Note that this particular issue was resolved not in the form of a bug fix, but simply by users solving the problem themselves and then reporting to the community how they did it. Notice in this case, how many other community members viewed this issue and responded to it.

Note also the amount of time that elapsed between the first instance of this particular problem and its final resolution—not much! Typically, an issue is more or less quickly resolved depending on the nature of the issue and how widely it is being experienced among other users.

9.5 EVOLVING INTO A DEMOCRATIC MERITOCRACY

Some new FOSS projects that promise to have broader impact may experience community pressure to expand. These projects can transition to a more mature phase that might be called the *democratic meritocracy* phase. This transition cannot be made without having a strong community and an active development process in place.

The democratic meritocracy phase is an ideal form of governance for a FOSS project, in the sense that the project is governed by a democratic process whose participants are representatives from the meritocracy of contributors. Typically, the sponsors of these types of projects are non-profit foundations, where the board of directors is also selected by the membership.

Any democratic process involves politics. A FOSS project that makes this transition has to invest energy into bringing transparency to the governing process so that the majority of its developers and users are happy. Probably the best example of such a community is the `Debian community`, which has thousands of voluntary developers and users and a very mature process of voting and selecting project leaders.

However, the Debian 100% voluntary model and maturity level are not easy to replicate. A lot of time is required to manage the political process, and it takes years to get this right. A more expedient alternative for reaching a high level of rigor and maturity for a project would be to fund a core development group that is selected from the leading core contributors (based again on merit).

In this section, we explore the transition of a young open source project with benevolent dictator governance into a mature project. That is, we ask the basic question, "How can an emerging FOSS project be rationally governed?"

9.5.1 Incubation

The formation of a vibrant community of users and developers marks a critical stage in an open source project's life that we call *incubation*. An open source project that fails to incubate risks becoming inactive and dying. This section discusses the process of incubation and what it takes for a young open source project to pass successfully through the incubation stage into a mature and democratic meritocracy.

The purpose of incubation is to establish a self-sustaining FOSS project. Both the Eclipse Project and the Apache Software Foundation have established *incubators* that invite open source projects with complementary goals to become members.

Both of these incubators accept new projects, provide guidance and support to help new projects develop their own collaborative communities, educate new developers in the principles of collaborative development, and propose to their boards the promotion of such products to "mature" status. (For more discussion, see `https://incubator.apache.org/policy/incubation.html`.)

However, membership in either of these two incubators is appropriate only for open source projects that develop middle-level software—that is, software for software developers, rather than non-technical persons. These biases exclude many important software projects, especially those which are developed for humanitarian organizations as end users.

So in this discussion, we use the term *incubation* in a more general sense, rather than implying membership in either the Eclipse or the Apache incubation process. Two key activities govern how successfully an open source project can pass through its incubation phase and become healthy and sustainable for the long run: building a vibrant community and establishing a viable bug tracking process.

Essential to successful incubation is the emergence of strong and sustainable communities of developers and users. At its beginning, a CO-FOSS project has only a single (lead) developer, a sponsoring client, and a single user. As the code base evolves, a core group of developers and a handful of "bleeding edge" users become actively involved. This is the case, for example, with the *OED* project discussed earlier in this book.

How does a project transform this fragile community into one that has a significant number of users and developers, where users are actively reporting issues and developers are actively contributing bug fixes and enhancements?

An active Web presence, including easily accessible developer and user forums and project wiki, are valuable catalysts that encourage the growth of sustainable user and developer communities.

Attracting new users and developers to a FOSS project requires active recruiting, not just passive "openness." For example, a certain amount of professional and social networking should become directly associated with the project. The lead developer(s) especially must make efforts to recruit promising new developers.

An active and engaged user and developer community is an essential measure of successful project incubation. (Absence of such a community suggests that the software is too single-client oriented to become more generally useful.) Growing a user community takes time, of course, but once it is established, that community can become self-sustaining.

9.5.2 Organization

Abstractly, Jensen [23] characterizes FOSS project organization as a socio-technical interaction network, or STIN for short. STINs are always in flux; they are self-organizing networks of activities, people, and tools, and often all these parts are geographically distributed around the world.

More concretely, mature FOSS projects tend to have three main organizational distinctions from proprietary projects. These are:

1. **Self-organizing** They allow participants to find their own level and project activity with which to become engaged, based on their interests and skills. Proprietary project leaders assign each participant to a project activity.

2. **Egalitarian** They openly invite contributions from everyone. Proprietary projects are hierarchically organized and closed in this regard.

3. **Meritocratic** They organize their work around public discussions, and decisions about future directions are based on merit. Proprietary projects organize around the results of private discussions among project leaders, and decisions about future directions are highly influenced by cost and profit.

The emergence of three inter-related communities seems to be a vital part of the transition from incubation to maturity. A mature FOSS project is often organized around three distinct roles that community members can play:

Users know and use the software actively. Many users provide feedback to developers (contributors and committers) when they find bugs or other difficulties when using the software. They also suggest new features that could improve the software's usability or applicability.

Contributors are users who also contribute bug fixes and minor features to the software, but don't have the right to alter the code base itself. Contributions can also be in the form of documentation, administrative support, and testing.

Committers are developers who review user contributions and install them in the code base. In this activity, the committer ensures that the code base keeps its integrity—i.e., that the new features are correctly implemented and that the bugs are actually fixed.

Communication among users, contributors, and committers should be open and transparent. It should encourage inclusiveness and diversity of opinion, separate from the influence of any company or sponsoring organization.

While a FOSS community itself is quite fluid, research has shown that most users are, in fact, passive users. Jensen [23] calls these users "free-riders," since they give nothing back in return for the privilege of using the software. Too many free-riders can kill a project.

Users who do provide feedback to the developers do so in an entirely voluntary spirit. Feedback typically occurs through the software's user forum, which is prominently accessible at the project's Web site.

Some users actually become *contributors* by providing bug fixes or suggestions for new features in the form of code patches. A user who is also a programmer can, in fact, do this since the open source code base is freely available for anyone to read. The determination of whether or not a contributor's suggestions are accepted and become part of the code base, however, is made by a committer.

Promotion to *committer* status is done on the merits of a person's collected contributions to the project over time. In this sense, the contributions become a portfolio of work that can be evaluated to assess the merits of that person's case for assuming the responsibilities of a committer.

Who decides on the promotion of a contributor to committer status? This is often done by a core project leadership group. In Apache this group is called the *Project Management Committee* (*PMC* for short), which has committers who oversee the project's organization. The PMC promotes a contributor to the role of committer strictly on the merits of his/her contributions to the project. This certification process is thus peer-to-peer and publicly documented (rather than closed).

The overall role of the PMC is to ensure that the community is behaving and governing itself in a manner that is consistent with the objectives of making the project successful. This includes operational, legal, and procedural oversight on all software releases.

While most users and contributors are volunteers, many committers can be paid employees of the project. What particular skills are required to attain committer status? Generally, an applicant's portfolio contains two types of contributions: those that illustrate technical competence and those that exhibit social skills.

> On the technical side, a portfolio should demonstrate that the applicant has programming and software architecture skills, documentation writing skills, and a general understanding of how software systems are built and maintained.
>
> On the social skills side, a portfolio should demonstrate good collaboration, reading and writing skills (e.g., effective use of e-mail and discussion forums), leadership skills, and an understanding and adaptability to various peer and sub-culture behaviors.

9.5.3 Task-Specific Roles

In addition to maintaining the code base, the participants in a mature FOSS project play roles that accomplish a number of other important project tasks. Those who play these roles are all committers, and many are members of the PMC. Here is a summary of these roles and their respective activities:

> The **project leader** maintains the project's release plan and current status, and moderates the developer forum.
>
> The **expert user** maintains the software's actors, use cases, requirements, and user roles.
>
> The **lead developer** maintains the software architecture.
>
> Other **developers** maintain the user interface design, domain classes, database design, code base, unit test suite, build package, and build schedule.
>
> **Testers** maintain bug reports and the user forum.
>
> **Writers** maintain on-line help text.

Bug Marshalls oversee opened bugs, filter them, and pass them on to developers.

Release Managers oversee the packaging and releasing of new versions of the software for general public consumption.

We also note that it is not unusual for an individual to play two or more of these roles simultaneously, depending on his/her particular skills and interests. The FOSS project itself publishes detailed guidelines for each of these tasks.

9.5.4 Oversight

The relationships among the levels of participation in the Sahana project are summarized in Figure 9.8. These levels represent the organization of a single mature FOSS project, identifying the various roles that individual participants play. This organization is very loose, and membership at any level is based solely on the merits of a person's contributions to the project.

However, a successful FOSS project eventually reaches a point in its life where an additional role is required. This role often emerges as the project becomes (part of) a non-profit organization whose investors actively set future directions for the project.

Figure 9.8 shows this fifth role as the Board of Directors. Unlike other roles, the Board is populated mainly by persons outside the community of developers, as indicated by the arrow in the upper-right corner of the figure.

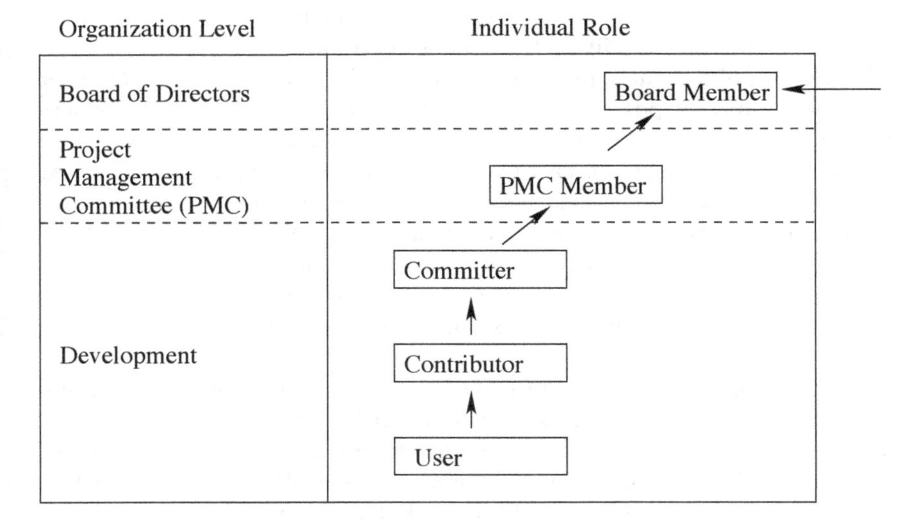

FIGURE 9.8 Organizational levels in the *Sahana* project.

The Board's job is to promote the adoption and growth of the software that the project is developing. The Board actively engages with partners from the private sector, academic institutions, and the public sector. The Board also establishes mechanisms for supporting the project, evaluating its success, making licensing decisions, and addressing general development and implementation issues.

Mature FOSS projects are often sponsored by non-profit foundations. For example, *Sahana* is sponsored by the Sahana Foundation. Because the *Sahana* project is in the humanitarian sector which has a lot of non-governmental and non-profit organizations, it is most naturally run by a non-profit foundation rather than a for-profit organization.

Through its Board of Directors, the Sahana Foundation engages with private sector, academic institutions, and public sector partners in promoting the adoption and support of *Sahana*. Such a Foundation is typically a legal entity established to manage the finances, legal issues and expense reimbursements for its core developers.

9.5.5 Decision Making and Conflict Resolution

The committers in a mature FOSS project are highly motivated and skilled individuals. Inevitably, rivalries and conflicts will arise because of individual differences in values and priorities. It is healthy for this high level of rivalry to occur as long as individuals are working on different parts of the code base.

Policy, procedural, and technical decision-making in a mature FOSS project aims to be fully transparent to all community members. Absence of transparency, for example when Board members make off-line decisions, can lead to demoralization among other community members—especially those who are invested more heavily in the collaborative process.

When a complex policy, procedural, or technical issue arises, a member of the PMC typically posts the issue on a public discussion forum, where it is debated and either ratified or rejected by consensus or majority vote. Here, *consensus* means that at least two other committers support a particular solution and no other committers post strong disagreements.

For example, *Sahana* uses a *lazy consensus* process. Voting is done with +1 denoting "for a motion," -1 denoting "against to a motion," and 0 denoting abstention. With a lazy consensus process, a 72-hour time frame is given for a decision. If there are no negative votes during that period, the motion passes.

A lazy consensus process is practical in a voluntary community where only a fraction of the participants actively contribute to any particular vote. Depending on the nature of the decision, the vote is made at either the committer, the PMC, or the Board level. For Board or PMC decisions, a poll is first run to gather community feedback, and this serves as additional input for the final decision.

Sometimes, of course, achieving consensus on a contentious issue is not possible. Often conflicts arise during discussions about community infrastructure, technical direction, expectations about developer roles, or interrelationships among roles.

These kinds of *conflicts* can be resolved by a process involving a small PMC made up of prominent members of the community. The PMC has the job of ensuring fairness throughout the community by solving persistent disputes.

The reputation of the individual PMC members carries some weight in conflict resolution, even though they may not vote unanimously on any single issue. Thus, the role of the Board is more that of a mediator that assists community members to resolve conflicts among themselves.

9.5.6 Domain Constraints

It is also important to understand the domain of users and organizations that the project is serving. Some domains may force additional constraints and priorities to be established.

Sometimes the application domain places special constraints on the governance of a mature FOSS project. Consider, for example, the humanitarian FOSS domain, where the software is targeted to help people recover from a disaster or longer-term condition that is debilitating to a significant population.

These systems often operate in a constrained setting—one that is not experienced in other FOSS development situations. For example, *Sahana* works in the domain of humanitarian response and disaster management. In this domain the following considerations must be factored into the software development process:

Telecoms and Internet access is either down or intermittently available.

Bandwidth is often at a premium so every character counts.

Power can go out at any time or not be available.

Any central data center or infrastructure might have been affected by the disaster.

People have little time to get familiar with new systems.

Off-the-shelf systems often have to be customized for the requirement or risk not capturing aspects of gathered data.

Local developers have very little time to learn and support the system.

There are many existing legacy systems and loads of spreadsheets with valuable data.

Data will come to you with different levels of granularity, validity, and redundancy. Such data often need to be cleaned up.

Considering such constraints, the governance of a FOSS project must adopt additional conventions that do not usually apply to other types of projects. Developers need to be sure that their project has a clear purpose and applicability, avoids complex user workflows, is debuggable, and uses operating system and database architectures that ensure scalability.

9.5.7 FOSS Project Foundations

Finally, we note that a mature FOSS project can either remain as a free-standing entity or seek membership in a larger (umbrella) organization to ensure continuity and sustainability.

As a free-standing entity, the project has complete autonomy over all its activities. For example, it may incorporate itself as a not-for-profit foundation or simply maintain a project Web site as the single point of contact for all its community members. As a free-standing entity, it may choose from among several alternative open source licensing arrangements for distribution and development of its code base. One example is the *OED project*, which at this writing is beginning the process of becoming a non-profit foundation.

For another example, the Sahana Software Foundation promotes free and open source software solutions for disaster and emergency management. The Foundation took over the governance and management of the *Sahana* software project from the Lanka Software Foundation in October 2009. By that time, *Sahana* had become the leading open source disaster management system worldwide, having been used by dozens of countries following natural and man-made disasters.

However, remaining a free-standing entity comes with certain risks. Perhaps the most significant risk is that the project may become dormant or even die from inactivity. This can occur, for example, when the number of contributors and committers becomes too low in relation to the number of "free-riding" users. Other risks include legal exposure and financial stress.

As an alternative, a FOSS project can become part of a larger umbrella organization that is dedicated to sustaining such projects over the long term. For example, the `Apache Software Foundation` provides support for the Apache Community of Open Source software projects, which provide software products for the public good. Apache projects are defined by collaborative consensus based processes, an open, pragmatic software license and a desire to create high quality software that leads the way in its field. We consider ourselves not simply a group of projects sharing a server, but rather a community of developers and users.

9.6 SUMMARY

In this chapter, we have examined the transition of a new CO-FOSS project into a more mature FOSS product that can serve the needs of more clients who share the same software need. The first step in this transition is the hand-off of the product from the instructor to the professional developer who will become responsible for bug fixing and hosting the software going forward.

Beyond the hand-off, we have also discussed how a new project can mature into a viable FOSS product for other organizations to adapt and reuse. Elements of maturity include identifying new clients, appropriate licensing choices, hosting alternatives, and self-assessment. Maturity also usually includes the creation of a community and developer forum and strategies for bug fixing and feature enhancement.

The governance of a mature FOSS project comes with additional considerations. These include the creation of a foundation, the identification of task-specific roles, an oversight organization, and a strategy for resolving conflicts and making decisions about project hosting and releasing code.

9.7 MILESTONE 9

1. After browsing through the projects registered on a project hosting service, such as GitHub or SourceForge, identify one project in each of the following categories and briefly explain why you categorized it that way:

   a. Dormant, CO-FOSS

   b. Vibrant, CO-FOSS

   c. Incubating

   d. Vibrant, mature FOSS

   e. Dormant, mature FOSS

2. Use the maturity model discussed in this chapter to assess your CO-FOSS project. Include a measurement of the size of its user and developer community, the frequency and currency of its bug reporting, and other criteria suggested by the model.

3. Discuss the steps required to transfer your project to a more expansive hosting service, such as GitLab, Bitbucket, or SourceForge.

4. For your CO-FOSS project, discuss its current licensing type (e.g., MPL or GPL) and the reasons why that particular type was chosen. Reconsidering your project's possible future uses, reevaluate that licensing choice and consider changing it to accommodate such future uses.

5. What steps are needed to establish your project's first release and begin to grow a larger developer and user community?

9.8 ENDING THE JOURNEY

The book is about a journey. The journey begins with an idea for a new CO-FOSS project, it continues with a realization of the idea in the form of a new open source software product, and it ends with the delivery and continued support of that product. Essential to the successful completion of such a journey is its defining *triad* (recall Figure 1):

1. An instructor and student team willing to imagine teaching and taking a course that will create such a product,

2. A non-profit client with a need and a desire to participate in developing that product, and

3. A professional developer willing to install and support that product going forward.

Having taken this journey several times since 2008, we know that it can provide an extraordinary experience for the instructor, the students, the client, and the developer. We also know that it can result in a concrete product that materially benefits the client and provides resources for future CO-FOSS projects to reuse.

We are hopeful that many more instructors, students, clients, and developers will form their own triads and embark on this journey. We hope that they will have similarly rewarding experiences, and that the products of their efforts will have significant impact in their own communities.

Bibliography

[1] http://en.wikipedia.org/wiki/capability˙maturity˙model˙integration.

[2] http://en.wikipedia.org/wiki/gnu˙general˙public˙license.

[3] http://en.wikipedia.org/wiki/linux˙kernel.

[4] https://en.wikipedia.org/wiki/unified˙modeling˙language.

[5] http://www.gnu.org/philosophy/free-sw.html.

[6] http://www.hostingadvice.com/how-to/nginx-vs-apache/.

[7] BBC. UK government backs open source. *Online*, February 2009.

[8] Grant Braught, John McCormick, James Bowring, Quinn Burke, Barbara Cutler, David Goldschmidt, Mukkai Krishnamoorthy, Wesley Turner, Steven Huss-Lederman, Bonnie MacKellar, and Allen Tucker. A multi-institutional perspective on h/foss projects in the computing curriculum. *ACM Transactions on Computing Education*, 18(2):1–31, July 2018.

[9] Alistair Cockburn. *Crystal Clear: A Human-Powered Methodology for Small Teams*. Addison-Wesley, 2005.

[10] E. F. Codd. A relational model of data for large shared data banks. *Communications of the ACM*, 13(6):377–387, June 1970.

[11] European Commission. Open innovation 2.0 conference, June.

[12] Festival Latinoamericano de Instalacion de Software Libre. https://flisol.info/flisol2017.

[13] FLOSS Definition. https://en.wikipedia.org/wiki/alternative˙terms˙for˙free˙software#floss.

[14] Free Software Foundation. https://en.wikipedia.org/wiki/gnu˙general˙public˙license.

[15] Free Software Foundation. http://www.fsf.org/licensing/licenses/.

[16] Mozilla Foundation. http://www.mozilla.org/about/.

[17] Martin Fowler. *Refactoring: Improving the Design of Existing Code.* Addison-Wesley, 2000.

[18] GNU. http://www.gnu.org/gnu/initial-announcement.html.

[19] Standish Group. https://www.infoq.com/articles/standish-chaos-2015.

[20] Jim Hamerly, Tom Paquin, and Susan Walton. Freeing the source: The story of mozilla. *Open Sources: Voices from the Open Source*, pages 197–206, 1999.

[21] Gijs Hillenius. Amsterdam to make openoffice and firefox default on city desktops. *Online*, April 2009.

[22] Gijs Hillenius. Fr: Gendarmerie saves millions with open desktop and web applications. *Online*, 2009.

[23] Chris Jensen and Walt Scacchi. Governance in open source software development projects. In Pär gerfalk, Cornelia Boldyreff, Jesus M. Gonzalez-Barahona, Gregory R. Madey, and John Noll, editors, *Open Source Software: New Horizons*, volume 319, pages 130–142, Heidelberg, May 2010. Springer.

[24] Android License. https://source.android.com/source/licenses.

[25] Bonnie MacKellar, Mihaela Sabin, and Allen Tucker. *Bridging the academia-industry gap in software engineering: A client-oriented open source software projects course*, pages 373–394. IGI Global, 2014.

[26] Netcraft. https://news.netcraft.com/archives/2018/01/19/january-2018-web-server-survey.html.

[27] OSI. http://www.opensource.org/licenses.

[28] Bruce Perens. http://slashdot.org/articles/99/02/18/0927202.shtml.

[29] Bruce Perens. Open sources: Voices from the open source revolution. *The Open Source Initiative*, pages 171–188, 1999.

[30] Federal Source Code Policy. https://sourcecode.cio.gov/#fn17.

[31] The Humanitarian FOSS Project. http://hfoss.org/.

[32] Peter H. Salus. *The Daemon, the Gnu, and the Penguin: How free and open source software is changing the world.* Reed Media Services, 2008.

[33] Desktop Browser Market Share. https://www.netmarketshare.com/.

[34] Red Hat Linux Market Share. https://www.gartner.com/doc/reprints?ct=150106&id=1-26vhvsw&st=sb.

[35] Sourceforge. https://sourceforge.net/projects/ampps/.

[36] Richard Stallman. http://www.gnu.org/gnu/manifesto.html.

[37] Richard Stallman.
http://www.gnu.org/philosophy/use-free-software.html.

[38] Richard Stallman. *The GNU Operating System and the Free Software Movement.* O'Reilly, 1999.

[39] Richard Stallman. Why 'open source' misses the point of free software. *Communications of the Association for Computing Machinery*, 52(6):31–33, June 2009.

[40] Zdnet Survey. http://www.zdnet.com/article/its-an-open-source-world-78-percent-of-companies-run-open-source-software/.

[41] Jenifer Tidwell. *Designing Interfaces 2e: Patterns for Effective Interaction Design.* O'Reilly, 2010.

[42] Linus Torvalds. *The Linux Edge.* O'Reilly, 1999.

[43] Allen Tucker, Ralph Morelli, and Chamindra de Silva. *Software Development: An Open Source Approach.* CRC Press, Boca Raton, Florida, 2011.

[44] David A. Wheeler.
http://www.dwheeler.com/essays/floss-license-slide.html.

[45] Robert Young. Giving it away: How red hat software stumbled across a new economic model and helped improve an industry. *Open Sources: Voices from the Open Source Revolution*, pages 113–126, 1999.

Index

acceptance testing, 239
agile development, 4
analyst, 101
anonymized database, 69
assertion, 141
asynchronous communication, 105
audience, 262

big marshall, 317
Bitbucket, 75
board of directors, 317
branch, 77
bug tracking, 27

client review, 163, 252
client-centered, 37
client-developer meetings, 93
client-server, 19, 55
closed issue, 202
cloud computing, 20
CO-FOSS, xxix, 5, 19
code reuse, 65, 135
code-sharing repository, 72
coding standards, 129
cohesion, 60
collision, 176
commit, 76, 118
committer role, 315
contributor role, 315
controller, 210
copyright notice, 131
coupling, 60
CREATE query, 172
cross-site scripting, 220
CRUD, 182
custom software, 7

database, 168
database module, 62
database permission model, 186
database security, 63
DELETE query, 173, 185
demo version, 270
democratic meritocracy, 312
design document, 67
developer, 101, 316
development tools, 109
discussion forum, 72
document-sharing, 72
documentation standards, 81
domain analysis, 45
domain class, 61
DROP query, 173

Eclipse IDE, 74

face-to-face communication, 72
failure, 144
feedback survey, 279
foreign key, 175
forum, 308
FOSS, xxx, 6, 10
foundations, 320

getter function, 138
GitHub, 74
GNU Savannah, 74
good documentation, 263
GPL, 11

H/FOSS, 18
hand-off, 288
help pages, 85
help table of contents, 274
HTML form, 222, 224

incubation, 24, 313
INSERT query, 173, 182

integrated development environment (IDE), 73, 114
integration testing, 152
issue tracking, 163, 202, 240, 252

JDBC, 190
jQuery, 221, 232

key, 175

layered architecture, 58
layering principle, 187
layering violation, 244
lazy consensus, 318
lead developer, 316
licensing, 293
licensing notice, 131
live version, 78
localization, 264
locking, 177

maturity, 296
milestone, 90
milestones, 108
mini-lecture, 87
model, 210
Model-View-Controller (MVC), 209

normalization, 173
NPFI, 288

observer, 102
on-line help, 271
open issue, 202
ORM, 190

password encryption, 216
PDO, 190
persistence, 168
pluggable architecture, 57
primary key, 175
project hosting, 74, 294
project leader, 316
project wiki, 108
proprietary software, 9
pull, 76

query, 172

refactoring, 243
relation, 173
relational database, 168
release manager, 317
repository, 76
responsive, 234
reuse, 130
roles, 101

sandbox database, 243
sandbox version, 69, 78
SELECT query, 173, 182
self-assessment, 281
self-assessments, 97
SEQUEL, 168
serial development, 3
session, 222
software maintenance, 304
software maturity, 22, 23
software platform, 112
Sourceforge, 74
SQL injection, 219
staging server, 56, 78, 109
Structured Query Language (SQL), 168
synchronous communication, 107
system access security, 64

table, 169
task management tool, 72
task model, 50
team leader, 101
technician, 102
test-driven development (TDD), 82
test suite, 141
tester, 316
tests suite, 82
to-do lists, 108

Unified Modeling Language (UML), 52
unique key, 127
unit test, 82, 141
unit testing, 6

UPDATE query, 173, 184
use case, 50
use case diagram, 53
user, 102
user interface, 63
user manual, 263
user role, 49, 315
user story, 49
user training, 278

version control system (VCS), 75, 117
view, 210

web framework, 21
web server, 21
writer, 316
writing quality, 264
writing style, 262
writing tone, 262